# DYNAMICS OF ROTOR-BEARING SYSTEMS

TITLES OF RELATED INTEREST

*Boundary element methods in elastodynamics*
G. D. Manolis & D. E. Beskos

*Boundary element methods in solid mechanics*
S. L. Crouch & A. M. Starfield

*A concise introduction to engineering economics*
P. Cassimatis

*Distribution-free tests*
H. R. Neave & P. B. Worthington

*Planning and design of engineering systems*
R. Warner & G. Dandy

*The finite element method in thermomechanics*
T.-R. Tsu

*Intelligent planning*
R. Wyatt

*Marine geotechnics*
H. Poulos

*Numerical methods in engineering and science*
G. deV. Davis

*Plastic design*
P. Zeman & H. M. Irvine

*Structural dynamics*
H. M. Irvine

*Theory of vibration*[*]
W. T. Thomson

[*] Originally published by, and available within North America from, Prentice-Hall.

# DYNAMICS OF ROTOR-BEARING SYSTEMS

**M. J. Goodwin**

*Principal Lecturer*
*Department of Mechanical and Computer-Aided Engineering*
*Staffordshire Polytechnic*

London
**UNWIN HYMAN**
Boston   Sydney   Wellington

© M. J. Goodwin, 1989
This book is copyright under the Berne Convention. No reproduction without permission. All rights reserved.

Published by the Academic Division of
**Unwin Hyman Ltd**
15/17 Broadwick Street, London W1V 1FP, UK

Unwin Hyman Inc.,
8 Winchester Place, Winchester, Mass. 01890, USA

Allen & Unwin (Australia) Ltd,
8 Napier Street, North Sydney, NSW 2060, Australia

Allen & Unwin (New Zealand) Ltd in association with the
Port Nicholson Press Ltd,
Compusales Building, 75 Ghuznee Street, Wellington 1, New Zealand

First published in 1989

---

**British Library Cataloguing in Publication Data**

Goodwin, M. J.
   Dynamics of rotor-bearing systems.
1. Rotors. Dynamics
I. Title
621.8'11
ISBN 0-04-621032-6

---

**Library of Congress Cataloging-in-Publication Data**

Goodwin, M. J., 1955–
   Dynamics of rotor-bearing systems.
Includes bibliographies and index.
1. Rotors—Dynamics.  2. Rotors—Bearings.  I. Title.
TJ1058.G66 1989    621.8    89-5626
ISBN 0-04-621032-6 (alk. paper)

---

Typeset in 10 on 12 point Times by
Mathematical Composition Setters Ltd, Salisbury,
Printed in Great Britain by the University Press, Cambridge

To Diane

# Contents

| | | |
|---|---|---|
| Acknowledgements | *page* | xiii |

**1 Introduction**    1

1.1 The importance of rotating machinery in the modern world    1
1.2 Rotor-bearing interaction    2
1.3 Causes and effects of vibration    2

**2 Bearing systems**    5

2.1 Introduction    5
2.2 Rolling element bearings    5
2.3 Hydrodynamic oil-lubricated journal bearings    16
2.4 Gas bearings    30
2.5 Squeeze-film bearings    48
2.6 Hydrostatic bearings    57

**3 Single-mass rotor dynamics**    69

3.1 Introduction    69
3.2 Flexible shaft in rigid bearings    69
3.3 Symmetrical rigid shaft in flexible anisotropic bearings    75
3.4 Symmetrical rigid shaft in flexible anisotropic bearings with damping and cross-coupling    77
3.5 Asymmetrical flexible shaft in flexible anisotropic bearings with damping and cross-coupling    83
3.6 Effects of flexible foundations    91
3.7 Gyroscopic effects    95
3.8 Aerodynamic effects    98

**4 Systems with many degrees of freedom**    102

4.1 Introduction    102
4.2 Method of influence coefficients    102
4.3 Transfer matrix method    107
4.4 Mechanical impedance (and receptance) methods    113
4.5 Dynamic stiffness matrix method    121
4.6 Effects of flexible bearings and pedestals    127
4.7 Magnetic effects    134

| | | |
|---|---|---|
| 4.8 | Dunkerley's formula | 138 |
| 4.9 | Rayleigh's method | 140 |

## 5  Torsional vibrations — 145

| | | |
|---|---|---|
| 5.1 | Introduction | 145 |
| 5.2 | Simple systems with one or two rotor masses | 145 |
| 5.3 | MDOF systems – transfer matrix methods | 150 |
| 5.4 | Geared systems and branching | 154 |
| 5.5 | Damping in torsional systems | 160 |
| 5.6 | Calculation of torsional natural frequencies and forced response using the method of mechanical impedances | 163 |

## 6  Instability in rotating machines — 167

| | | |
|---|---|---|
| 6.1 | Introduction | 167 |
| 6.2 | Oil whirl | 167 |
| 6.3 | Resonant whip | 177 |
| 6.4 | Internal friction | 178 |
| 6.5 | Effect of rotor polar asymmetry | 181 |
| 6.6 | Instability due to aerodynamic forces | 185 |
| 6.7 | Effects of a pulsating torque | 187 |

## 7  Balancing — 189

| | | |
|---|---|---|
| 7.1 | Introduction | 189 |
| 7.2 | Rigid rotors whose initial imbalance is known | 189 |
| 7.3 | Rigid rotors with an unknown imbalance | 197 |
| 7.4 | Modal balancing of long flexible rotors with an unknown imbalance | 204 |
| 7.5 | Balancing long flexible rotors with an unknown initial imbalance using the method of influence coefficients | 209 |
| 7.6 | Balancing tolerances | 211 |

## 8  Measuring bearing impedances — 215

| | | |
|---|---|---|
| 8.1 | Introduction | 215 |
| 8.2 | Static force method | 217 |
| 8.3 | Use of an electromagnetic (or other) vibrator | 222 |
| 8.4 | Use of centrifugal forces | 229 |
| 8.5 | Transient methods – measurement on a running machine | 237 |

## 9  Measurements and diagnostics — 248

| | | |
|---|---|---|
| 9.1 | Introduction | 248 |

| | | |
|---|---|---|
| 9.2 | Signal measurement and display | 248 |
| 9.3 | Shaft imbalance | 252 |
| 9.4 | Misalignment, pre-loaded shaft | 253 |
| 9.5 | Rubs | 257 |
| 9.6 | Loose components | 260 |
| 9.7 | Shaft cracks | 261 |
| 9.8 | Rolling element bearing faults | 262 |
| 9.9 | Faults in gears | 266 |
| 9.10 | Protection against spurious signals | 267 |
| 9.11 | Monitoring strategies and standards | 271 |

Index 278

# Acknowledgements

We are grateful to the following individuals and organizations who have kindly given permission for the reproduction of copyright material (figure numbers in parentheses):

M. F. White and the *Journal of Applied Mechanics* (2.1, 2.4); FAG Kugelfischer and R. Oldenbourg Verlag, reproduced by permission from P. Eschmann, L. Hasbargen & K. Weigand, *Ball and Roller Bearings Theory, Design and Application*, © 1985 John Wiley & Sons Ltd (2.2, 2.3, 2.5, 2.6, 2.7, 2.8); Figures 2.11, 2.15, 2.18, 6.2, 9.9 reproduced from A. Cameron, *Basic Lubrication Theory* (1981), by permission of Ellis Horwood Ltd; J. S. Rao (2.14, 2.19, Table 7.1); McGraw-Hill Book Company (2.16); A. A. Raimondi and the American Society of Lubrication Engineers (2.17, 2.20, 2.21, 2.22, 2.23, 2.24); M. Chandra & R. Sinkason and Elsevier (2.25); American Society of Lubrication Engineers (2.29); E. P. Garguilo and the American Society of Mechanical Engineers (2.31); B. C. Majumdas and the American Society of Mechanical Engineers; N. S. Rao and the American Society of Mechanical Engineers (2.33, 2.34); E. J. Hahn and the American Society of Mechanical Engineers (2.36, 2.38); S. S. Kossa & R. A. Cookson, reproduced by permission from Cookson & Kossa, *International Journal of Mechanical Science* **21**, © 1979 Pergamon Press PLC; American Society of Mechanical Engineers (2.41, 2.43, 2.44); Society of Tribologists & Lubrication Engineers (2.42); Figure 3.6 reproduced from Smith, *Journal Bearings in Turbomachinery* (1969), by permission of Chapman & Hall; N. F. Rieger, reproduced from Rieger, *Lecture 7 – Geared Drive Systems* (notes presented at course 'Rotordynamics 2 – Problems in Turbomachinery', Udine 1985) by permission of C.I.S.M. (5.11); R. L. Eshleman and the American Society of Mechanical Engineers (6.12a); R. Holmes, reproduced from Holmes, *The measurement of oil film stiffness and damping* (1982), by permission of the Council of the Institution of Mechanical Engineers (8.4); R. Nordmann, reproduced from Nordmann, *Vibrations in Rotating Machinery* (1980), by permission of the Council of the Institution of Mechanical Engineers (8.11); P. G. Morton and the *GEC Journal of Science and Technology* (8.13); Bently Nevada (9.8a(i & ii), 9.8b); R. A. Collacott (9.8a(iii)); Hewlett-Packard Company (9.10c, 9.16, 9.17); Figure 9.14 reproduced from E. Downham, *Vibration in Rotating Machinery: Malfunction Diagnosis – Art & Science* (1976) by permission of the Council of the Institution of Mechanical Engineers; Figure 9.19 reproduced from E. Downham and R. Woods *The Rationale of Monitoring*

ACKNOWLEDGEMENTS

*Vibration on Rotating Machinery in Continuously Operating Process Plant (1971)*, by permission of the American Society of Mechanical Engineers; Verein Deutscher Ingenieure (9.20); Sound and Vibration Acoustical Publications Inc; IRD Mechanalysis (9.22).

# DYNAMICS OF ROTOR-BEARING SYSTEMS

# 1 Introduction

## 1.1 The importance of rotating machinery in the modern world

Rotating machinery, one of the most important classes of machinery, is used extensively throughout the industrialized world. Its uses are extremely diverse: in power stations, marine propulsion systems, aircraft engines, machine tools, automobiles, medical equipment, household accessories, and many other applications. Indeed, it is difficult to think of many types of machine that do not include rotating components in one form or another.

Rotating plant is often used in situations where its correct functioning is of crucial importance. Failure of the machine components in applications such as aeroengines, turbogenerators, military equipment, space satellites and others, may put human life in jeopardy and cost six-figure sums to repair. Small wonder that manufacturers, operators and governments are willing to devote a great deal of effort to ensure that plant is reliable and personnel are well trained.

It is because of the wide range of applications of rotating machinery, particularly in situations where its reliable operation is paramount, that rotor dynamics has evolved as a subject area worthy of study in its own right. Before the twentieth century, rotating shafts and bearings were not considered to interact with one another; some interest was shown in rotor balance, and bearings were important because they could reduce power loss, but they were rarely considered to interact in any system-like manner. In the early years of the twentieth century, however, machine design developed to a stage where plant was required to operate at high speed and, in many cases, above one or more critical speeds. The importance of shaft flexibility was then realized, and the influence of the bearings and foundations on the dynamic behaviour of the system better understood (although it remains a subject for further research).

Because of the part played by rotating machinery in the modern world, and the sophistication of the mathematical analyses required in the design of many machines, a study of rotating machines forms an important part of many undergraduate mechanical engineering courses, and is the central theme of many postgraduate studies. Furthermore, a detailed understanding is usually required by personnel employed in research, design, maintenance, and other areas of industry. The aim of this book is partly to review the general principles of operation of rotating machinery, thus satisfying the requirements of the undergraduate, but mainly to guide the graduate who continues to work with rotating machines in gaining a deeper understanding of the subject. The book should be particularly useful as a specialist reference text for personnel employed in industry.

## 1.2 Rotor-bearing interaction

A fundamental understanding of the lateral vibrations of rotating machines was first conceived in the early twentieth century, much of the pioneering work being attributed to Jeffcott (1919). Shortly afterwards, bearing flexibility began to be recognized by machine designers as being important, and it was observed that even in oil-film bearings there could be enough flexibility associated with the oil film to reduce the machine critical speeds significantly below those values calculated on the assumption of 'pinned' supports (Stodola 1925). Later, further investigations were to allow also for flexibility of the bearing pedestals and machine foundations themselves (Smith 1933).

It is well known by present-day engineers concerned with rotor dynamics that bearings play an important part not only in determining machine critical speeds but also in the imbalance response and stability characteristics. Indeed, there is a consensus view that rotating components and bearings should be treated together as a system, rather than as individual components which have little effect upon each other's operation. In fact, the relationship between the two is so close that most machine operators use vibration data collected at the bearings to infer information about the condition of the machine in general, and in many cases bearing–pedestal vibration.

It is within the above context that this book has been written, one of the aims being to clarify the dependence of both rotor and bearing dynamic characteristics upon each other. Chapter 2 is devoted to a study of different types of bearing, and to their operating characteristics. It explains the operation of some of the more common bearing types available for selection by the designer, and shows how each might be modelled in terms of spring and damper elements. The latter characteristic of bearings is so important in the modelling of the complete system that great efforts have been made by researchers to measure such spring and damping coefficient values; Chapter 8 reviews some of the measurement techniques that have been used. Bearings can also play an important part in determining a machine's stability, and Chapter 6 discusses some situations where the bearing characteristics are crucial in this context.

## 1.3 Causes and effects of vibration

Vibration in rotating machinery causes high levels of noise and wear, together with increased rotor-bending stresses which lead to a lower machine fatigue life. Machine vibrations are themselves an indicator of the wellbeing of the machine, and can provide an early warning of the development of fault conditions; in fact most rotating machine health-monitoring programmes include measurement of machine vibrations. The most common source of vibration is imbalance; all rotating machines have some residual imbalance because balancing procedures are imperfect, hence some vibration is always present during operation. For this reason, designers estimate the accuracy of their balancing procedures and so are able to determine the likely levels of machine vibration before construction. Chapters 3 and 4 discuss analytical and numerical methods used by designers when performing such calculations; Chapter 3 deals mainly with analytical methods that may be used with relatively simple machines where the shaft carries a single mass rotor; in many cases these calculations can be performed by hand. Such calculations may also be used as a starting point in the design of more complicated machines, prior to the application of more accurate numerical methods, the more common of which are discussed in Chapter 4. While Chapters 3 and 4 deal with the effects of imbalance, Chapter 7 explains how its magnitude can be reduced. It describes several approaches to machine balancing which may be used in different circumstances. In addition to lateral vibrations, machine torsional vibrations are also worthy of attention at the design stage in many machine types. If machines are operated at a speed coincident with a torsional natural frequency, then excessive torsional oscillations are likely to set in and could lead to failure of the machine. In Chapter 5, theoretical techniques used to calculate machine torsional natural frequencies are explained; once again these vary in complexity and involve hand calculations for very simple systems, but may require numerical methods for more complicated systems where several vibration modes need to be investigated.

In addition to imbalance, there are other reasons why rotating plant may vibrate excessively. Under some conditions it is possible for vibrations to be excited from within, where no external forcing on the system is apparent. Such machines are said to be unstable, and a number of circumstances under which unstable machine vibrations are likely to set in are well known. In many cases the mechanism causing such vibrations is not completely understood, but some of the mathematical approaches towards investigating the more common types of instability are described in Chapter 6. In other cases the vibration may not be associated with imbalance or instability, but may be related to some other fault conditions; Chapter 9 describes some of the other more well documented fault conditions and explains how they may be recognized and rectified.

## References

Jeffcott, H. H. 1919. The lateral vibration of loaded shafts in the neighbourhood of a whirling speed. The effect of want of balance. *Phil. Mag.* series 6 **57**, 304f.

Smith, D. M. 1933. The motion of a rotor carried by a flexible shaft in flexible bearings. *Proc. Roy. Soc.* A **142**, 92f.

Stodola, A. 1925. Kritische Wellenstörung infolge der Nachgiebigkeit des Ölpolslers im Lager (Critical shaft perturbations as a result of the elasticity of the oil cushion in the bearings) *Schweizerische Bauzeitung* **85**, no. 21, May.

# 2 Bearing systems

## 2.1 Introduction

All rotating machinery is supported by one or more bearings which play a vital part in determining the behaviour of the rotating system under the action of both static and dynamic loads. In order to design the system properly it is necessary that the designer should be aware of the possible types of bearing which could be used with the system under consideration, and of the performance characteristics associated with each bearing type. The aim of this chapter is to discuss the bearing types which are most frequently used and, where appropriate, to present the theory of operation for the bearing.

In this chapter several different theoretical approaches are demonstrated. That used for each bearing type is the one which is most appropriate or is most commonly used. However, the reader should not necessarily regard any of the methods described as being exclusive to any particular bearing type. The main thrust of the chapter is orientated towards the operation of bearings under the action of dynamic loads; in particular the stiffness and damping of different bearing types are discussed, since it is these characteristics which have a major influence on the overall system dynamics, as is explained later in Chapters 3, 4 and 6.

This chapter deals initially with rolling element bearings, focusing on bearing stiffness. The next section deals with oil-film hydrodynamic bearings and discusses static and dynamic operating characteristics; the 'short bearing' and 'long bearing' approximations are described, and the bearing stiffness and damping coefficients are introduced. The remainder of the chapter discusses the characteristics of gas bearings, squeeze-film bearings, and hydrostatic bearings. In each case a description of the mode of operation of the bearing is given together with an appropriate mathematical treatment.

## 2.2 Rolling element bearings

### 2.2.1 Introduction

Rolling element bearings are probably the most common type of bearing used in machinery. They are relatively compact, can transmit heavy loads of various forms, and can easily be installed and serviced. Rolling element bearings can be obtained in a number of different forms, each offering an improved performance under particular operating conditions. Table 2.1 summarizes the various bearing types available, together with their relative merits.

Bearings are selected on the basis of both the magnitude and direction of the loading which they are required to carry, and also on the basis of the

**Table 2.1** Types of rolling element bearing

| Bearing type | Suitable for load type | Other remarks |
| --- | --- | --- |
| Deep groove ball bearings | Radial and/or axial | Available with shields and seals. Available in double row form to accommodate higher radial loads |
| Angular contact bearings | Larger axial loads than deep groove ball bearing. Normally axial load at least as large as radial load | Available in 'matched pairs' to provide accurate axial positioning of shaft |
| Four-point (duplex bearing) | High load-carrying capacity. To be used when axial load greater than radial load. | Available in double row form |
| Self-aligning bearings | Lower radial and axial load-carrying capacity than deep groove ball bearing | Can accommodate large amounts of misalignment |
| Cylindrical roller bearings | Large radial loads and very light axial loads | Separate inner and outer rings. Available in 'matched assembly' form for tight control over internal clearance |
| Spherical roller bearings | Very high radial loads and light axial loads | Non-separable. Operates at lower speed than cylindrical roller. Can accommodate misalignment |
| Tapered roller bearings | Can support very large radial and axial loads | Lower operating speed than angular contact bearing. Provides a very rigid shaft mounting |
| Needle roller bearings | Can support large radial loads at speeds similar to those of cylindrical roller bearings | Very small bearing outer diameter. Can be used without one or both rings to save space provided the seatings are surface treated. |
| Thrust ball bearings | Axial loads only (can be in either direction) | Axial load must *always* be present. Can function with a small amount of misalignment when used with a spherical seating washer. Available in cylindrical and needle roller form for very large loads |

speed at which they are required to operate. Instructions for selecting bearings are contained in most manufacturer's catalogues, and it is beyond the scope of this book to repeat such information. Instead, this section deals with the subject of bearing elastic deformation and the evaluation of bearing stiffness. The bearing stiffness plays a particularly important role in the determination of the overall rotor-bearing system dynamic characteristics, but values of bearing stiffness are not usually documented in manufacturer's selection catalogues. Damping in rolling element bearings is very small, but may be enhanced by the use of squeeze-films; this is discussed later in section 2.5.

## 2.2.2 Bearing elastic deformation

Before the stiffness of rolling element bearings can be calculated, consideration must first be given to the amount by which a loaded shaft, supported by the bearing, can be displaced from its concentric position by a given load. This displacement depends upon the elastic deformation of the bearing races and of the rolling elements themselves; the elastic deformation of these components depends upon the size of the bearing and on the bearing internal clearance.

The bearing internal clearance is an important factor because it determines the size of the stressed area of the races. In Figure 2.1, for a bearing with a significant positive internal clearance $c_r$, there is a relatively small area of interference between the rolling elements and the outer race when a given load is applied to the bearing. For zero internal clearance the size of this contact area is increased and occupies 180° of included angle of the bearing; for negative internal clearance, i.e. pre-load, this contact area is greater still. For a given load, the size of the contact area determines the magnitude of the stresses in the bearing components, and so determines the amount of elastic deformation.

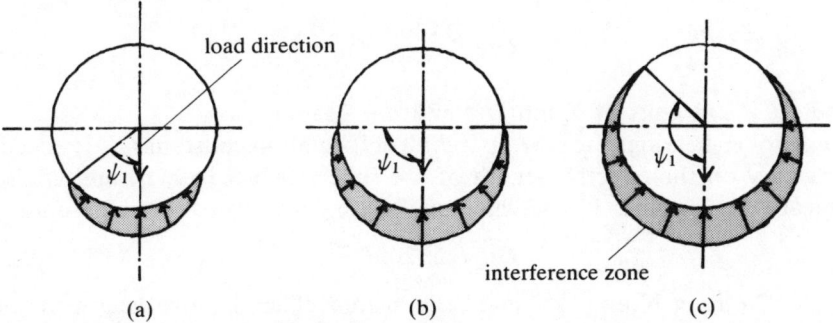

**Figure 2.1** Load zones in rolling element bearings. The shaded area indicates the load zone where there is interference between rolling elements and raceways. The load direction is vertical, (a) $c_r > 0$, $\psi_1 < 90°$; (b) $c_r = 0$, $\psi_1 = 90°$; (c) $c_r < 0$, $\psi_1 > 90°$.

The deformation of the whole bearing under a given load may be evaluated after first considering that at the location of a single rolling element which has a known load applied to it. In the case of a ball-bearing the contact area between the rolling element and the race is zero when no load is applied; 'point contact' is said to occur. For two bodies with point contact, made of the same material and subjected to a compressive load $F$, Hertzian contact theory may be developed (Eschmann et al. 1985) to show that the total deformation $\delta$ is given by

$$\frac{\delta}{d} = c_\delta \left(\frac{F}{d^2}\right)^{2/3} \qquad (2.1)$$

where $d$ is the diameter of the rolling element, and $c_\delta$ is a deformation constant which depends upon the material properties and the geometry of the surfaces. For steel ball, angular contact and thrust ball bearings the Hertzian theory yields values of $c_\delta$ as shown in Figure 2.2.

In the case of rolling element bearings elastic deformation can take place at both contact between inner race and rolling element, and also at contact between outer race and rolling element. The total deformation at the location of one rolling element is therefore given by

$$\frac{\delta}{d} = (c_\delta(\text{inner}) + c_\delta(\text{outer})) \left(\frac{F}{d^2}\right)^{2/3} \qquad (2.2)$$

which may be rewritten as

$$F = C_\delta \, \delta^{3/2} \qquad (2.3)$$

where $C_\delta = d^{1/2}/(c_\delta(\text{inner}) + c_\delta(\text{outer}))^{3/2}$. From Figure 2.2 it can be seen that the value of $C_\delta$ will depend mainly on the bearing's curvature ratio $\varkappa$. If its dependence on the bearing size and contact angle is neglected then values of $C_\delta$ may be obtained directly from Figure 2.3 or, when the curvature ratios at inner and outer races are very similar and less than 0.1, which is usually the case with deep groove ball-bearings, from the approximation

$$C_\delta = \frac{34\,300}{\varkappa^{0.35}} d^{1/2} \qquad (2.4)$$

where $C_\delta$ is in units of $N.mm^{-3/2}$ and $d$ is in mm.

For roller-bearings the corresponding deformation constant $C_{\delta L}$ is dependent only on the effective length of the rollers themselves, $l_e$, and is also indicated in Figure 2.3. The value of $C_{\delta L}$ may also be calculated from

$$C_{\delta L} = 26\,200 \, l_e^{0.92} \qquad (2.5)$$

where $C_{\delta L}$ is in $N.mm^{-1.08}$ and $l_e$ is in mm. The effective length of the rollers is that which is actually in contact with the raceway, usually this is the actual roller length minus the roller corner radii. The relationship between the deformation constant, the applied compressive load and the elastic deformation at a single roller is given by Eschmann et al. (1985) as

$$F = C_{\delta L}\, \delta^{1.08} \quad (2.6)$$

This is different to that indicated by Equation (2.3) because, of course, in roller-bearings there is 'line contact' between the rolling elements and the bearing races, not 'point contact'.

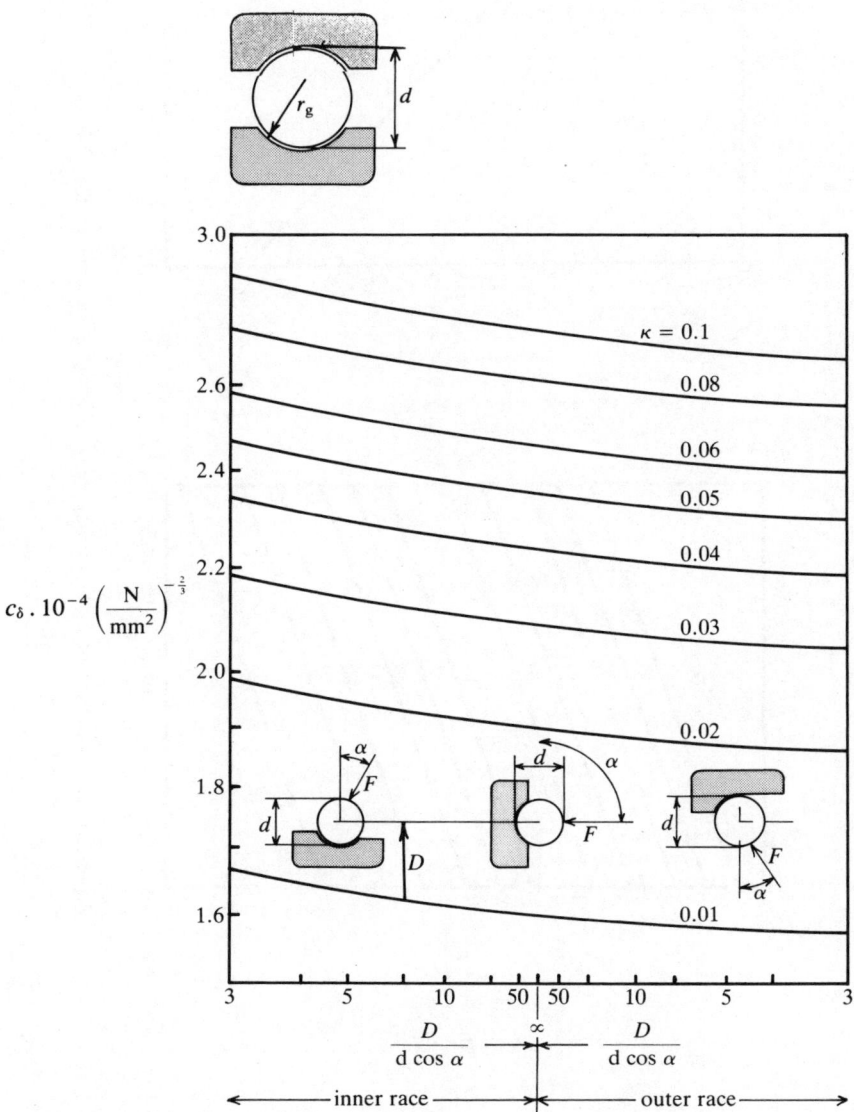

**Figure 2.2** Variation of $C_\delta$ with bearing geometry. The curvature ratio, $\varkappa = (2r_g - d)/d$.

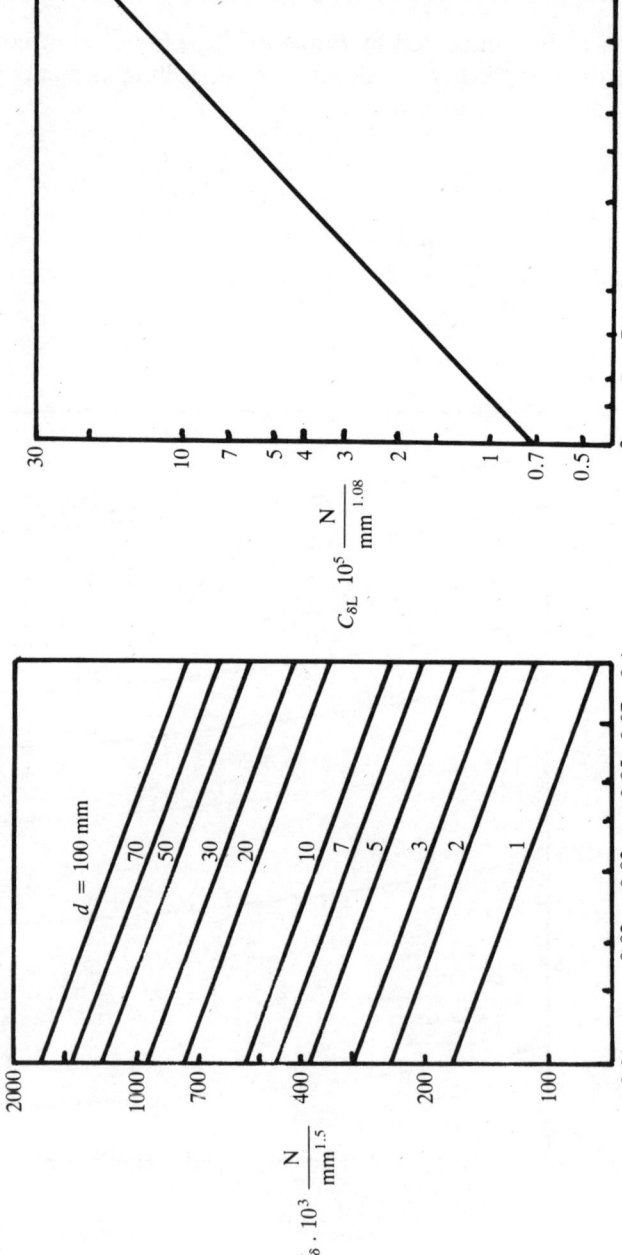

**Figure 2.3** left- Variation of $C_\delta$ with curvature ratio $\varkappa$; right- Variation of $C_{\delta L}$ with roller length $l_e$.

The above relationships indicate the elastic deformation at one rolling element under the action of a single compressive force applied at the element. In real bearings where more than one element is in compression, the effects at each element can be allowed for as follows. In Figure 2.4 the bearing inner race is displaced from the concentric position by a distance $x_m$, part of which consists of the radial internal clearance $c_r$. The elastic deformation of the bearing in the direction of the applied radial load is then given by

$$x = x_m - c_r \tag{2.7}$$

and the elastic deformation at any other rolling element position is given by

$$x = x_m \cos \psi - c_r \tag{2.8}$$

Setting the deformation to zero in Equation (2.8) indicates that only rolling elements at angular orientations within $\pm \psi_1$ to the load direction carry any significant loading, and that the load zone is defined by

$$\psi_1 = \cos^{-1} \frac{c_r}{x_m} \tag{2.9}$$

For a given bearing, with known angular positions of the rolling elements, a displacement of the inner race $x_m$ may be assumed and thus the resulting elastic deformation at each rolling element calculated from Equation (2.8). The contact forces at each element may then be evaluated from Equation (2.3) and added as vectors to give the net radial force $F_r$ applied to the bearing in order to produce the assumed displacement $x_m$.

The above process may be shortened somewhat if the fraction of the net radial load applied that is transmitted through the rolling element directly in line with the applied load is known. If this force $F_m$ can be established then the resulting inner race displacement $x_m$ may be calculated directly, using

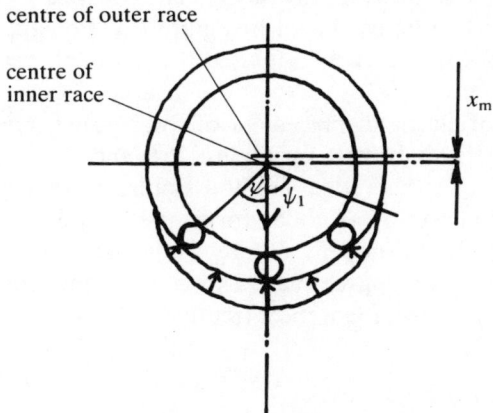

**Figure 2.4** Displacement $x_m$ of bearing inner race under load.

Equation (2.3) for ball bearings, as

$$x_m = \left(\frac{F_m}{C_\delta}\right)^{2/3} + c_r \qquad (2.10)$$

where the term $c_r$ has been added to allow for the take-up of the internal clearance. The corresponding expression for roller-bearings, using Equation (2.6) is

$$x = \left(\frac{F_m}{C_{\delta L}}\right)^{1/1.08} + c_r \qquad (2.11)$$

It is noteworthy that expressions similar to Equations (2.10) and (2.11) were developed by Palmgren (1959), who neglected the effects of bearing clearance and geometry, as

$$x_m = 4.36 \times 10^{-8} \left(\frac{F_m^2}{d}\right)^{1/3} \qquad (2.12)$$

for ball-bearings, and

$$x_m = 3.06 \times 10^{-10} \frac{F_m^{0.9}}{l_e^{0.8}} \qquad (2.13)$$

for roller-bearings, where $F_m$ is in N and $d$ and $l_e$ are in m. Before Equation (2.10) can be applied, however, the maximum compressive force, $F_m$, on a single element must first be determined. The relationship between $F_m$ and the net radial force applied, $F_r$, is usually written in the form

$$F_m = \frac{k}{n} F_r \qquad (2.14)$$

where $n$ is the number of rolling elements and $k$ is a number which depends upon the applied load, the number of rolling elements, the deformation constant $C_\delta$ and the bearing clearance. The variation of $k$ with these quantities is found to be as shown in Figure 2.5. For approximate calculations a value of $k = 5$ is frequently used for both ball bearings and roller bearings.

The variation of elastic displacement of the bearing inner race, $x_m - c_r$, with applied load is indicated by Figures 2.6 and 2.7. Investigations by White (1979), using a theoretical method similar to that described above, have also yielded curves of a similar form; the theoretical force–displacement relationships published by White were also very similar to those inferred by Palmgren's approximate expressions and, for ball bearings, similar to those given by a theoretical treatment described by Harris (1966).

### 2.2.3 Bearing stiffness

Once the relationship between the applied radial force and the bearing

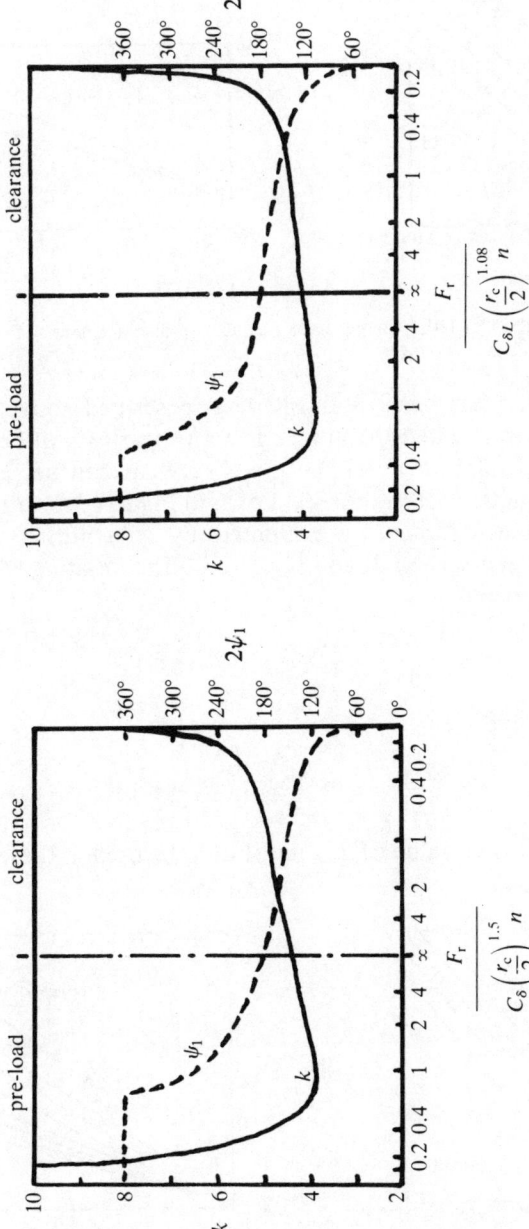

**Figure 2.5** Variation of $k$ and $\psi_1$ with bearing parameters, left- for ball bearings; right- for roller bearings.

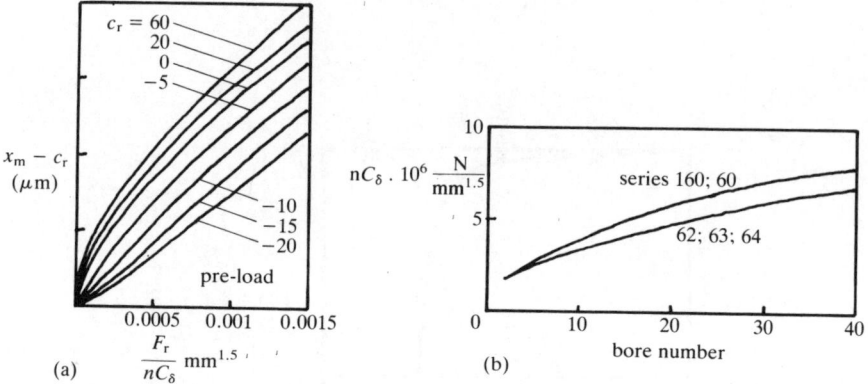

**Figure 2.6** Variation of ball bearing inner race elastic displacement with applied load.

elastic deformation has been established, as described above, the bearing stiffness can be calculated. Rolling element bearing stiffness is highly non-linear with displacement of the inner race, as can be seen from the variation in the gradient of the curves shown in Figures 2.6 and 2.7. Strictly, the bearing stiffness $K$ can be evaluated by substituting $F = F_m$ and $\delta = x_m - c_r$ into Equations (2.3) and (2.6) and differentiating with respect to displacement to give

$$K = \frac{dF_r}{dx_m} = 1.5 \frac{n}{k} C_\delta (x_m - c_r)^{0.5} \quad (2.15)$$

for ball bearings and

$$K = \frac{dF_r}{dx_m} = 1.08 \frac{n}{k} C_{\delta L} (x_m - c_r)^{0.08} \quad (2.16)$$

for roller bearings. The value of $x_m$ that should be used in these expressions

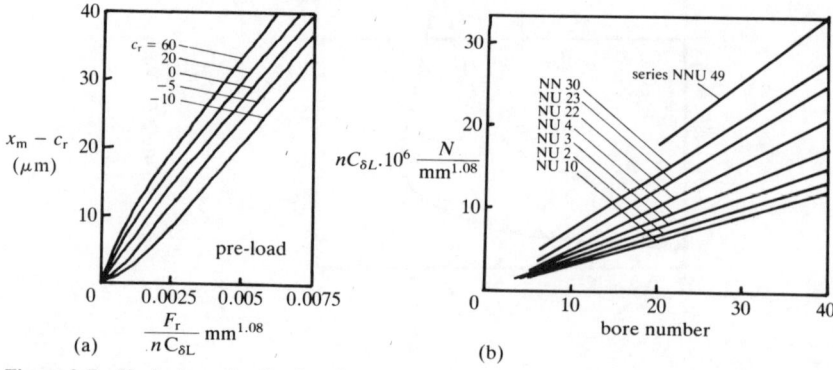

**Figure 2.7** Variation of roller bearing inner race elastic displacement with applied load.

should be that which is caused to occur as a consequence of the steady load which is applied to the bearing.

If large amplitudes of vibration are expected to occur at the bearings, such that the bearing stiffness non-linearity becomes a significant factor in machine response calculations, then an effective stiffness can be defined as the ratio force amplitude/displacement amplitude. When the rotating load is greater than the steady load the bearing clearance can be allowed for by subtracting the radial clearance from the displacement amplitude, since only the bearing elastic displacement must be used in stiffness calculations. The bearing effective stiffness is then given by

$$K = \frac{F_{\text{rotating}}}{x_m - c_r} \qquad (2.17)$$

Stiffness coefficients evaluated in this way have been shown to agree well with experimental measurements (White 1979).

The discussion above has made no reference to the effects of movement of the rolling elements around the raceways on the bearing force–displacement relationship and stiffness. In fact, although the number and location of the rolling elements in the load zone changes as the bearing is rotated, the displacement and stiffness of the bearing are little affected, and so rotation of the bearing need not be allowed for in stiffness calculations.

The comparative stiffnesses of different radial bearing types are shown in Figure 2.8. More detailed information concerning bearing types other than ball and roller can be found in Eschmann *et al.* (1985).

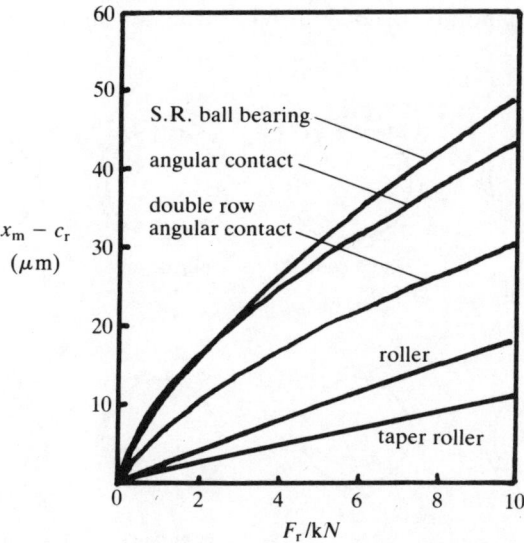

**Figure 2.8** Comparative stiffness of different types of rolling element bearing.

## 2.3 Hydrodynamic oil-lubricated journal bearings

### 2.3.1 Description of operation

Journal bearings consist of a circular section length of shaft (the journal) rotating inside a bearing bush which is nominally circular. The journal diameter is usually 99.8–99.9% of that of the bush, and the clearance space between the two is partially filled by the lubricating fluid. At zero rotational speed, under a steady load, the journal rests at the bottom of the clearance space; when it rotates, however, oil is dragged along by the journal due to the oil viscosity and a thin film of oil is built up between the journal and bush so that under normal operating conditions there is no surface-to-surface contact. When designing for steady loads the problem is to ensure that there is sufficient film clearance to take up the effects of surface undulations and likely changes in load during the machine's life whilst ensuring that the bearing is not over-designed and is not prone to self-excited vibrations (see Chapter 6). With these factors in mind, most journal bearings operate at steady-state eccentricity ratios (journal eccentricity/radial clearance) of about 0.6–0.7.

Journal bearings can have a significant effect on a machine's vibration characteristics. The oil film behaves like a complicated arrangement of springs and dampers and so influences the machine critical speeds and imbalance response. Moreover, bearing fluid film forces can cause rotor instability which results in serious levels of self-excited vibration (e.g. oil-film whirl).

The most common forms of journal bearing are shown in Figure 2.9. The

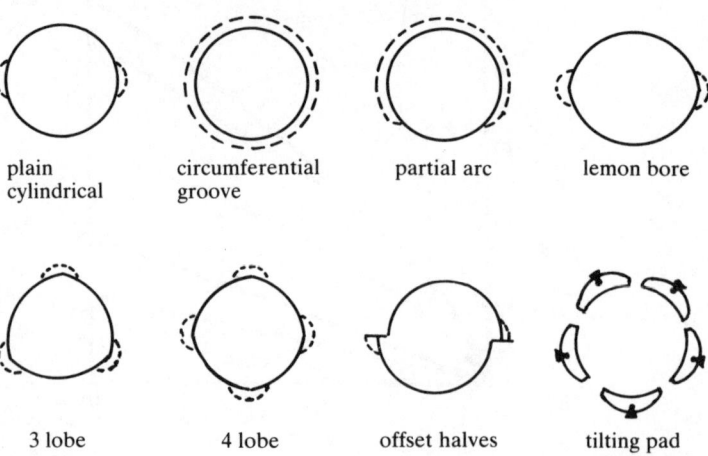

Figure 2.9 Some frequently used types of journal bearing geometry.

simplest to manufacture, and probably the most common, is the plain cylindrical bearing. The oil inlet port (and outlet when present) is frequently on the bearing horizontal centre line and may be either circular ('hole entry') or in the form of an axial groove type of recessed pocket to aid oil distribution. The plain bearing is able to carry very heavy loads except in the direction towards the oil feed pocket. An alternative form of the plain cylindrical bearing is the circumferential groove bearing in which the surface is effectively divided into two half-bearings by a circumferential oil distribution groove. This type of bearing is frequently used when the direction of the load may change significantly because there are no oil supply pockets to restrict the sense in which the load is applied. However, the magnitude of the load which can be applied is less than half of that of the plain bearing. Another variation of the plain bearing is the partial arc bearing where all but the lower part of the bearing bush bore is relieved. Typically, 60–120° of included arc of surface are left to carry the load whose direction is virtually constant. The advantage obtained by relieving most of the bush bore is to reduce friction losses in the bearing whilst retaining most of the bearing load-carrying capacity.

The dependence of the machine vibration characteristics on the bearing has lead to the development of several non-circular-bore hydrodynamic bearings. The lemon-bore bearing is effectively two partial arc bearings mounted in opposition to each other, and is manufactured by removing material from the interface between the upper and lower halves of what was originally a circular bearing. The resulting bearing has a load-carrying capacity which is comparable with that of the partial arc and which is more resistant to oil-film whirl. The stiffness and damping in the horizontal direction are relatively low however. The three-lobe and four-lobe bearings are further variations which provide even greater resistance to oil-film whirl. They have relatively large oil-film stiffness and damping values which are dependent on the orientation of the lobes, but their load-carrying capacity is only about half of that for the partial arc bearing. Another form of cylindrical bearing is that used with offset halves; this type is able to support loads which are slightly less than those carried by the partial arc bearing and has high oil-film stiffness and damping. It also has a very high resistance to oil-film whirl but is suitable for machines which rotate in one direction only. However, of all the journal bearing types available, only the tilting pad bearing is 100% resistant to oil-film whirl. This bearing type contains a number of bearing arcs, each of which is independently mounted on a fulcrum whose axis of rotation runs parallel with that of the journal. The pads are thus able to find their own equilibrium alignment position during bearing operation. Unfortunately, this bearing type is also one of the most costly to manufacture and has a load-carrying capacity which is only about half of that of the plain bearing. For this reason tilting pad bearings are generally used only when oil-film whirl would otherwise cause a problem.

## 2.3.2 Reynolds' equation, and approximate solutions

The characteristics of hydrodynamic bearings can be derived theoretically by correctly modelling the bearing fluid film. The equation which does this was derived by Reynolds (1886) and is:

$$\frac{\partial}{\partial x}\left[\frac{\rho h^3}{\mu}\frac{\partial p}{\partial x}\right] + \frac{\partial}{\partial y}\left[\frac{\rho h^3}{\mu}\frac{\partial p}{\partial y}\right] = 6\left[U\frac{\partial}{\partial x}(\rho h) + 2\frac{\partial}{\partial t}(\rho h)\right] \quad (2.18)$$

where $x$ locates the distance around the bearing circumference of the point under consideration (measured from some arbitrary datum) and $y$ is the position in the axial direction; $\rho$ is the lubricant density and $\mu$ its dynamic viscosity, $h$ is the film clearance, $p$ the lubricant pressure, $U$ the tangential velocity of the journal surface, and $t$ time. The derivation of Equation (2.18) is described in several sources, for example Cameron (1981). For the purposes of mathematical modelling, the lubricant film may be considered as though it were unwrapped from around the journal as indicated in Figure 2.10; the justification for doing this is that the film clearance is very small compared with the journal diameter and so the effects of curvature can be ignored for the purposes of evaluation of the lubricant pressure variation.

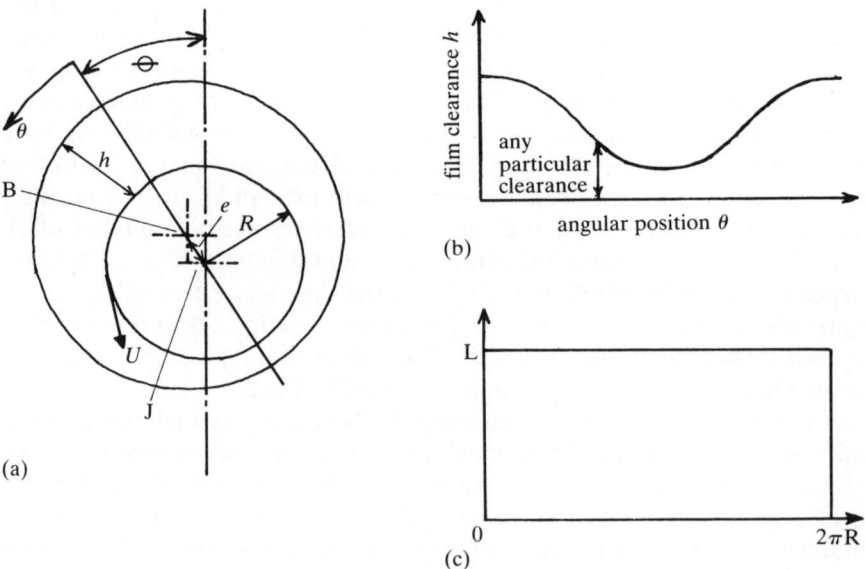

**Figure 2.10** Diagram of a journal bearing, and the variation of clearance around the circumference. (a) Axial view of bearing. B is the bearing centre, and J the journal centre. R is the journal radius, and $\theta$ the angular position around the bearing measured from JB produced, as shown and ⊖ is the attitude angle; (b) Variation of film clearance, h, with angular position $\theta$; (c) Plan view of oil-film unwrapped from around the journal. The vertical axis is the axial sense, the horizontal axis the circumferential sense, bearing length is L.

## JOURNAL BEARINGS

In most applications the lubricant density does not vary substantially throughout the oil film; even though its viscosity might vary, a constant effective viscosity based on the lubricant thermal balance (ESDU 1966) may be used for calculations. (This is not the case for gas bearings; see section 2.4.) For such cases the terms $\rho$ and $\mu$ may be taken outside the differential signs in Equations (2.18). Furthermore, for steady-state operation only, the film clearance at any one point does not change with time and so $\partial h/\partial t$ is equal to zero. Making these simplifications enables Equation (2.18) to be rewritten as

$$\frac{\partial}{\partial x}\left[h^3 \frac{\partial p}{\partial x}\right] + \frac{\partial}{\partial y}\left[h^3 \frac{\partial p}{\partial y}\right] = 6\mu U \frac{\partial h}{\partial x} \qquad (2.19)$$

Equation (2.19) describes the variation of lubricant pressure in both the axial and circumferential directions. An approximate solution for $p$ may be obtained by using either the Ocvirk (1952) 'short bearing' approximation (where pressure variation in the circumferential direction is assumed to be negligible compared with that in the axial direction, and so $\partial p/\partial x$ is set to zero), or the 'long bearing' approximation (where the converse applies and $\partial p/\partial y$ is set to zero). Both such approximations enable closed-form solutions of Equation (2.19) to be obtained, provided that appropriate boundary conditions are selected to enable evaluation of the constants of integration (see Figure 2.11). These approximate solutions are described in detail by Cameron (1981) and Pinkus and Sternlicht (1961). An application of the short bearing approximation to squeeze-film bearing analysis is also discussed in Section 2.5.

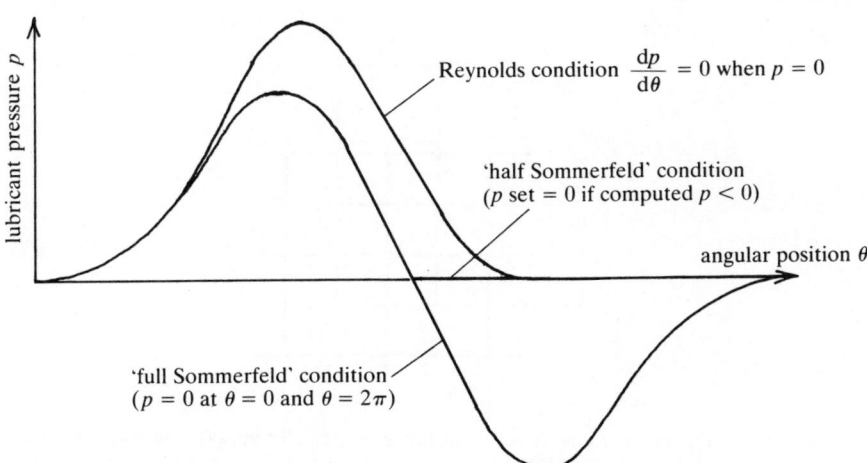

**Figure 2.11** Boundary conditions used in journal bearing analysis.

## 2.3.3 Finite bearings

Real bearings are neither infinitely long nor infinitely short; most have a length/diameter ratio in the range 0.5–1.5, so although the approximate solutions described above may be used in preliminary design calculations, final design studies will often require a more realistic analysis. For a more general solution of the Reynolds equation it is necessary to revert to a numerical method. By far the most common numerical technique used in hydrodynamic bearing analysis is the finite difference method, because of its relative simplicity and because of the ease with which it can be adopted to suit most bearing geometries. An outline of this method is given below.

In the first instance the problem should be reduced to a two-dimensional one by considering the oil film to be unwrapped from around the journal. The oil film is then divided into a number of sections of finite size by describing it in terms of mesh nodes, as indicated in Figure 2.12. In between each of the nodes described in Figure 2.12(a) a number of further points are

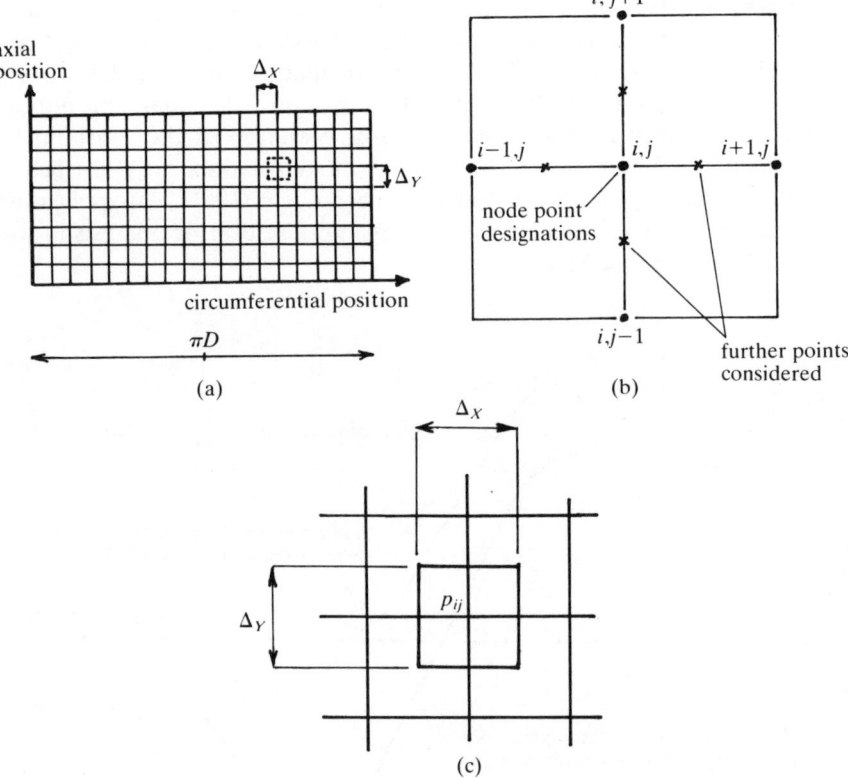

**Figure 2.12** Finite difference mesh for modelling a bearing oil-film, (a) finite difference mesh superimposed on plan view of oil-film; (b) any node of coordinates $i$, $j$ and surrounding nodes of significance; (c) the shading indicates the area over which pressure $p_{ij}$ is considered to act.

considered, for example those surrounding the node $i, j$ are shown in Figure 2.12(b). Since the Reynolds equation describes the behaviour of the lubricant at any location in the fluid film, it can be written for every specific node contained within the finite-difference mesh. For example, Equation (2.19) may be written for the node $i, j$ as

$$h_{i,j+1/2}^3 \frac{p_{i,j+1} - p_{i,j}}{\Delta X} - h_{i,j-1/2}^3 \frac{p_{i,j} - p_{i,j-1}}{\Delta X} + h_{i+1/2,j}^3 \frac{p_{i+1,i} - p_{i,j}}{\Delta Y}$$

$$- h_{i-1/2,j}^3 \frac{p_{ij} - p_{i-1,j}}{\Delta Y} = 6\mu U \frac{h_{ij+1/2} - h_{i,j-1/2}}{\Delta X} \quad (2.20)$$

where the derivatives have been written in finite-difference form. Equation (2.20) may then be rewritten in the form

$$p_{i,j} = K_1 + K_2 p_{i+1,j} + K_3 p_{i-1,j} + K_4 p_{i,j+1} + K_5 p_{i,j-1} \quad (2.21)$$

where $K_1$, $K_2$, etc. are constants whose values are known for every node.

Equation (2.21) may then be written for every node in the mesh, except for those where the lubricant pressure is already known (for example at the nodes representing the ends of the bearing where the lubricant pressure must be equal to the ambient pressure). These equations can then be solved simultaneously to provide the values of lubricant pressure at every mesh node.

Alternatively, an iterative method of solution may be employed such as successive relaxation where the value of the pressure at each node is determined successively according to the most up-to-date values of the terms on the right-hand side of Equation (2.21) (initially, all pressures might be set to zero). This evaluation of pressures is repeated for all nodes several times over until the change in the value of lubricant pressure at any node is no greater than a small fraction of 1% of the lubricant pressure.

At this stage some mention of boundary conditions is merited (see also Figure 2.11). At first consideration it may appear to be difficult to impose the correct Reynolds conditions (also known as the Swift–Stieber conditions, after Swift 1932 and Stieber 1933) because of the uncertainty of where the lubricant pressure becomes zero; it appears that only the half-Sommerfeld conditions (also known as the Gumbel conditions, 1921) can be imposed by setting any negative pressures to zero, after a solution has been found. However, if a second iteration procedure, or solution of simultaneous equations, is carried out with these pressures set to zero then different nodes will be found to have sub-zero pressures. This process may be repeated until there is no change in the lubricant film edge position. In the case of iterative procedures the process can be speeded up by assigning any sub-zero pressures to be zero as and when they are evaluated. When this process is completed it is found that, because the Reynolds equation is a continuous function, the final pressure distribution corresponds to the Reynolds boundary conditions with the constraint of $\partial p/\partial \theta = 0$ at the trailing edge of the lubricant film automatically catered for.

Once the lubricant pressures have been evaluated for a particular journal attitude angle and eccentricity, the corresponding forces provided by the lubricant on the journal may be evaluated by integrating numerically all of the elemental forces associated with each node. For example, each node pressure is considered to contribute to force on the journal over an area $\Delta x \times \Delta y$ surrounding it, when integrating using a trapezium rule (see also Figure 2.12(c)). Thus, if the node is situated at some angle $\psi$ to the horizontal, as shown in Figure 2.13, then the lubricant force on the journal in the horizontal direction (left to right) is

$$F_h = \sum_{j=1}^{m} \sum_{i=1}^{n} p_{ij} \cos \psi_{ij} \, \Delta x \, \Delta y \qquad (2.22)$$

and that in the vertical direction (upwards) is

$$F_v = \sum_{j=1}^{m} \sum_{i=1}^{n} p_{ij} \sin \psi_{ij} \, \Delta x \, \Delta y \qquad (2.23)$$

where there are $n$ axial node positions and $m$ circumferential node positions.

Strictly, the above expressions for the lubricant forces acting on the journal should also include allowance for the tangential friction forces, whose magnitude may be evaluated as shown below; in most cases, however, the friction forces are small when compared with the pressure forces. For bearings designed to carry vertical loads only (for example gravity loads) the relationship between eccentricity ratio $\varepsilon$ and journal attitude angle $\phi$ may be determined by investigating different values of $\phi$ for a given value of $\varepsilon$ until the value of $F_h$ is found to be zero; this 'trial and error' method enables corresponding values of $\varepsilon$, $\phi$, and Sommerfeld number $S$ to be found. The steady-state journal loci for plain cylindrical bearings and their relationship

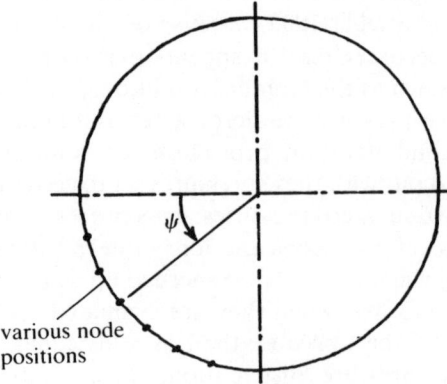

**Figure 2.13** Some node positions around the bearing circumference.

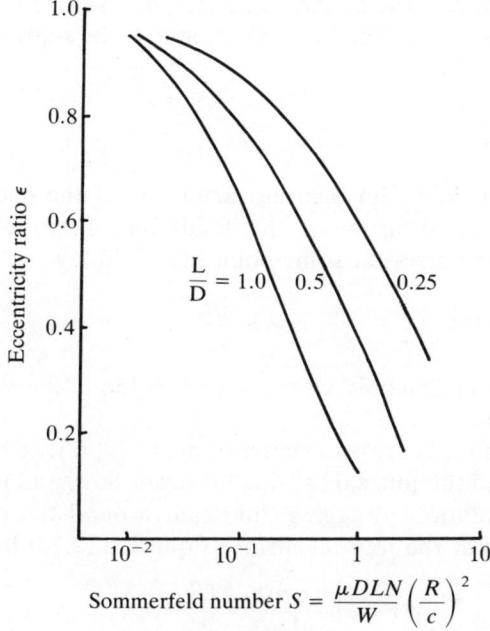

**Figure 2.14** Variation of bearing eccentricity ratio with Sommerfeld number.

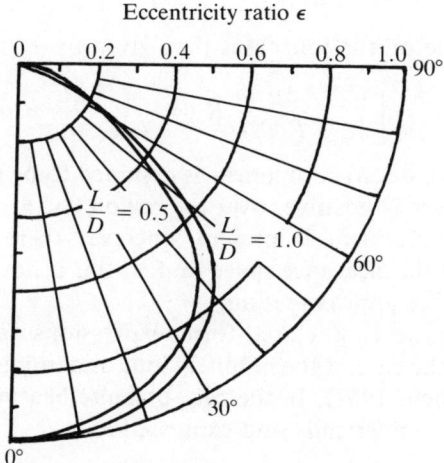

**Figure 2.15** Variation of bearing eccentricity ratio with journal attitude angle for plain cylindrical bearings of length/diameter ratio 0.5 and 1.0.

with the Sommerfeld number is shown in Figures 2.14 and 2.15, where $L$ is the bearing length, $D$ is the bearing diameter, $W$ the load on the bearing, $R$ the bearing radius, $c$ the radial clearance, and $N$ the shaft rotational speed in rev. sec.$^{-1}$

## 2.3.4. Friction

Friction forces in fluid film bearings arise out of the energy dissipation associated with the shearing of the fluid film. From elementary fluid mechanics the shear stress at some point in a fluid layer is given by

$$\tau = \mu \frac{du}{dz} \qquad (2.24)$$

where $\mu$ is the fluid dynamic viscosity and $du/dz$ is the velocity gradient across the fluid film.

In a hydrodynamic bearing the latter term arises partly from the variation in pressure around the journal causing lubricant flow, and partly as a result of the journal rotation 'dragging' lubricant around the clearance space. Thus, for a point at the journal surface Equation (2.24) becomes

$$\tau = \mu \left[ \frac{1}{\mu} \frac{dp}{dx} \frac{h}{2} + \frac{U}{h} \right] \qquad (2.25)$$

$$\underbrace{\phantom{xxxxxxx}}_{\text{pressure-induced term}} \quad \underbrace{\phantom{xxxxxxx}}_{\text{velocity-induced term}}$$

The total drag force on the journal is then given by

$$F_d = \int_{-L/2}^{+L/2} \left[ \int_0^{\pi+d} \frac{dp}{R d\theta} \frac{h}{2} R \, d\theta + \int_0^{2\pi} \frac{\mu U}{h} R \, d\theta \right] dy \qquad (2.26)$$

where the pressure-induced component is assumed to be present only when the lubricant pressure is positive, over the region $0 < \theta < \pi + d$, whilst the velocity-induced component is present wherever there is lubricant, for example all around the clearance space (in fact this is not strictly true but it provides a reasonable approximation).

It is possible to develop closed-form expressions for the integral in Equation (2.26) in the case of the infinitely long and infinitely short bearings (Pinkus and Sternlicht 1961). In the case of finite bearings the integration may be carried out numerically and expressed as

$$F_d = \sum_{i=1}^{n} \sum_{j=1}^{l} \frac{h_{ij}}{2} \left( \frac{dp}{dx} \right)_{ij} \Delta x \, \Delta y + \sum_{i=1}^{n} \sum_{j=1}^{l} \frac{\mu R U}{h_{ij}} \Delta x \, \Delta y$$

$$+ \sum_{i=1}^{n} \sum_{j=l}^{m} \frac{KRU}{h_{ij}} \Delta x \, \Delta y \qquad (2.27)$$

where $l$ is the node column corresponding to the edge of the pressurized section of lubricant film. Equation (2.27) is a slight variation on Equation (2.26) in that the last term of (2.27) now allows for the break up of the lubricant film into streamers in the cavitated portion of the oil film (see Figure 2.16) where $K$ is a constant whose value is less than unity. This feature of real lubricant films is not allowed for in Equation (2.26). The magnitude of $K$ may be determined empirically, and is the concern of current research. The variation of bearing friction with Sommerfeld number is shown for plain cylindrical bearings in Figure 2.17.

### 2.3.5 Lubricant flow rate

In the case of both the infinitely short and infinitely long bearing there is a lubricant pressure variation in the axial direction for $0 < \theta < \pi$ causing a flow of lubricant out of the ends of the bearing. This flow rate, allowing for both ends of the bearing, is given by basic fluid mechanics theory as

$$Q = 2 \int_0^\pi \frac{h^3}{12\mu} \frac{dp}{dy} R \, d\theta \qquad (2.28)$$

In the case of the finite bearing the integration in Equation (2.28) is usually carried out numerically; using a trapezium rule for integration, Equation (2.28) may then be expressed as

$$Q = 2 \sum_{j=1}^{m} \frac{h_{ij}^3}{12\mu} \left(\frac{dp}{dy}\right)_{ij} \Delta x \qquad (2.29)$$

for the finite-difference mesh in Figure 2.12.

The variation of lubricant flow with Sommerfeld number for a plain cylindrical bearing is shown in Figure 2.18. The effect of various design parameters on lubricant flow and other performance characteristics of journal bearings may also be found in ESDU (1966).

### 2.3.6 Dynamic characteristics

Hydrodynamic bearing lubricant films are themselves flexible, that is to say that when a dynamic load is applied to the bearing, as a consequence of imbalance for example, the journal is caused to orbit about the static equilibrium position which it would otherwise take up when carrying only a static load. The effective stiffness and damping associated with this flexibility of the lubricating film have a most significant effect on system critical speeds, on forced response and on system stability. For these reasons it is important to be able to evaluate the bearing dynamic characteristics at the design stage.

The approach which is usually adopted is to consider there to be some small displacements $dx$ and $dy$ of the journal away from its static equilibrium position, in the horizontal and vertical directions respectively.

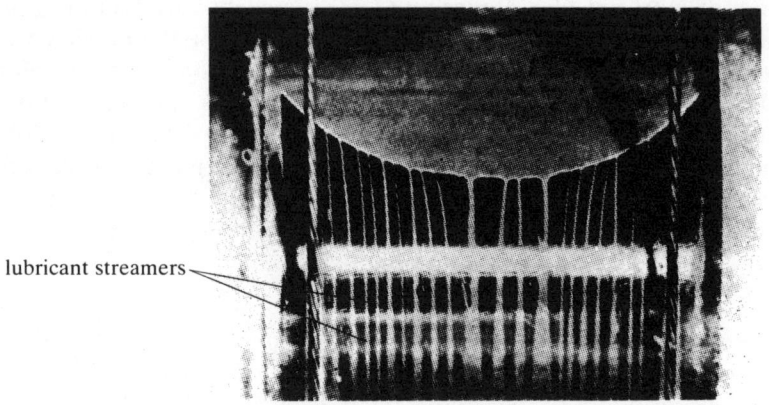

**Figure 2.16** Break-up of lubricant film into streamers in the cavitated region.

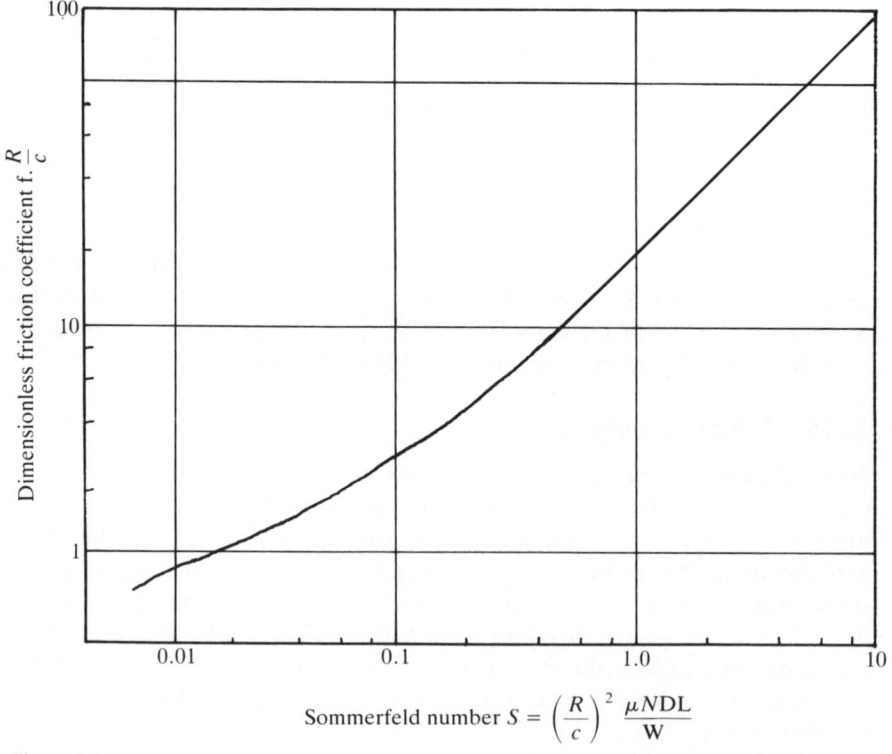

**Figure 2.17** Variation of friction coefficient with Sommerfeld number for a 360° oil-film and bearing length/diameter ratio 1.0.

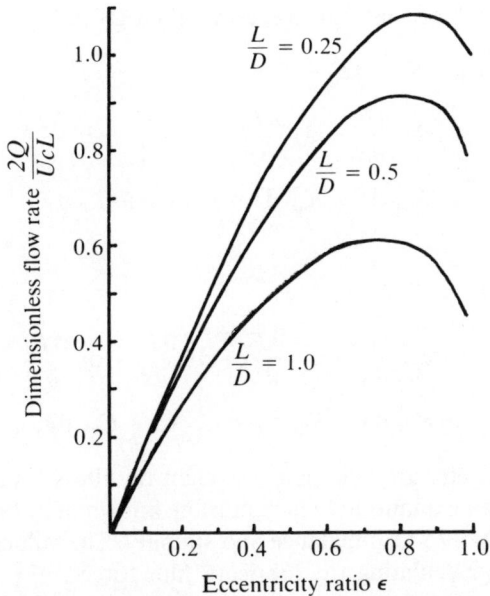

**Figure 2.18** Variation of bearing lubricant flow rate $Q$ with eccentricity ratio for a plain cylindrical bearing. The curves are for a 360° oil-film with Reynolds boundary conditions.

The journal centre is also assumed to possess velocities $d\dot{x}$ and $d\dot{y}$ in the horizontal and vertical directions respectively, whereas under the action of only a steady load its velocity would be zero. The effects of these changes in the state of the journal on the lubricant film forces are then evaluated. When there is no dynamic load on the journal, the lubricant steady-state film forces in the horizontal and vertical directions could be expressed as functions of the journal displacements, $x$ and $y$, and velocities, $\dot{x}$ and $\dot{y}$, away from the bearing centre; that is

$$F_x = f_1(x, y, \dot{x}, \dot{y}) \tag{2.30}$$

$$F_y = f_2(x, y, \dot{x}, \dot{y}) \tag{2.31}$$

where $\dot{x}$ and $\dot{y}$ are both zero. When there are changes in these displacements and velocities as described above, the new values of lubricant film force will be $F_x + dF_x$ and $F_y + dF_y$ respectively. These forces may also be expressed using Equations (2.30) and (2.31) as a four-variable Taylor series, neglecting small terms, as

$$F_x + dF_x = f_1(x, y, \dot{x}, \dot{y}) + dx \frac{\partial F_x}{\partial x} + dy \frac{\partial F_x}{\partial y} + d\dot{x} \frac{\partial F_x}{\partial \dot{x}} + d\dot{y} \frac{\partial F_x}{\partial \dot{y}} \tag{2.32}$$

$$F_y + dF_y = f_2(x, y, \dot{x}, \dot{y}) + dx \frac{\partial F_y}{\partial x} + dy \frac{\partial F_y}{\partial y} + d\dot{x} \frac{\partial F_y}{\partial \dot{x}} + d\dot{y} \frac{\partial F_y}{\partial \dot{y}} \tag{2.33}$$

The changes in the lubricant film forces are then given by

$$dF_x = F_x + dF_x - F_x$$

$$= dx\frac{\partial F_x}{\partial x} + dy\frac{\partial F_x}{\partial y} + d\dot{x}\frac{\partial F_x}{\partial \dot{x}} + d\dot{y}\frac{\partial F_x}{\partial \dot{y}}$$

$$= K_{xx}\,dx + K_{xy}\,dy + C_{xx}\,d\dot{x} + C_{xy}\,d\dot{y} \qquad (2.34)$$

and

$$dF_y = F_y + dF_y - F_y$$

$$= dx\frac{\partial F_y}{\partial x} + dy\frac{\partial F_y}{\partial y} + d\dot{x}\frac{\partial F_y}{\partial \dot{x}} + d\dot{y}\frac{\partial F_y}{\partial \dot{y}}$$

$$= K_{yx}\,dx + K_{yy}\,dy + C_{yx}\,d\dot{x} + C_{yy}\,d\dot{y} \qquad (2.35)$$

where $K_{xx}, K_{xy}, \ldots$, etc. are known as the eight oil-film stiffness and damping coefficients. It is these quantities which must be known in order to calculate the overall system response to imbalance and stability. The values of $K_{xx}$ and $K_{yx}$ may be found by calculating the lubricant film forces, $F_{x1}$ and $F_{y1}$ when a small displacement $dx$ is imposed on the journal, with $dy$, $d\dot{x}$ and $d\dot{y}$ all set to zero. The process is then repeated with a displacement of $-dx$ to calculate forces $F_{x2}$ and $F_{y2}$. The stiffness coefficients $K_{xx}$ and $K_{yx}$ are then given by

$$K_{xx} = \frac{F_{x1} - F_{x2}}{2\,dx} \qquad (2.36)$$

and

$$K_{yx} = \frac{F_{y1} - F_{y2}}{2\,dx} \qquad (2.37)$$

Similarly, the values of the other stiffness and damping coefficients may be evaluated by imposing small variations in $dy$, $d\dot{x}$ and $d\dot{y}$ on the journal, and investigating their effect on the lubricant film forces.

It should be noted that when proceeding in the manner described above care should be taken when formulating Equation (2.18) in finite difference form, as the quantity $\partial h/\partial t$ is no longer zero when $d\dot{x}$ or $d\dot{y}$ are non-zero, and the appropriate value of $U$ is also dependent on $d\dot{x}$ and $d\dot{y}$.

Figure 2.19 shows the variation of dimensionless bearing stiffness and damping coefficients with eccentricity ratio for a 120° plane bearing. Solutions for several other types of bearing have also been carried out by Lund et al. (1965) and by Holmes (1960). In addition to the 'spring' and 'damping' coefficients introduced above, some authors have suggested the use of 'inertia' coefficients (Lund 1966) and 'moment' coefficients (Rao and Mukherjee, 1977), although such coefficients usually have only a small effect on the bearing dynamics.

## JOURNAL BEARINGS

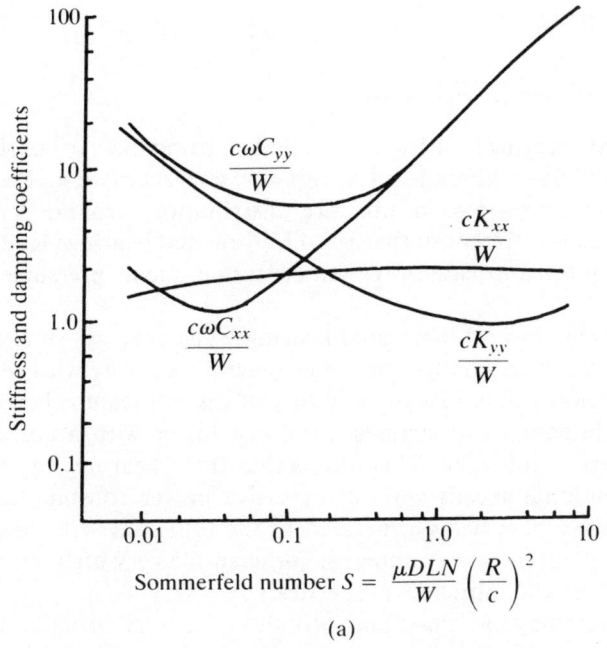

(a)

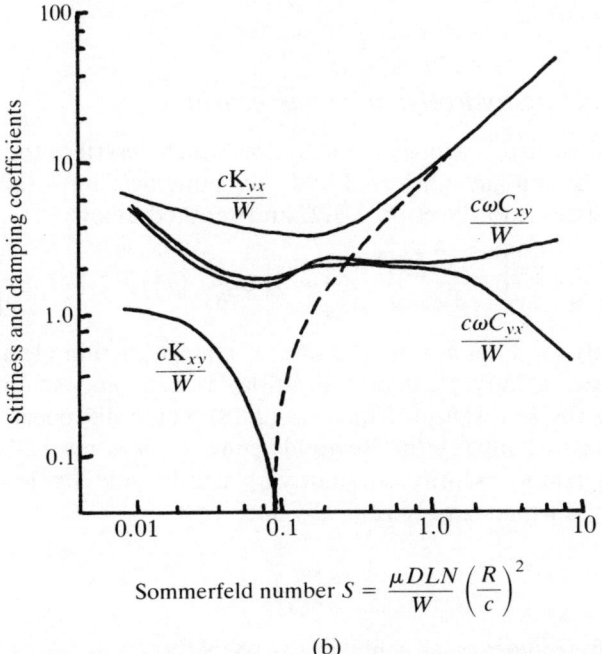

(b)

**Figure 2.19** Variation of stiffness and damping coefficient for a plain cylindrical bearing, $L/D = 1.0$, with Sommerfeld number, (a) direct coefficients; (b) cross-coupling coefficients.

## 2.4 Gas bearings

### 2.4.1 Description of operation

Gas-lubricated bearings operate on similar principles to oil-lubricated bearings, but have different performance characteristics. Because the lubricant is compressible, a pressure distribution around the bearing clearance which is different to that for oil-lubricated bearings is set up; and, because the lubricant cannot cavitate, sub-ambient pressures may be significant.

In comparison with oil-lubricated bearings, gas bearings of similar sizes are able to carry much smaller loads and generally operate with smaller film clearances. However, because the viscosity of the lubricant is far below that of oil, the lubricant shear stresses are much lower with a corresponding reduction in friction torque. This means that these bearings can operate at very high rotational speeds without excessive power consumption and so without excessive heat being generated in the lubricant. Gas bearings can also operate in extreme environments, for example very high temperatures, without affecting the lubricant properties.

Gas bearings may be classified into three distinct groups: self-acting hydrodynamic bearings, externally pressurized aerostatic bearings or porous wall bearings.

### 2.4.2 Self-acting hydrodynamic gas bearings

As with oil-lubricated bearings, the equation which describes the behaviour of the lubricant in a gas-lubricated hydrodynamic bearing is the Reynolds equation as discussed in Section 2.3.2. and restated below:

$$\frac{\partial}{\partial x}\left[\frac{\rho h^3}{\mu}\frac{\partial p}{\partial x}\right] + \frac{\partial}{\partial y}\left[\frac{\rho h^3}{\mu}\frac{\partial p}{\partial y}\right] = 6\left[U\frac{\partial}{\partial x}(\rho h) + 2\frac{\partial}{\partial t}(\rho h)\right] \quad (2.18)$$

Under steady load conditions the rate of change of film clearance with respect to time, at any particular location, is zero and so under these circumstances the last term of Equation (2.18) would disappear.

In the process of solving the Reynolds equation it is normally assumed that the gas dynamic viscosity does not vary significantly, i.e. $\mu$ = constant, and that the gas expands polytropically such that

$$\frac{p}{p_a} = \left(\frac{\rho}{\rho_a}\right)^n \quad (2.38)$$

where the suffix a indicates an ambient gas property.

Frequently isothermal conditions are assumed, so that $n = 1$. Also, since the net force on the journal will be dependent on the variation of the gas pressure from the ambient pressure it is helpful to write $p$ as

$$p = p_a + p' \tag{2.39}$$

where $p'$ is the increase of gas pressure above the ambient pressure, $p_a$. Substituting for $p$ from Equation (2.39) into (2.18) and for $\rho$ from (2.38) into (2.18) gives, for steady journal operating conditions:

$$\frac{\partial}{\partial x}\left[\left(1+\frac{p'}{p_a}\right)^{1/n}\frac{h^3}{\mu}\frac{\partial p}{\partial x}\right] + \frac{\partial}{\partial y}\left[\left(1+\frac{p'}{p_a}\right)^{1/n}\frac{h^3}{\mu}\frac{\partial p}{\partial y}\right]$$

$$= 6U\frac{\partial}{\partial x}\left[\left(1+\frac{p'}{p_a}\right)^{1/n}h\right] \tag{2.40}$$

which can be expressed in non-dimensional form as

$$\frac{\partial}{\partial \bar{x}}\left[(1+\lambda\bar{p})^{1/n}\bar{h}^3\frac{\partial \bar{p}}{\partial \bar{x}}\right] + \left(\frac{D}{L}\right)^2\frac{\partial}{\partial \bar{y}}\left[(1+\lambda\bar{p})^{1/n}\bar{h}^3\frac{\partial \bar{p}}{\partial \bar{y}}\right]$$

$$= 12\frac{\partial}{\partial \bar{x}}[(1+\lambda\bar{p})^{1/n}\bar{h}] \tag{2.41}$$

where

$$\bar{x} = x/D, \quad \bar{y} = \frac{y}{L}, \quad \bar{h} = \frac{h}{c}, \quad \bar{p} = \frac{c^2 p}{\mu U R}$$

and $\lambda = \mu U R / c^2 p_a$.

Equation (2.41) is the same as Equation (2.18), although it is now expressed in non-dimensional form, except for the inclusion of the $\lambda\bar{p}$ term. When $\lambda$ tends to zero the equations become identical. The term $\lambda$ is known as the 'compressibility number' and describes how significant the lubricant compressibility is on the bearing characteristics.

Equation (2.41) (or 2.18) may be solved analytically or numerically to determine the pressure distribution within the bearing lubricant and hence the oil-film forces acting on the journal. Analytical solutions generally involve simplifications of the equation and so are only approximate; the more popular methods include perturbation methods and the 'linearized $ph$ solution' where the product of pressure and film thickness is treated as the dependent variable. These solutions are discussed in more detail by Pinkus and Sternlicht (1961). For more accurate solutions it is necessary to revert to numerical methods (Raimondi 1961, Elrod and Malanoski 1960).

Solution of Equation (2.41) for finite bearings using, for example, the method of finite differences could now proceed as described earlier for oil-lubricated bearings. The finite-difference representation of the differential equations would, of course, be slightly different to allow for the lubricant compressibility term. The boundary conditions which must be satisfied are that $p' = 0$ at $x = \pm L/2$ and that $p(0, y) = p(\pi D, y)$.

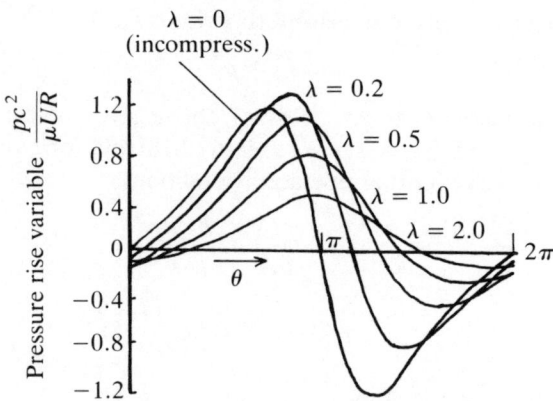

**Figure 2.20** Variation of lubricant pressure $p$, around the circumference of a gas journal bearing; $L/D = 1.0$. Curves are shown for values of the compressibility number $\lambda$ ranging from 0 to 2.0.

Figure 2.20 shows the variation in pressure distribution around the circumference of a gas bearing, for various values of compressibility number, for a bearing having a ratio $L/D = 1$. Figures 2.21, 2.22, 2.23 and 2.24 show the corresponding eccentricity ratio for a given specific static load ratio and compressibility number, and the variation of friction force, attitude angle, and steady-state locus with compressibility number, respectively.

Typical bearing dynamic characteristics are shown in Figure 2.25, in terms of the eight bearing stiffness and damping coefficients introduced in

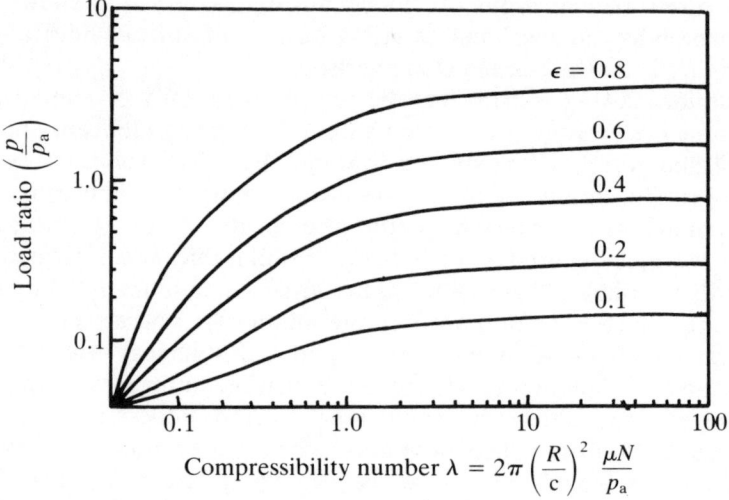

**Figure 2.21** Variation of bearing eccentricity ratio $\varepsilon$ with load ratio $p/p_a$ and compressibility number $\gamma$, $L/D = 1.0$.

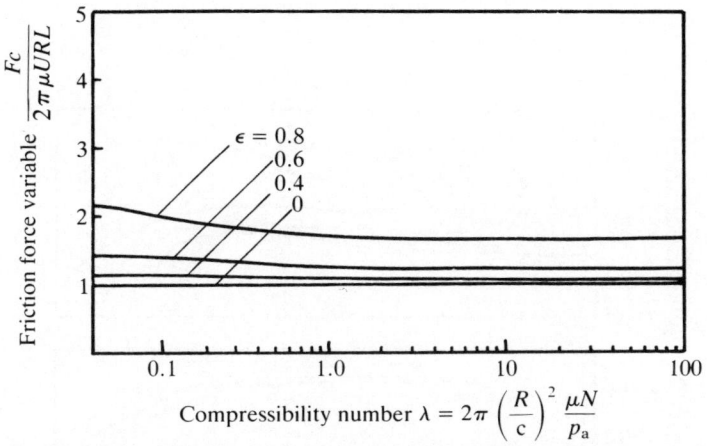

**Figure 2.22** Variation of bearing friction force variable with compressibility number $\lambda$ and eccentricity ratio $\varepsilon$; $L/D = 1.0$.

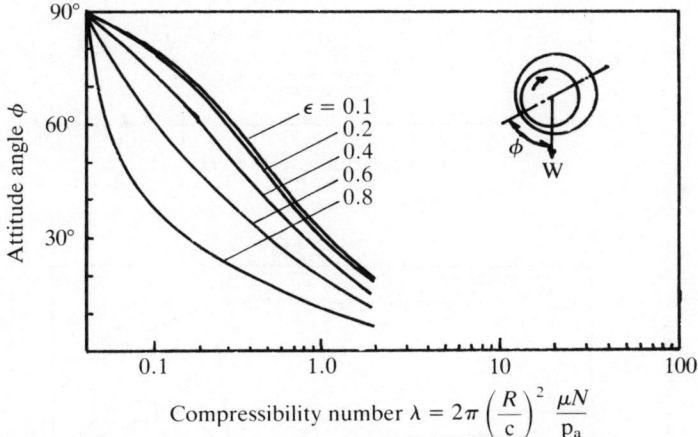

**Figure 2.23** Variation of bearing attitude angle $\phi$ with compressibility number $\lambda$ and eccentricity ratio $\varepsilon$; $L/D = 1.0$.

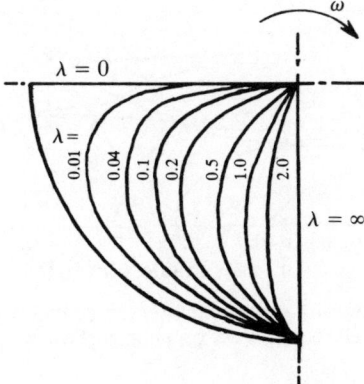

**Figure 2.24** Steady-state loci of gas journal bearings with various compressibility numbers $\lambda$; $L/D = 1.0$.

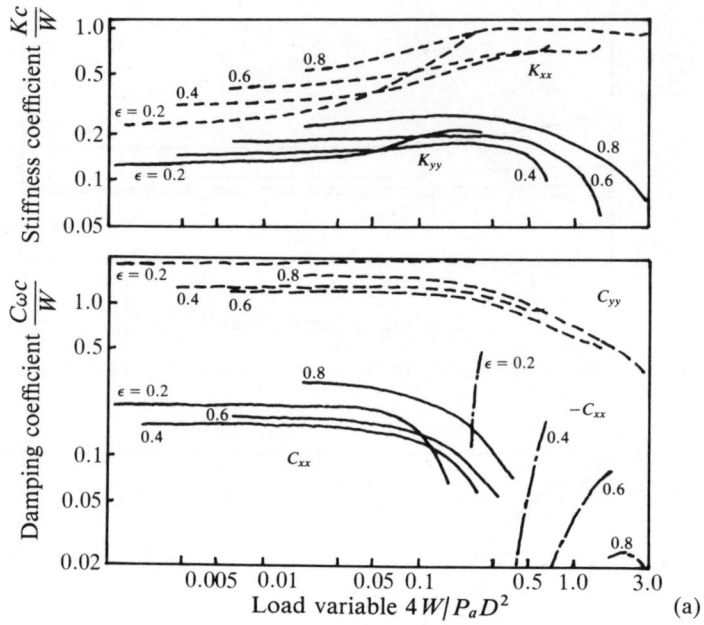

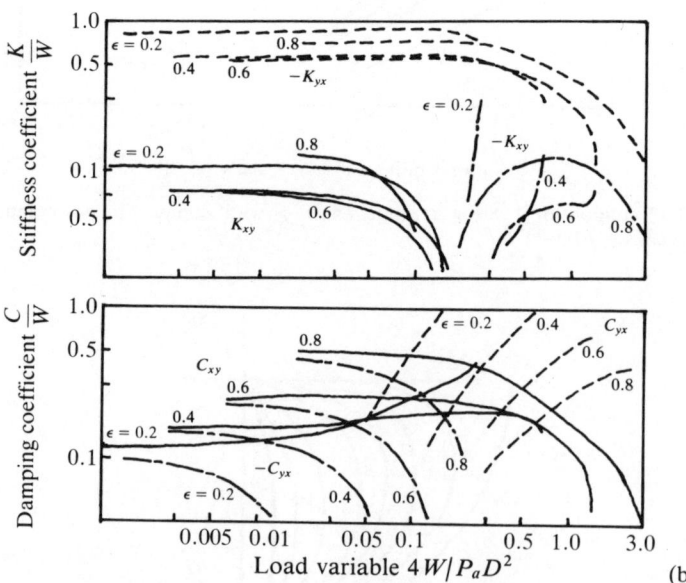

**Figure 2.25** Variation of bearing stiffness and damping coefficient with load and operating parameters, (a) direct coefficients; (b) cross-coupling coefficients.

Section 2.3. It is noteworthy that these coefficients are dependent upon the frequency of vibrations because for high frequencies the lubricant compressibility will be significant. This is because, at high frequencies, as the two bearing surfaces move towards each other there will be a tendency for the lubricant to compress as well as to be squeezed out of the clearance space. For very low relative velocities of the two surfaces towards one another the operation may be likened to steady-state loading and the compressibility effect will be less significant.

### 2.4.3 *Externally pressurized gas bearings*

Gas journal bearings may be supplied with lubricant from a pressurized supply via feed ports located at various positions along the bearing length. Two common arrangements are those where the bearing bush contains either one or two circumferential rows of feed holes, as shown in Figure 2.26. Externally pressurized gas bearings have the advantages of self-acting bearings together with an ability to carry much larger loads with greater lubricant film thicknesses. External pressurization is also used in combination with hydrodynamic action to increase bearing stiffness (Lund 1964, Grassam and Powell 1964).

A number of different theoretical treatments of externally pressurized gas bearings, varying in their accuracy and complexity, have been suggested by several authors. The approach which follows is one of the simpler methods;

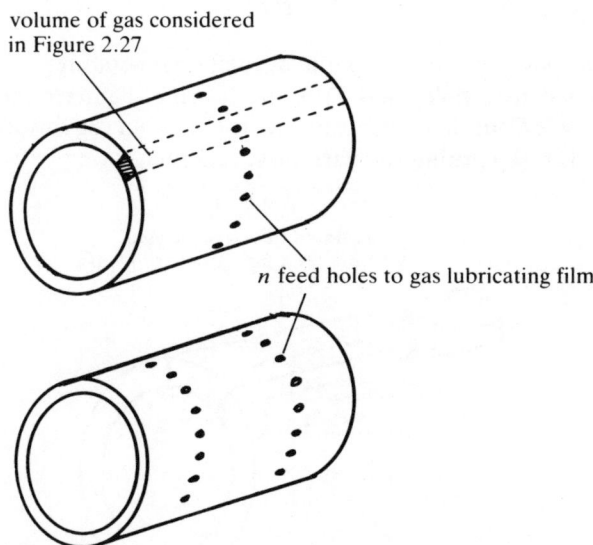

**Figure 2.26** Two common arrangements of lubricant feed holes in externally pressurized gas bearings.

it involves considering the gas flow path from inlet feed hole to the bearing edge to be composed of a number of slots (equal to the number of feed holes) as shown in Figure 2.27. The method has been found to predict bearing load capacities and stiffnesses to within 10% in most cases, for bearings of $L/D$ ratio less than 2 (Shires and Pantall 1963). A similar method is also described in the text by Fuller (1984), for circular thrust bearings. Comprehensive design charts for eternally pressurized journal bearings are published in the MTI Gas Bearing Design Manual (1972).

Considering the equivalent flow slot corresponding to one bearing feed hole shown in Figure 2.27, and treating flow along the slot (i.e. the bearing clearance space) as flow between flat plates, the mass flow rate of lubricant is

$$\dot{m} = -\frac{\rho w h^3}{12\mu}\frac{dp}{dy} \qquad (2.42)$$

where $h$ is the clearance, $w$ is the slot width and where $dp/dy$ is the lubricant pressure gradient in the axial direction. Before this equation can be further developed it must be remembered that the lubricant density is a function of pressure. Isothermal conditions are normally justified on the basis that any heat generated is conducted away by the bearing materials and so the lubricant temperature change along the bearing is negligible. Under these conditions

$$\rho = \frac{p}{RT} \qquad (2.43)$$

where $R$ is the gas constant and $T$ the absolute temperature.

If there are $n$ feed holes and $D$ is the bearing diameter then $w$ can be substituted by $\pi D/n$; then substituting for $\rho$ from Equation (2.43) into Equation (2.42), separating the variables, and integrating between the inlet

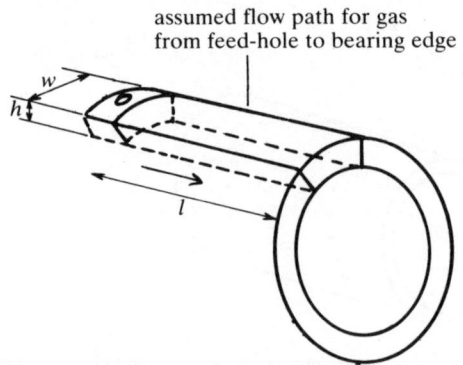

**Figure 2.27** Flow path for gas, from feed hole to bearing edge.

pocket where the pressure is $p$ and the ends of the bearing where it is $p_a$, gives

$$p_p^2 - p_a^2 = \frac{24\mu \dot{m}nRTl}{\pi D h^3} \qquad (2.44)$$

Equation (2.44) can be developed to give the total lubricant mass flow rate out of both ends of a concentric 360° journal bearing as

$$\dot{m} = \frac{(p_p^2 - p_a^2)\pi D h_o^3}{12RTl} \qquad (2.45)$$

Considering the inlet feed hole to behave as a nozzle where the inlet pressure is the supply pressure, $p_s$, and the pressure downstream is the static pressure in the throat, $p_p$, then these two pressures are related by

$$\frac{p_p}{p_s} = \left[1 - \frac{(\gamma-1)}{2}\left(\frac{v}{c_s}\right)^2\right]^{\gamma/(\gamma-1)} \qquad (2.46)$$

where $\gamma$ is the gas ratio of specific heats, $v$ is the gas velocity at the throat, and $c_s$ is the speed of sound at the supply conditions. Re-arranging Equation (2.46) gives

$$v = \left[\left(\frac{2c_s^2}{\gamma-1}\right)\left(1 - \left(\frac{p_p}{p_s}\right)^{(\gamma-1)/\gamma}\right)\right] \qquad (2.47)$$

and the mass flow rate through the inlet hole is

$$\dot{m} = C_d \rho_t A v \qquad (2.48)$$

where $C_d$ is the nozzle discharge coefficient, $\rho_t$ is the lubricant density at the throat, and $A$ is the throat cross-sectional area. Assuming isentropic expansion such that

$$\rho_t = \rho_s \left(\frac{p_s}{p_s}\right)^{1/\gamma} \qquad (2.49)$$

and representing the speed of sound at the supply point as

$$c_s = (\gamma R T_s)^{1/2} \qquad (2.50)$$

and substituting Equations (2.47), (2.49) and (2.50) into (2.48) gives

$$\dot{m} = C_d A \rho_s (2RT_s)^{1/2} \left[\frac{\gamma}{\gamma-1}\left\{\left(\frac{p_p}{p_s}\right)^{2/\gamma} - \left(\frac{p_p}{p_s}\right)^{(\gamma+1)/\gamma}\right\}\right] \qquad (2.51)$$

The value of the discharge coefficient $C_d$ varies with pressure ratio $p_p/p_s$ but is usually about 0.8 for $p_p/p_s$ less than about 0.7.

The value of the throat area $A$ is governed by the type of feed hole as shown in Figure 2.28. The two types are the same when $h = d/4$; if $h$ is less than this the inlet behaves as an annular orifice unless a feed pocket is included as in Figure 2.28(a). Simple orifices are usually preferred because

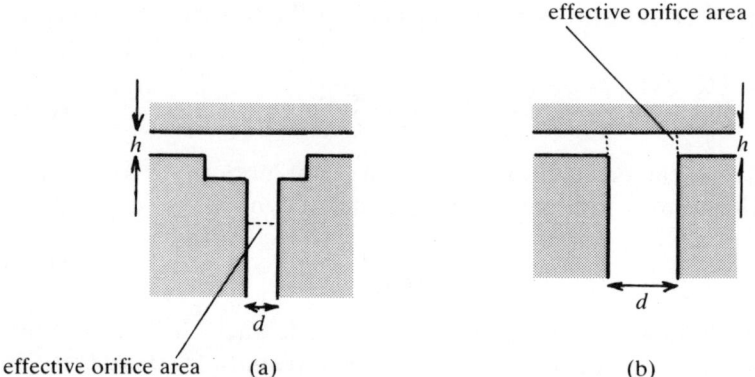

**Figure 2.28** Inlet orifice types in externally pressurized gas bearings. $h$, film clearance; $d$, diameter of orifice; (a) simple orifice with feed pocket; (b) annular orifice.

the throat area is independent of local film thickness and so results in a lubricant film having a higher static stiffness.

If the bearing has two rows of feed holes then the flow through one feed hole must be the same as that through one equivalent slot. Equating expressions for mass flow rate from Equations (2.44) and (2.51) gives

$$p_p^2 - p_a^2 = \frac{24\mu RTlC_d A\rho_s(2RT_s)^{1/2}}{wh^3}\left[\frac{\gamma}{\gamma-1}\left\{\left(\frac{p_p}{p_s}\right)^{2/\gamma} - \left(\frac{p_p}{p_s}\right)^{(\gamma+1)/\gamma}\right\}\right] \quad (2.52)$$

This equation can now be solved to determine the lubricant pressure at inlet to the bearing clearance using an iterative method, or alternatively using an approximation suggested by Shires and Pantall (1963). Once $P_p$ has been established then the mass flow rate for one slot can be calculated using Equation (2.44). Equation (2.44) can then be written in general form expressing the lubricant pressure at any point in the bearing clearance as:

$$p = \left[\left\{\frac{24\mu \dot{m}nRTx}{\pi Dh}\right\} + p_a^2\right]^{1/2} \quad (2.53)$$

where $x$ is the distance from the bearing edge to the point under consideration.

The pressure profile may then be integrated along the length of the slot, and summed vectorially for each slot since each slot has a different clearance, to give the total load supported by the bearing.

The bearing load calculated on the basis of the above theory makes no allowance for the effects of dispersion of the gas as it exits the feed hole (i.e. instead of flowing immediately in the axial direction), nor does it make any allowance for the effects of non-axial flow due to the difference in pressures in the imaginary axial slots which make up the bearing clearance. The effects of dispersion may be allowed for by a load dispersion coefficient given by Powell (1970) and defined as

$$C_{\text{ld}} = \frac{\text{actual load with dispersion}}{\text{theoretical load without dispersion}}$$

$$= 0.89 \left(\frac{dn}{D}\right)^{0.21} \left(\frac{Ln}{D}\right)^{0.42} \left(\frac{p_p}{p_a}\right)^e \quad (2.54)$$

where

$$e = 0.0505 \left(\frac{\pi D}{nd}\right)^{0.379} \left(\frac{\pi D}{nl}\right)^{0.758}$$

The effects of non-axial flow may be similarly allowed for by a further empirically derived coefficient given by Powell (1970) as

$$C_{\text{af}} = \frac{\text{actual load with non-axial flow}}{\text{theoretical load without non-axial flow}}$$

$$= 0.315 \left[\frac{\cosh(6.36l/D) - 1}{\sinh(6.36l/D)} + \tanh\left\{6.36 \frac{L - 2l}{D}\right\}\right] \left[\frac{L - l}{D}\right]^{-1}$$

In the above two formulae $L$ is the bearing length, $D$ the bearing diameter, $d$ the nozzle diameter, $n$ the number of inlets, and $l$ the distance between the inlet holes and the bearing edge.

The remainder of this section deals with the determination of the dynamic stiffness and damping coefficients for externally pressurized gas bearings. The most rigorous of analyses would invariably involve some numerical means of integrating the Reynolds equation to produce a realistic lubricant pressure profile; some examples are discussed in Szeri (1980). However, in many cases a satisfactory set of parameter values may be obtained using an approximate method. The method described below is a linearized first-order perturbation method similar to that described by Mullan & Richardson (1964). The method is suitable for cases where journal eccentricity is not excessive.

If the journal is considered to be concentric then the first term on the right-hand side of the Reynolds equation (2.18) is zero, since $\partial(\rho h)/\partial x = 0$, and Equation (2.18) may then be written in dimensionless form as

$$\frac{\partial}{\partial \theta} \bar{p} \bar{h}^3 \frac{\partial \bar{p}}{\partial \theta} + \frac{\partial}{\partial y} \bar{p} \bar{h}^3 \frac{\partial \bar{p}}{\partial y} = \sigma \frac{\partial}{\partial t'} (\bar{p} \bar{h}) \quad (2.56)$$

where

$$\rho = p/RT, \bar{p} = p/p_a, \bar{h} = h/c, \bar{y} = y/R, t' = \omega t \quad \text{and} \quad \sigma = 12\mu\omega R/p_a c^2$$

where $\sigma$ is the 'squeeze number' and is a function of the compressibility number – see Section 2.4.2. The other symbols are as defined in Section

2.4.2. The boundary conditions which must be applied are

$$\bar{p}(\theta,\bar{y}) = \bar{p}(\theta+2\pi,\bar{y}), \qquad \bar{p}(\theta, L/D) = 1, \qquad \dot{m}_{\text{out}} = \dot{m}_{\text{in}} \qquad (2.57)$$

The first two conditions above are necessary because of the geometry of the bearing and because of the ambient pressure at the bearing edge. The third condition may be written in an alternative form by equating that part of the axial flow towards one end of the bearing, which comes from a small circumferential length $w$ of a circumferential feed groove, to half of the inlet lubricant flow over that circumferential distance. The flow out of the bearing over such a region is given by Equation (2.42), so that the third boundary condition may be written as

$$-p\frac{dp}{dy}\frac{wh^3}{12\mu RT} = \frac{1}{2}\frac{\partial Q_{\text{in}}}{\partial \theta}\,d\theta \qquad (2.58)$$

where $\partial Q_{\text{in}}/\partial\theta$ is the inlet flow per unit bearing arc. Setting $d\theta = 2w/D$ and non-dimensionalizing gives

$$-\bar{h}^3\bar{p}\frac{d\bar{p}}{d\bar{y}} = \frac{6\mu RT_s}{c^3 p_a^2}\frac{\partial Q_{\text{in}}}{\partial\theta} \qquad (2.59)$$

or, when the journal and bearing are concentric,

$$-\bar{h}^3\bar{p}\frac{d\bar{p}}{d\bar{y}} = \bar{\dot{m}} \qquad (2.60)$$

where $\bar{\dot{m}}$ is a dimensionless total inlet flow for the bearing. Its value can be calculated by using Equation (2.51) (summed for each feedhole) and dividing by $2\pi$ to give a value for $\partial Q_{\text{in}}/\partial\theta$ in Equation (2.59).

The linearization process is performed by assuming only small displacements of the journal from the equilibrium position so that only small changes in the lubricant pressure distribution take place. If the change in film clearance is such that

$$\bar{h} = 1 + \varepsilon \cos\theta \qquad (2.61)$$

where the terms are as defined in Section 2.3, and the subsequent lubricant pressures are

$$\bar{p} = \bar{p}_0 + \delta\bar{p} \qquad (2.62)$$

where $\bar{p}_0$ is the pressure at $\varepsilon = 0$, then Equations (2.61) and (2.62) may be substituted into Equation (2.56), and the following two equations obtained by comparing zeroth- and first-order powers of $\varepsilon$ on each side of the equation (and neglecting the products of small terms):

$$\frac{\partial^2(\bar{p}_0\,\delta\bar{p})}{\partial\theta^2} + \frac{\partial^2(\bar{p}_0\,\delta\bar{p})}{\partial y^2} = \frac{\sigma}{p}\left[\frac{\partial}{\partial t'}(\bar{p}_0\,\delta\bar{p}) + \bar{p}_0^2\cos\theta\,\frac{\partial\varepsilon}{\partial t'}\right] \qquad (2.63)$$

$$\frac{\partial^2\bar{p}_0^2}{\partial\bar{y}^2} = 0 \qquad (2.64)$$

GAS BEARINGS

Equation (2.64) may now be integrated, subject to the boundary conditions, to yield an expression for $\bar{p}_0$, while Equation (2.63) may be made linear in $\bar{p}_0 \, \delta\bar{p}$ provided an approximation is made, that is replacing $\sigma/p_0$ on the right-hand side by its average value $\sigma/p_{0\,\text{av}}$. This equation may now be solved for $\bar{p}_0 \, \delta\bar{p}$, using numerical or other methods, whereupon division by $\bar{p}_0$ yields the dynamic pressure $\delta\bar{p}$. The resulting force is collinear with the journal displacement (no cross-coupling). Some typical values of bearing dynamic stiffness and damping calculated in this way are shown in Figures 2.29 and 2.30.

### 2.4.4 Externally pressurized porous gas journal bearings

In some types of rotating machine the bearing material is manufactured from a porous material which helps to ensure an even distribution of lubricant at the bearing surface. The porous bearing also has an inherent capability to damp small oscillations. The porous material itself acts as a device which controls the flow of gas so that no additional flow restrictor is necessary to provide compensation. In effect, the bearing material takes the place of the orifice or capillary used in other types of bearing with an externally pressurized lubricant supply (see also Section 2.6). A diagram of a porous bearing is shown in Figure 2.31. For porous bearings the high load-carrying capacity and high stiffness characteristics are enhanced at higher rotational speeds because hydrodynamic action complements the effects of external pressurization. For low loads and moderate speeds the hydrodynamic action is insignificant. The following theoretical treatment

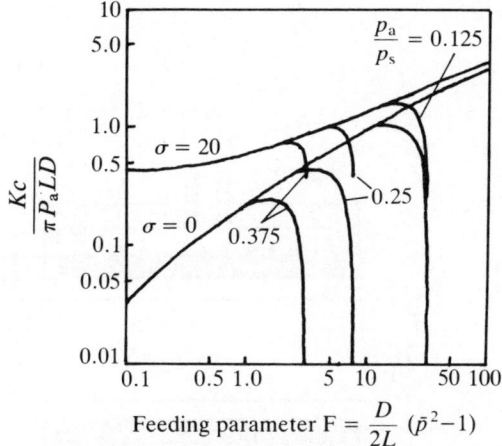

**Figure 2.29** Variation of bearing stiffness $K$ with feeding parameter for externally pressurized gas bearings. $L$, bearing length; $D$, bearing diameter; $\bar{p}$, pressure downstream of orifice/ambient pressure, for a concentric journal. $L/D = 1.0$, ratio of specific heats $\gamma = 1.1$.

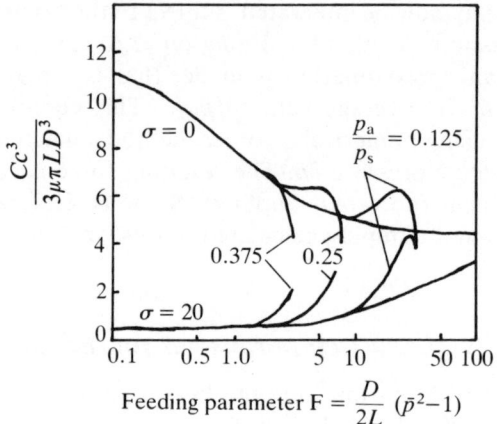

**Figure 2.30** Variation of bearing damping $C$ with feeding parameter. Other parameters are as defined in Figure 2.29.

therefore neglects the hydrodynamic action. Further, the analysis is only suited to bearings operating with an eccentricity ratio less than 0.5, for the sake of further simplification.

The porous material which goes to make up the bearing bush is usually relatively thin, and the ends may be sealed, so that the gas may be assumed to flow only in the radial direction through the material. In such a situation it may be assumed that the gas obeys Darcy's law which for steady-state conditions is

$$\frac{\partial^2 \bar{p}'^2}{\partial \bar{z}^2} = 0 \qquad (2.65)$$

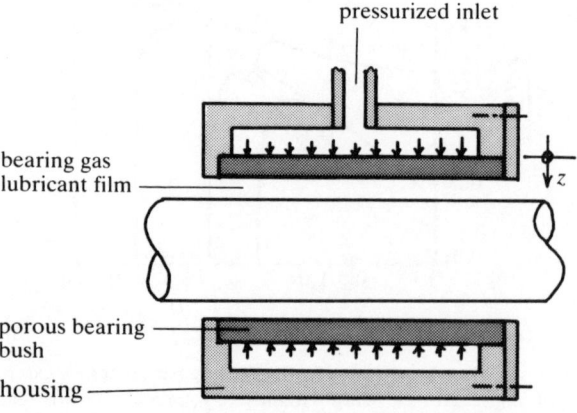

**Figure 2.31** Diagram of a porous gas journal bearing.

where $\bar{p}' = p'/p_a$ is the non-dimensional pressure in the porous bush, the suffix 'a' indicating ambient pressure, and $\bar{z} = z/t$ is the non-dimensional distance in the radial direction, $t$ being the wall thickness of the porous bush. For porous bearings the non-dimensional Reynolds equation for the lubricant film becomes modified to allow for additional inflow of lubricant in the radial direction from the porous bush. For steady-state conditions this is now written as

$$\frac{\partial}{\partial \theta}\left[\bar{h}^3 \frac{\partial \bar{p}^2}{\partial \theta}\right] + \left(\frac{D}{L}\right)^2 \frac{\partial}{\partial \bar{y}}\left[\bar{h}^3 \frac{\partial \bar{p}^2}{\partial \bar{y}}\right] = \Lambda\left[\frac{\partial \bar{p}'^2}{\partial \bar{z}}\right] \quad (2.66)$$

where the prime indicates that the pressure term refers to that for lubricant in the porous bearing bush and $\Lambda$ is a dimensionless feeding parameter given by

$$\Lambda = \frac{3KD^2}{c^3 t} \quad (2.67)$$

where K is the permeability of the porous bush. The boundary conditions to the problem are that when

$$\bar{z} = 0, \quad \bar{p}' = \bar{p}_s, \text{ the dimensionless supply pressure,}$$
$$\bar{z} = 1, \quad \bar{p}' = \bar{p}$$
$$\bar{p}(\theta, y) = \bar{p}(\theta + 2\pi, y) \quad (2.68)$$

and

$$\bar{p}(\theta, \pm 1) = 1$$

The lubricant film may be linearized by using a first-order perturbation method. If the non-dimensional bearing clearance is assumed to be given by $\bar{h} = 1 + \varepsilon \cos \theta$, where $\varepsilon$ is the eccentricity ratio and is not excessive, then the lubricant pressures may be assumed to vary in proportion to $\varepsilon$ and may be expressed as

$$\bar{p} = \bar{p}_0 + \varepsilon \bar{p}_1, \quad \bar{p}' = \bar{p}'_0 + \varepsilon \bar{p}'_1 \quad (2.69)$$

in the bearing clearance and porous medium respectively. Substituting these modified expressions for clearance and pressure into Equations (2.65) and (2.66) and neglecting terms involving $\varepsilon^2$ and higher orders, gives the following equations:

$$\frac{\partial^2 {\bar{p}'_0}^2}{\partial \bar{z}^2} + 2\varepsilon \frac{\partial^2}{\partial \bar{z}^2}(\bar{p}'_0 \bar{p}'_1) = 0$$

$$\frac{\partial}{\partial \theta}\left[\frac{\partial}{\partial \theta}(\bar{p}_0^2 + 2\varepsilon \bar{p}_0 \bar{p}_1) + 3\varepsilon \cos\theta \frac{\partial \bar{p}_0^2}{\partial \theta}\right] + \left(\frac{D}{L}\right)^2 \frac{\partial}{\partial y}$$

$$\times \left[\frac{\partial}{\partial y}(\bar{p}_0^2 + 2\varepsilon \bar{p}_0 \bar{p}_1) + 3\varepsilon \cos\theta \frac{\partial \bar{p}_0^2}{\partial y}\right] = \Lambda\left[\frac{\partial {\bar{p}'_0}^2}{\partial \bar{z}} + \frac{\partial}{\partial \bar{z}}(2\varepsilon \bar{p}'_0 \bar{p}'_1)\right]$$

$$(2.70)$$

Remembering that $\bar{p}_0$ does not change with respect to $\theta$ for the concentric journal and comparing coefficients of terms of $\varepsilon^0$ and $\varepsilon^1$ on each side of each equation gives

$$\frac{\partial \bar{p}_0'^2}{\partial \bar{z}} = 0 \tag{2.71}$$

$$\frac{\partial^2}{\partial \bar{z}^2}(\bar{p}_0' \bar{p}_1') = 0 \tag{2.72}$$

$$\left(\frac{D}{L}\right)^2 \frac{\partial^2 \bar{p}_0^2}{\partial \bar{y}^2} - \Lambda \frac{\partial \bar{p}_0'^2}{\partial \bar{z}} = 0 \tag{2.73}$$

$$\frac{\partial^2}{\partial \theta^2}(\bar{p}_0 \bar{p}_1) + \left(\frac{D}{L}\right)^2 \frac{\partial^2}{\partial y^2}(\bar{p}_0 \bar{p}_1) + 1.5\left(\frac{D}{L}\right)^2 \cos\theta \frac{\partial^2 \bar{p}_0^2}{\partial y^2} = \Lambda \frac{\partial}{\partial \bar{z}}(\bar{p}_0' \bar{p}_1') \tag{2.74}$$

The boundary conditions associated with these variations in clearance and pressure are, when

$$\begin{aligned}
\bar{z} &= 0, & \bar{p}_0' &= \bar{p}_s, & \bar{p}_1' &= 0 \\
\bar{z} &= 1, & \bar{p}_0' &= \bar{p}_0, & \bar{p}_1' &= \bar{p}_1 \\
& & \bar{p}_1(\theta, y) &= \bar{p}_1(\theta + 2\pi, y) &
\end{aligned} \tag{2.75}$$

and

$$\bar{p}_0(\theta, y = \pm 1) = 1, \qquad \bar{p}_1(\theta, y = \pm 1) = 0$$

Integrating Equations (2.71) and (2.72) once using the boundary conditions in (2.75), and then substituting for $\partial \bar{p}_0'^2/\partial \bar{z}$ and $\partial(\bar{p}_0' \bar{p}_1')/\partial \bar{z}$ in Equations (2.73) and (2.74) gives

$$\left(\frac{D}{L}\right)^2 \frac{\partial^2 \bar{p}_0^2}{\partial \bar{y}^2} - \Lambda(\bar{p}_0^2 - \bar{p}_s^2) = 0 \tag{2.76}$$

$$\frac{\partial^2}{\partial \theta^2}(\bar{p}_0 \bar{p}_1) + \left(\frac{D}{L}\right)^2 \frac{\partial^2}{\partial \bar{y}^2}(\bar{p}_0 \bar{p}_1) + 1.5\left(\frac{D}{L}\right)^2 \cos\theta \frac{\partial^2 \bar{p}_0^2}{\partial \bar{y}^2} - \Lambda(\bar{p}_0 \bar{p}_1) = 0 \tag{2.77}$$

Equation (2.76) may then be integrated twice using the boundary conditions $\bar{p}_0 = \bar{p} = 1$ at $\bar{y} = 2y/L = 1$ and $\partial \bar{p}_0/\partial \bar{y} = 0$ at $\bar{y} = 0$ (since the pressure distribution must be symmetrical about the bearing centre line) to give

$$\bar{p}_0^2 = \frac{(1 - \bar{p}_s^2)\cosh(\bar{y}\sqrt{\Lambda}\,L/D)}{\cosh(\sqrt{\Lambda}\,L/D)} + \bar{p}_s^2 \tag{2.78}$$

Similarly, Equation (2.77) may be integrated using the boundary conditions $\bar{p}_0 \bar{p}_1 = 0$ at $\bar{y} = 1$ and $\partial(\bar{p}_0 \bar{p}_1)/\partial \bar{y} = 0$ at $\bar{y} = 0$. This equation may be solved numerically or, alternatively, a closed-form analytical expression is given by

Majumdar (1978) as

$$\bar{p}_0\bar{p}_1 = \frac{1.5\Lambda(1-\bar{p}_s^2)\cos\theta}{\cosh(\sqrt{\Lambda}\,L/D)}$$

$$\times \left[\frac{\cosh(\sqrt{1+\Lambda}\,\bar{y}L/D)}{\cosh(\sqrt{1+\Lambda}\,L/D)}\{\cosh(\sqrt{1+\Lambda}\,L/D)-\cosh(\sqrt{\Lambda}\,L/D)\}\right.$$

$$\left.-\cosh(\sqrt{1+\Lambda}\,\bar{y}L/D)-\cosh(\sqrt{\Lambda}\,\bar{y}L/D)\right] \quad (2.79)$$

The change in lubricant pressure due to the journal eccentricity may now be evaluated as $\varepsilon p_1 = \varepsilon \bar{p}_1 p_a = \varepsilon \bar{p}_0 \bar{p}_1 p_a/\bar{p}_0$ using Equations (2.78) and (2.79). Then the load supported by the bearing is given by

$$W = -2\int_0^{L/2}\int_0^{2\pi}\varepsilon p_1 R\cos\theta\,d\theta\,dy \quad (2.80)$$

as in the case of oil-film bearings discussed earlier.

The flow of gas from both ends of the bearing is given by integrating Equation (2.68), where $\rho = p/RT$ and $h = c(1+\varepsilon\cos\theta)$, to give

$$\dot{m} = \frac{-2c^3}{12\mu RT}\int_0^{2\pi} p\frac{dp}{dy}(1+\varepsilon\cos\theta)^3 R\,d\theta \quad (2.81)$$

This flow rate does not change substantially with bearing eccentricity ratio. For the concentric position Equation (2.81) may be integrated to give

$$\dot{m} = \frac{-2c^3 p_a^2}{24\mu RT(L/D)}\int_0^{2\pi}\frac{\partial(\bar{p}_0^2)}{\partial\bar{y}}\,d\theta \quad (2.82)$$

which may be evaluated after substituting for $\partial\bar{p}_0/\partial\bar{y}$ using Equation (2.78).

Figure 2.32 shows the variation of load with bearing feed parameter, for a porous bearing with $L/D=1$, and $L/D=2$; computed using the above analysis.

A similar analytical approach may be used to determine the bearing dynamic characteristics, as used by Rao (1977). Equation (2.65) is modified, to allow for time-dependent effects, to become

$$\frac{\partial^2 \bar{p}'^2}{\partial\bar{z}^2} - 2\sigma\gamma\frac{\partial p'}{\partial t'} = 0 \quad (2.83)$$

where $\gamma = \eta c^2 t^2/(12KR^2)$ is the non-dimensional porosity of the bearing material, and $\eta$ is the material porosity and $\sigma$ is the squeeze number defined below. Similarly, the Reynolds equation (2.66) is written to include the squeeze-film term and becomes

$$\frac{\partial}{\partial\theta}\left[\bar{h}^3\frac{\partial\bar{p}^2}{\partial\theta}\right] + \left(\frac{D}{L}\right)^2\frac{\partial}{\partial\bar{y}}\left[\bar{h}^3\frac{\partial\bar{p}^2}{\partial\bar{y}}\right] = \Lambda\frac{\partial\bar{p}'^2}{\partial\bar{z}} + 2\sigma\frac{\partial(\bar{p}\bar{h})}{\partial t'} \quad (2.84)$$

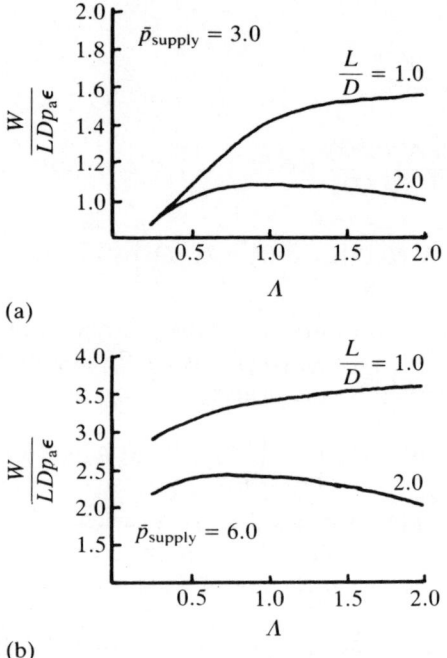

**Figure 2.32** Variation of load variable, $W/LD\,p_a$, $\varepsilon$, with feed parameter, $2D\Lambda$, (a) $\bar{p}_s = 3.0$; (b) $\bar{p}_s = 6.0$.

where $t' = \omega t$ is dimensionless time, and $\sigma = 12\mu\omega R^2/(p_a c^2)$ and is also dimensionless. The subsequent linearization and perturbation procedure results in equations which are the same as those for the steady loading case except that allowance is made for sinusoidal variations in the perturbed parameters by writing

$$\bar{h} = 1 + \varepsilon e^{jt'} \cos \theta$$
$$\bar{p} = \bar{p}_0 + \varepsilon e^{jt'} \bar{p}_1 \qquad (2.85)$$
$$\bar{p}' = \bar{p}'_0 + \varepsilon e^{jt'} \bar{p}'_1$$

and Equations (2.71)–(2.74) now become

$$\frac{\partial^2}{\partial \bar{z}^2}(\bar{p}'_0{}^2) = 0 \qquad (2.86)$$

$$\frac{\partial^2}{\partial \bar{z}^2}(\bar{p}'_0 \bar{p}'_1) - j\sigma\gamma\bar{p}'_1 = 0 \qquad (2.87)$$

$$\left(\frac{D}{L}\right)^2 \frac{\partial^2}{\partial \bar{y}^2}(\bar{p}_0^2) - \Lambda \frac{\partial}{\partial \bar{z}}(\bar{p}'_0{}^2) = 0 \qquad (2.88)$$

$$\frac{\partial^2}{\partial \theta^2}(\bar{p}_0 \bar{p}_1) + \left(\frac{D}{L}\right)^2 \frac{\partial^2}{\partial \bar{y}^2}(\bar{p}_0 \bar{p}_1) + 1.5\left(\frac{D}{L}\right)^2 \cos\theta \frac{\partial^2}{\partial \bar{y}^2}(\bar{p}_0^2)$$

$$= \Lambda \frac{\partial}{\partial \bar{z}}(\bar{p}_0' \bar{p}_1') + j\sigma(\bar{p}_1 + \bar{p}_0 \cos\theta) \quad (2.89)$$

Proceeding with the analysis as before, and solving Equations (2.86) and (2.88), the expression obtained for $\bar{p}_0^2$ remains unchanged from that given in Equation (2.78). Also an approximate solution to Equation (2.87) (Rao 1977) is

$$\frac{\partial}{\partial \bar{z}}(\bar{p}_0' \bar{p}_1') = (j\sigma\gamma\bar{p}_0)^{1/2}\bar{p}_1 \quad (2.90)$$

Substituting for $\partial^2 \bar{p}_0^2/\partial \bar{y}^2$ and $\partial(\bar{p}_0'\bar{p}_1')/\partial \bar{z}$ from Equations (2.76) and (2.90) into Equation (2.89) gives

$$\frac{\partial^2}{\partial \theta^2}(\bar{p}_0 \bar{p}_1) + 1.5\Lambda \cos\theta(\bar{p}_0^2 - \bar{p}_s^2) + \left(\frac{D}{L}\right)^2 \frac{\partial^2}{\partial \bar{y}^2}(\bar{p}_0 \bar{p}_1)$$

$$= \Lambda(j\sigma\gamma\bar{p}_0)^{1/2}\bar{p}_1 + j\sigma(\bar{p}_1 + \bar{p}_0 \cos\theta) \quad (2.91)$$

$\bar{p}_s^2$ can now be substituted from Equation (2.78) and then Equation (2.91) can be solved to find $\bar{p}_1$ numerically, for example using finite differences as described earlier for oil-film journal bearings. The dynamic load supported by the bearing oil film is then given by

$$W_d = -2\int_0^{L/2}\int_0^{2\pi} p_1 R \cos\theta \, d\theta \, dy \quad (2.92)$$

which is related to the bearing stiffness $K$, the damping $C$, and the journal

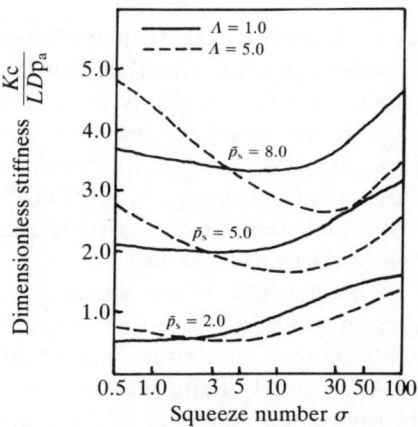

**Figure 2.33** Variation of porous bearing stiffness $K$ with squeeze number $\sigma$; $L/D = 1.0$, $\gamma = 1.0$.

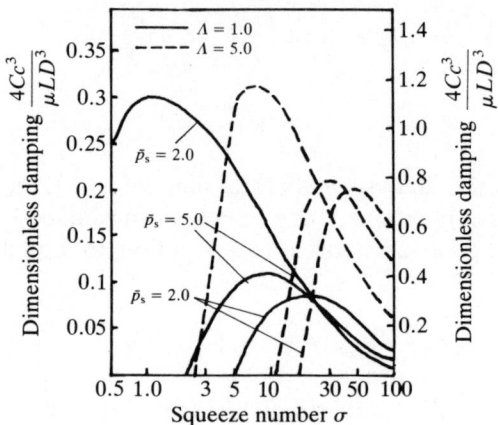

**Figure 2.34** Variation of porous bearing damping $C$ with squeeze number $\sigma$; $L/D = 1.0$, $\gamma = 1.0$. The left-hand vertical scale gives the values for $\Lambda = 5.0$, the right-hand scale for $\Lambda = 1.0$.

displacement $X$ by

$$W_d = (K + j\omega C)X \qquad (2.93)$$

The variations of bearing stiffness and damping with squeeze number obtained using this method are shown in Figures 2.33 and 2.34.

## 2.5 Squeeze-film bearings

### 2.5.1 Description of operation

Squeeze-film bearings are a special case of hydrodynamic journal bearing where the speed of rotation is zero. In other words, the hydrodynamic lubricant pressure forces are built up purely as a result of the action of squeezing the lubricant in the clearance space between the journal and bearing surfaces. This class of bearing has been used extensively in conjunction with rolling element bearings in turbomachinery, for example in the support of aero-engine compressor shafts. In these cases the rolling element bearings on which the shaft is mounted are installed with their outer race housing acting as the journal in the squeeze-film bearing, the two bearings in effect acting in series, as shown in Figure 2.35. The bearing is frequently supplied with oil via a circumferential oil inlet groove linked to an oil supply source. Alternatively there may be a number of feed holes spaced around the bearing circumference, midway along its length. The feed oil may or may not be pressurized. The oil then flows axially out of the bearing and is collected in a drain and returned to the supply

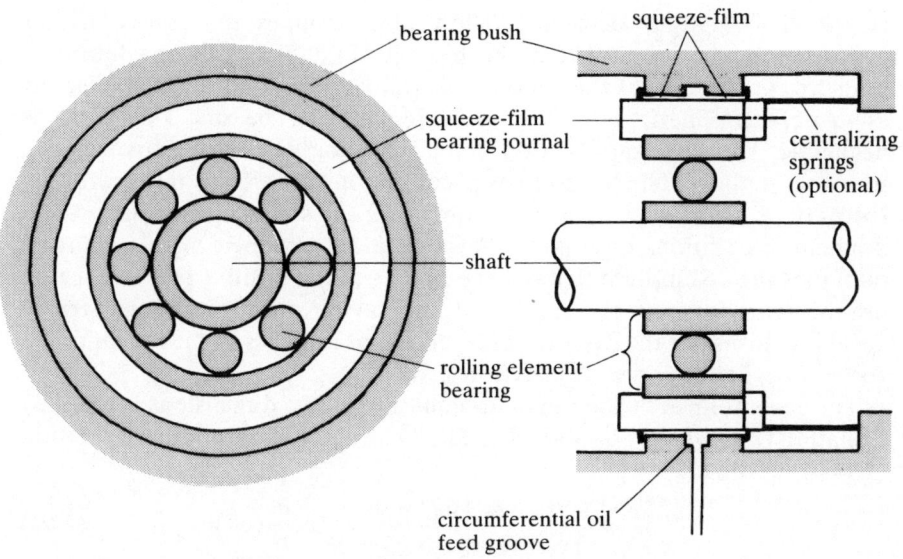

**Figure 2.35** Diagram of a squeeze-film bearing.

reservoir. The bearing may also be fitted with end seals to restrict the lubricant flow out of the bearing clearance. When squeeze-film bearings are included in a machine design they are provided for the purpose of increasing the effective rotor support damping with a view to reducing machine vibrations, especially when operating close to the machine's critical speed. Some designs incorporate centralizing springs which ensure that the shaft is nominally concentric within the squeeze-film bearing, thereby counteracting the effects of a gravity load on the shaft. In the absence of centralizing springs, however, the bearing is still capable of functioning satisfactorily, if correctly designed, and the rotor does not simply drop to the bottom of the clearance space as might be expected after initial consideration; this, of course, would be the case if there were no dynamic load acting. Because the squeeze film is highly non-linear, a correctly designed bearing exhibits a net steady-state lift force to counteract gravity loads and this allows bearing operation without surface-to-surface contact in the presence of a suitable dynamic load.

## 2.5.2 Operation with an unbalanced load

Squeeze-film bearings are generally designed with small length/diameter ratios, and so most researchers adopt the short bearing approximation of the Reynolds equation for the purposes of carrying out a theoretical analysis; see for example Mohan and Hahn (1974), Holmes and Dede

(1980). The analysis described below also assumes that the lubricant properties do not vary within the bearing fluid film, that the lubricant pressures at the ends of the bearing are atmospheric, and that the effect of any sub-atmospheric lubricant pressures on the bearing operation is negligible. This last simplification is imposed simply by setting any negative pressures in the oil film to zero before considering the effects of the resulting forces on the journal; it is in effect equivalent to using the half-Sommerfeld boundary conditions discussed in Section 2.3.2. Implicit in this approximation is the assumption that the extremities of the positive pressure region are the same for both the completed and cavitated films although strictly speaking this is not the case, the difference that it makes to the net fluid film force is negligible – see also Section 2.3.2.

The general form of the Reynolds equation in two dimensions is given by Equation (2.1) which, for a squeeze-film bearing, with zero journal rotation ($U = 0$), becomes

$$\frac{\partial}{\partial x}\left[\frac{\rho h^3}{\mu}\frac{\partial p}{\partial x}\right] + \frac{\partial}{\partial y}\left[\frac{\rho h^3}{\mu}\frac{\partial p}{\partial y}\right] = 12\frac{\partial}{\partial t}(\rho h) \qquad (2.94)$$

Assuming constant lubricant density $\rho$ and dynamic viscosity $\mu$, and invoking the short bearing approximation $\partial p/\partial x = 0$, this equation may be written as

$$\frac{\partial}{\partial y}\left[h^3\frac{\partial p}{\partial y}\right] = 12\mu\frac{\partial h}{\partial t} \qquad (2.95)$$

At any instant in time the journal of the squeeze-film bearing will have a radial velocity $\dot{e}$ where $e$ is the eccentricity, and a tangential velocity $e\dot{\phi}$, where $\phi$ is the attitude angle, as indicated in Figure 2.36. The term $\partial h/\partial t$ on

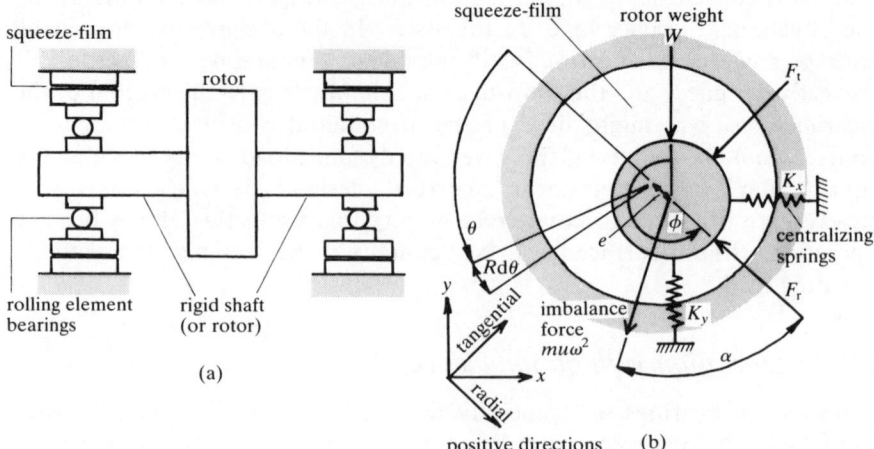

**Figure 2.36** Squeeze-film bearing representation for the purposes of analysis, (a) general system arrangement; (b) diagram of bearing system.

SQUEEZE-FILMS

the right-hand side of Equation (2.95) is the rate of increase of film clearance, and this may now be expressed in terms of the radial and tangential velocities as

$$\frac{\partial h}{\partial t} = e\dot{\phi} \sin \theta + \dot{e} \cos \theta \qquad (2.96)$$

so that Equation (2.95) now becomes

$$\frac{\partial}{\partial y} h^3 \frac{\partial p}{\partial y} = 12\mu(e\dot{\phi} \sin \theta + \dot{e} \cos \theta) \qquad (2.97)$$

Integrating Equation (2.97) twice gives

$$p = \frac{6\mu}{h^3} (e\dot{\phi} \sin \theta + \dot{e} \cos \theta) y^2 + C_1 y + C_2 \qquad (2.98)$$

whereupon applying the axial boundary conditions $p = 0$ at $y = \pm L/2$ (where $L$ is the bearing length and $y$ is measured from the mid-length plane) gives

$$p = \frac{6\mu}{h^3} \left[ y^2 - \frac{L^2}{4} \right] (e\dot{\phi} \sin \theta + \dot{e} \cos \theta) \qquad (2.99)$$

Equation (2.99) gives the nominal oil-film pressure at any particular location around the bearing. If at some location $\theta$ a small element of bearing surface whose area is $R \, d\theta \, dz$ is considered, then the fluid film force acting on this elemental area is

$$dF = pR \, d\theta \, dz \qquad (2.100)$$

acting radially inwards. The components of this force acting in the radial and tangential directions defined in Figure 2.36 are then

$$\begin{aligned} dF_r &= -pR \cos \theta \, d\theta \, dz \\ dF_t &= -pR \sin \theta \, d\theta \, dz \end{aligned} \qquad (2.101)$$

Then the net forces, in the radial and tangential directions, produced by all of the bearing lubricant film, are

$$\begin{aligned} F_r &= -R \int_{-L/2}^{L/2} \int_0^{2\pi} p \cos \theta \, d\theta \, dz \\ F_t &= -R \int_{-L/2}^{L/2} \int_0^{2\pi} p \sin \theta \, d\theta \, dz \end{aligned} \qquad (2.102)$$

However, Equations (2.102) make no allowance for the effects of lubricant film cavitation. As discussed earlier, when considering lubricant film forces all apparently negative pressures should be set to zero. Alternatively, the forces could be calculated by changing the limits of integration in Equations (2.102) so as to include only the positive pressures. From Equation (2.101) it can be seen that the lubricant pressure is positive provided that

$$(e\dot\phi \sin \theta + \dot e \cos \theta) < 0 \tag{2.103}$$

and so the transition from negative to positive pressure must occur when

$$e\dot\phi \sin \theta + \dot e \cos \theta = 0$$

that is when

$$\tan \theta_1 = \left[\frac{-\dot e}{e\dot\phi}\right] \tag{2.104}$$

Then the limits of integration must be between $\theta_1$ and $\theta_1 + \pi$, (the half-Sommerfeld boundary conditions). Further consideration reveals that this is the same as 90° to either side of the direction of the instantaneous velocity of the journal. The instantaneous forces in the radial and tangential directions are then given by

$$F_r = -R \int_{-L/2}^{L/2} \int_{\theta_1}^{\theta_1+\pi} p \cos \theta \, d\theta \, dz$$

$$F_t = -R \int_{-L/2}^{L/2} \int_{\theta_1}^{\theta_1+\pi} p \sin \theta \, d\theta \, dz \tag{2.105}$$

Substituting for $p$ from Equation (2.99) and integrating with respect to $z$, and writing $e = \varepsilon c$ and $h = c(1 + \varepsilon \cos \theta)$ where $\varepsilon$ is the eccentricity ratio and $c$ the bearing concentric radial clearance, gives

$$F_r = \frac{\mu R L^3}{c^2} \left[\varepsilon\dot\phi \int_{\theta_1}^{\theta_1+\pi} \frac{\sin \theta \cos \theta}{(1 + \varepsilon \cos \theta)^3} d\theta + \dot\varepsilon \int_{\theta_1}^{\theta_1+\pi} \frac{\cos^2 \theta}{(1 + \varepsilon \cos \theta)^3} d\theta\right]$$

and $\tag{2.106}$

$$F_t = \frac{\mu R L^3}{c^2} \left[\varepsilon\dot\phi \int_{\theta_1}^{\theta_1+\pi} \frac{\sin^2 \theta}{(1 + \varepsilon \cos \theta)^3} d\theta + \dot\varepsilon \int_{\theta_1}^{\theta_1+\pi} \frac{\sin \theta \cos \theta}{(1 + \varepsilon \cos \theta)^3} d\theta\right]$$

Thus the bearing radial and tangential fluid film forces may be evaluated for any instant in time provided that $\varepsilon$, $\dot\varepsilon$ and $\dot\phi$ are known. The remaining integrals in Equations (2.106) may be evaluated using numerical methods, for example using the trapezium rule, or in closed form analytically. The latter is a relatively cumbersome task but enables subsequent evaluation of the forces to be made much faster once the integral is known. The values of the closed-form solutions are given by Booker (1965).

Knowledge of the oil-firm forces enables the resulting journal orbits within the bearing to be predicted by solving the journal equations of motion. For example, for the rotor mounted on a rigid shaft shown in Figure 2.36(a) the equations of motion in the radial and tangential directions are

$$-F_r + mu\omega^2 \cos\alpha + K_x x \sin\phi - (W + K_y y)\cos\phi = m(\ddot{e} - e\dot{\phi}^2)$$
$$-F_t - mu\omega^2 \sin\alpha + K_x x \cos\phi + (W + K_y y)\sin\phi = m(e\ddot{\phi} + 2\dot{e}\dot{\phi})$$
(2.107)

where $m_u$ is the imbalance moment on the shaft and $\omega$ is the shaft angular velocity, $\alpha$ is the angle between the imbalance force and the eccentricity direction (refer to Figure 2.36). For any particular initial conditions the left-hand sides of Equations (2.107) are now known; it is only $\dot{e}$, $\ddot{e}$, $\dot{\phi}$ and $\ddot{\phi}$, i.e. the right-hand sides, which remain to be determined. One of the simplest forms of numerical solutions is the Euler method which is outlined below.

If the first of Equations (2.107) is taken as an example, where the left-hand side can be evaluated for any particular moment in time $t$ as described earlier, then writing the left-hand side of the equation as $N$ we may write

$$N_t = m\ddot{e} - me\dot{\phi}^2 \qquad (2.108)$$

Over a small time interval $\delta t$, the journal velocity in the radial direction may be approximated by

$$\dot{e} = \frac{e_{t+\delta t} - e_t}{\delta t} = z_t \text{ (say)} \qquad (2.109)$$

similarly $\ddot{e} = \dot{z}$ is given by

$$\dot{z} = \frac{z_{t+\delta t} - z_t}{\delta t}$$

$$= \frac{N}{m} + e\dot{\phi}^2 \qquad (2.110)$$

(using Equation 2.108). Then we may write

$$e_{t+\delta t} = e_t + \delta t z_t$$
$$z_{t+\delta t} = z_t + \delta t \left[\frac{N_t}{m} + e_t \dot{\phi}_t^2\right] \qquad (2.111)$$

Similarly, for the second of Equations (2.107) we may write

$$M_t = me\ddot{\phi} + 2m\dot{e}\dot{\phi} \qquad (2.112)$$

where $M$ is the known left-hand side. Then, developing in a manner as shown above, we may write

$$\phi_{t+\delta t} = \phi_t + \delta t \dot{\phi}_t$$
$$\dot{\phi}_{t+\delta t} = \dot{\phi}_t + \delta t \left[\frac{M_t}{me_t} - 2\frac{z_t}{e_t}\dot{\phi}_t\right] \qquad (2.113)$$

Using this form of representation it is possible to start with some assumed initial conditions to include assumed values of $\dot{\varepsilon}$, $\ddot{\varepsilon}$, $\dot{\phi}$ and $\ddot{\phi}$ as well as those of $\varepsilon$ and $\phi$ and then proceed to determine new values of each of these parameters at some small time interval later on. At this point new values on the left-hand side of Equations (2.107) must also be determined before the process is repeated once more. The process should be repeated until the journal orbit settles down to prescribe some unchanging path; this then is the whirl orbit which the journal and rigid rotor would follow under steady operating conditions. In some instances it may be found that such a steady-state solution is not possible, i.e. the journal orbit is forever changing even after a very large number of iterations; this represents an unstable system and is one which the machine designer should strive to avoid. Some examples of journal whirl orbits so calculated are shown in Figure 2.37.

### 2.5.3 Bistable operation

Squeeze-film bearings, if incorrectly designed, may be susceptible to bistable operation whereupon, depending on initial conditions, two or more

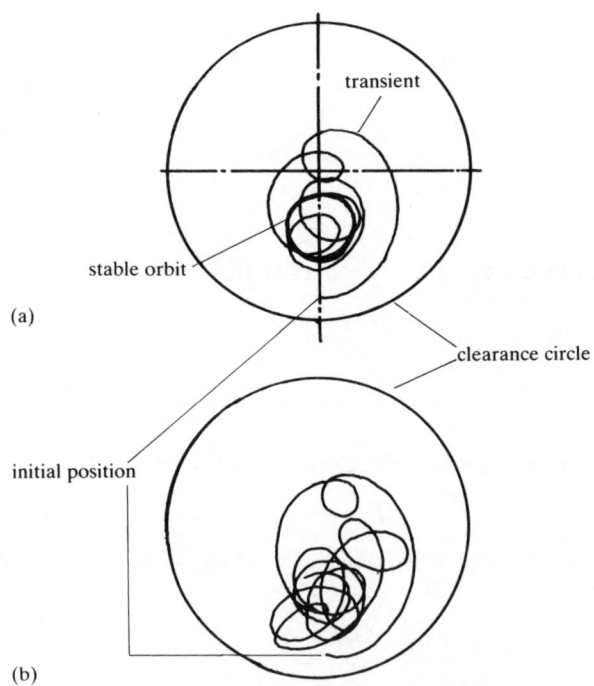

**Figure 2.37** Examples of computed squeeze-film bearing journal trajectories resulting in, (a) stable orbits; (b) unstable orbits.

modes of operation are possible for the same running conditions. One or more of these modes is usually accompanied by high levels of vibration amplitude and large bearing transmitted forces. This phenomenon is depicted in Figure 2.38 where the broken line shows the transmissibility response for a well-designed bearing. In contrast, the solid line shows a case where at a speed parameter of 2.6 (points aa on the diagram) transmissibility can be either 0.69 or 1.1. The speed at which this occurs is called a 'jump' speed since a sudden jump in vibration and transmissibility amplitudes is possible. At a higher speed parameter of 2.8 (points bb on the diagram) bistable operation ceases but reappears when speed parameter is 3.9 (points cc) when transmissibility can be either 0.48 or 1.95 (this is a 'jump down' speed if speed is increasing). Bistable bearing operation is a condition which the designer should strive to avoid because of the possibility of corresponding high vibration amplitudes. Several such modes of operation have been predicted theoretically although not all have been demonstrated experimentally. The subject is currently attracting much research.

## 2.5.4 The $2\pi$ film and pressurization effects

The analysis described in Section 2.5.3 makes use of the '$\pi$ film model', that is it is based on the assumption that the circumferential pressure distribution around the bearing is such that positive film pressures exist only for

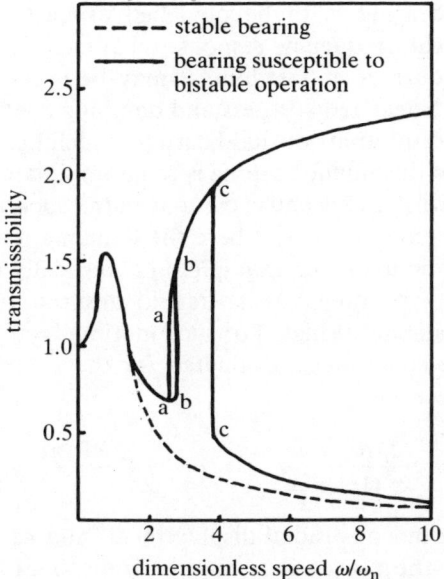

**Figure 2.38** Variation of force transmissibility with dimensionless rotational speed.

180° of bearing arc. If the supply pressure is raised sufficiently high it is possible to ensure that all parts of the bearing oil film have positive oil-film pressures at all times. In this situation the bearing lubricant film extends around the full circumference of the bearing – the '$2\pi$ film'. In order to model this situation theoretically the limits of integration in Equations (2.105) and (2.106) must remain at 0 and $2\pi$. Although negative pressures are now allowed for in the mathematics, this merely serves to represent the pressure throughout the film being raised by some value sufficient to raise the would-be negative pressures above zero. Because this pressure increase is the same throughout the film it has no other effects (the additional force so created on one side of the bearing cancels out that created on the opposite side).

The instantaneous value of radial fluid film force in a $2\pi$ film is always negligible, while that of the tangential force is approximately twice that for the corresponding $\pi$ film. This has an effect on the equivalent bearing stiffness and damping coefficients, as discussed in the following section.

Pressurized squeeze-film bearings operating with (uncavitated) $2\pi$ lubricant films have been shown to behave in a more linear manner than the $\pi$ films (and exhibit a frequency response which more closely resembles that of a system with linear supports). They have also been shown to be less susceptible to bistable operation.

### 2.5.5 *Equivalent linearized stiffness and damping coefficients*

If a squeeze-film bearing is to be modelled as part of the process of producing theoretical frequency response characteristics for a complex system with many degrees of freedom, it may be necessary to model the bearing in terms of linearized stiffness and damping coefficients, similar in nature to those used in other journal bearing modelling.

One way in which this might be done is to approximate the likely journal orbit locus to one with a concentric circular path. Such an approximation would be valid for circumstances where the dynamic load is substantially greater than the static load, for example on a vertically mounted rotor or for a horizontal rotor running close to its critical speed, or for squeeze-film bearings with centralizing springs. For such motion $\dot{\varepsilon} = 0$, $\dot{\phi} = \omega$ and $\theta_1 = \pi$. The resulting force components evaluated for the $\pi$ film from Equations (2.106) are then given by

$$F_\mathrm{r} = \frac{2\mu RL^3 \varepsilon \omega e}{c^3(1-\varepsilon^2)^2}, \qquad F_\mathrm{t} = \frac{\mu RL^3 \pi e \omega}{2c^3(1-\varepsilon^2)^{3/2}} \qquad (2.114)$$

Since $e$ is the instantaneous radial displacement, and $e\omega$ the instantaneous tangential velocity, these equations may be transposed to yield values of equivalent radial spring stiffness and tangential damping coefficient given by

$$K_r = \frac{F_r}{e} = \frac{2\mu RL^3 \varepsilon \omega}{c^3(1-\varepsilon^2)^2}$$

$$C_t = \frac{F_t}{e\omega} = \frac{\mu RL^3 \pi}{2c^3(1-\varepsilon^2)^{3/2}}$$
(2.115)

Inspection of the above equations for $K_r$ and $C_t$ reveals that their values do not change substantially provided that $\varepsilon$ is less than about 0.4, and so may be suitable for journal whirl orbits whose magnitude lies within this range. Equations (2.115) may, of course, be used to evaluate direct spring and damping coefficients in the $x$ and $y$ directions, while all cross-coupling coefficients would be zero.

In situations where it is not appropriate to assume concentric circular orbits of the journal in the bearing, and where stable whirl orbits similar to those shown in Figure 2.37(a) may result, some other form of analysis is more appropriate. In these cases the aim is to establish equivalent values of linearized oil-film spring and damping coefficients which result in a theoretical system response that is the same as that which occurs in practice. The derivation of such equivalent linearized coefficients is still the subject of much research, and coefficient values are still unavailable in closed-form expressions. Recent attempts have based the analysis on the relationship between the dynamic force and the resulting displacement amplitudes (Holmes and Dogan 1982) or on the energy stored and dissipated during a whirl cycle (Hahn 1984). An earlier analytical study by White (1972) has also yielded expressions for eight linearized coefficients suitable for stability analyses, but these too are only applicable when the journal orbit is predominantly circular and of constant eccentricity.

## 2.6 Hydrostatic bearings

### 2.6.1 Description of operation

Hydrostatic journal bearings differ from hydrodynamic bearings in that the lubricant pressure required to separate the bearing surfaces is supplied from some external pressure source and not as a result of journal rotation. The normal arrangement of a hydrostatic journal bearing is similar to that shown in Figure 2.39 where the bearing bush has a number of 'pockets' incorporated within it. The lubricant is supplied to the bearing through flow restrictors, which are normally capillary tubes or orifice plates, before entering the bearing pockets, whereupon its pressure will have reduced to $p_p$ below the supply pressure $p_s$. The lubricant then flows from the bearing pockets out over the bearing lands to a drain area where the pressure is

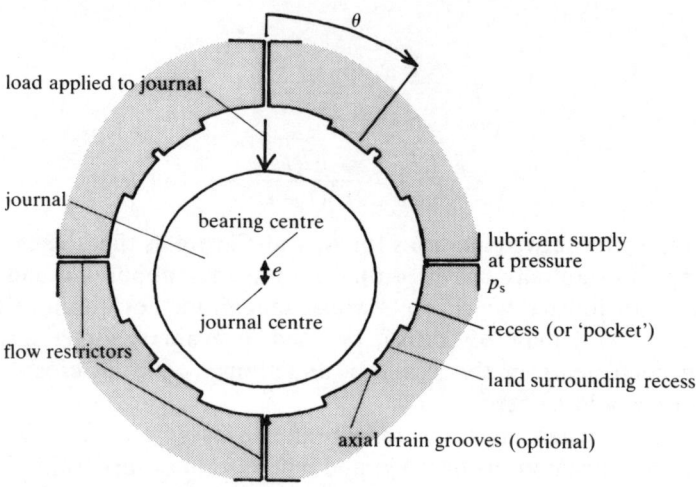

**Figure 2.39** Diagram of a hydrostatic journal bearing with four recesses.

negligible. It will be understood that the flow restrictor forms an important part of the bearing if consideration is given to the consequences of an increased load on one of the bearing 'pads'. Such a load increase would result in a smaller clearance at the pad lands, and hence to an increased resistance to lubricant flow. As a result of this, the pocket pressure would now increase so that a greater proportion of the supply pressure is expended on overcoming flow resistance at the bearing lands. Without the flow restrictor the bearing pocket pressure would always be equal to the supply pressure and no change in bearing load could be accommodated.

Hydrostatic bearings can be designed to have very high oil-film stiffness values and can support very large loads even at zero rotational speed. Because in most cases there is always full film lubrication, there is no wear of the bearing. Furthermore, frictional losses associated with the rotating journal are proportional to rotating speed and are therefore zero at zero speed, giving a low start-up resistance.

## 2.6.2 Steady-state operating characteristics

Theoretical treatments of hydrostatic bearings may take the form of simplified 'lumped parameter' types of analysis, where lubricant flow rates over the bearing lands are approximated in relatively simple expressions. Such types of analysis have been discussed by O'Donoghue *et al.* (1970) and Goodwin *et al.* (1983), amongst others. A similar approach may be used for hybrid bearings (Section 2.6.4). However, a more accurate mathematical model of such bearings may be obtained if a numerical method is used to solve the Reynolds equation for the bearing lubricant pressure distribution,

# HYDROSTATIC BEARINGS

finite difference solutions being the most popular. It is this approach that is described below.

The bearing surface may be modelled in the form of a finite-difference mesh as shown in Figure 2.40 where the boundary conditions are $p = 0$ at the bearing edges, and $p = p_n$ (for $n = 1, ..., 4$) at the bearing pockets; initially the pocket pressures are not known, of course, and these must first be determined. For a given journal position in the bearing, the clearance at each mesh node is known (as described in Section 2.3, for example). If unit pressure is assigned as the value of $p_1$, whilst $p_2$, $p_3$ and $p_4$ are set to zero together with the pressures at the edge nodes, the resulting pressure field may be determined by solving the Reynolds equation

$$\frac{\partial}{\partial \theta}\left[h^3 \frac{\partial p}{\partial \theta}\right] + \left(\frac{R}{L}\right)^2 \frac{\partial}{\partial y}\left[h^3 \frac{\partial p}{\partial y}\right] = 0 \qquad (2.116)$$

in finite-difference form as discussed in Section 2.3.3. It is noteworthy that for hydrostatic operation the effect of journal rotation is negligible, and so the velocity term is excluded from the Reynolds equation.

When the pressure field has been obtained, the resulting flow out of each bearing pocket may be evaluated numerically. For one mesh point $M$ on the edge of pocket 1 in Figure 2.40 the flow out is

$$Q = \frac{h^3}{12\mu}\left(\frac{dp}{dx}\right)\Delta y \qquad (2.117)$$

where $\Delta y$ is the distance between mesh points in the $y$ sense.

A similar expression should be evaluated for all mesh points around all pockets, so enabling calculation of the flow out of each pocket when unit pressure is the boundary condition at pocket 1. The resulting flow values so calculated are the flow coefficients for unit pressure at pocket 1, and may be designated $Q_{1,1}$, $Q_{1,2}$, $Q_{1,3}$ and $Q_{1,4}$. (All except $Q_{11}$ will, of course, be negative). By changing the boundary conditions so that unit pressure is next

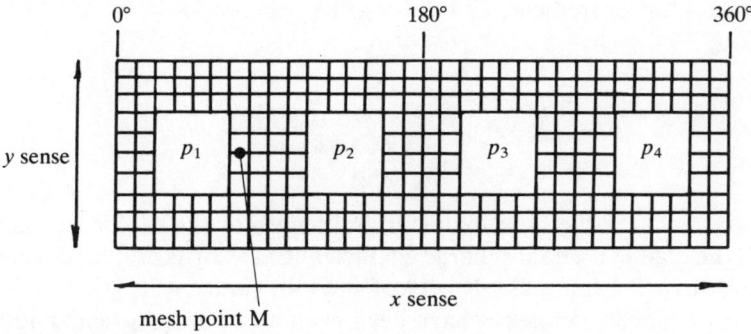

**Figure 2.40** Developed view of finite difference mesh modelling a four-recess hydrostatic bearing. $M$ is a mesh point.

set as the value only of $p_2$, then of $p_3$, and finally of $p_4$ the remaining flow coefficients in the flow matrix

$$\begin{bmatrix} Q_{1,1} & Q_{1,2} & Q_{1,3} & Q_{1,4} \\ Q_{2,1} & Q_{2,2} & Q_{2,3} & Q_{2,4} \\ Q_{3,1} & Q_{3,2} & Q_{3,3} & Q_{3,4} \\ Q_{4,1} & Q_{4,2} & Q_{4,3} & Q_{4,4} \end{bmatrix} \quad (2.118)$$

may be evaluated.

For any one bearing pocket, the net flow in under steady operating conditions is zero. In other words,

$$\begin{matrix} \text{flow in through} \\ \text{restrictor} \end{matrix} + \begin{matrix} \text{flow in over bearing} \\ \text{land from other} \\ \text{pockets} \end{matrix} - \begin{matrix} \text{flow out over bearing} \\ \text{land due to pocket} \\ \text{pressure} \end{matrix} = 0$$

(2.119)

For pocket number $i$, Equation (2.199) may be written, in the case of a capillary compensated bearing, as

$$K_c(p_s - p_i) - \sum_{j=1 \neq i}^{j=4} Q_{j,i} p_j - Q_{i,i} p_i = 0 \quad (2.120)$$

where $p_s$ is the bearing lubricant supply pressure and $K_c$ is the capillary restrictor coefficient. In the case of an orifice-compensated bearing, Equation (2.120) becomes

$$K_0(p_s - p_i)^{1/2} - \sum_{j=1 \neq i}^{j=4} Q_{j,i} p_j - Q_{i,i} p_i = 0 \quad (2.121)$$

Equation (2.120) (or 2.121) may be written for each bearing pocket and the resulting pocket pressures then determined by relaxation, as described by O'Donoghue *et al.* (1970), or by solving the pocket flow equations simultaneously, as described by Ghosh and Majumdar (1978).

The restrictor coefficients required in the analysis may be obtained by experimental measurement, or by using the expressions

$$K_c = \frac{3\pi d_c^4}{32l}$$

$$K_0 = \frac{3\pi C_d d_0 \mu}{(\rho/2)^{1/2}}$$

(2.122)

where $d_c$ and $l$ are the bore diameter and length of the capillaries, and $d_0$ and $C_d$ the diameter and discharge coefficient of the orifices; $\mu$ and $\rho$ are the dynamic viscosity and mass density of the lubricant.

When the pocket pressures have been evaluated for a particular journal position, the actual pressure distribution over the lands may then be determined since all boundary condition pressures are then known. The bearing load may then be evaluated by numerically integrating the lubricant pressure over all of the bearing surface, as described for journal bearings in Equations (2.10) and (2.11).

The lubricant flow rate may be determined numerically by summing the outflow from the ends of the bearing using the expression

$$Q = 2 \sum_{j=1}^{m} \frac{h_{i,j}^3}{12\mu} \left(\frac{\mathrm{d}p}{\mathrm{d}y}\right)_{i,j} \Delta x \qquad (2.123)$$

as in the case of hydrodynamic bearings, or by summing the flow through the restrictor elements into each of the $n$ bearing pockets

$$Q = \sum_{i=1}^{i=n} K_c(p_s - p_i) \text{ or } \sum_{i=1}^{i=n} K_o(p_s - p_i)^{1/2} \qquad (2.124)$$

The variation of bearing load capacity and lubricant flow rate with journal eccentricity, for a four-pocket bearing, is shown in Figures 2.41 and 2.42.

## 2.6.3 Dynamic characteristics

When evaluating the bearing characteristics under dynamic operation, consideration must be given to the effects of squeezing the lubricant between the journal and the bearing lands. The Reynolds equation must

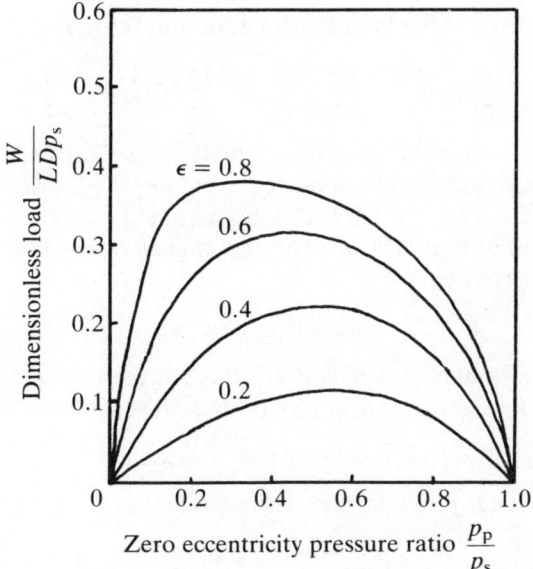

**Figure 2.41** Variation of load variable with pressure ratio and eccentricity ratio for a capillary compensated hydrostatic journal bearing.

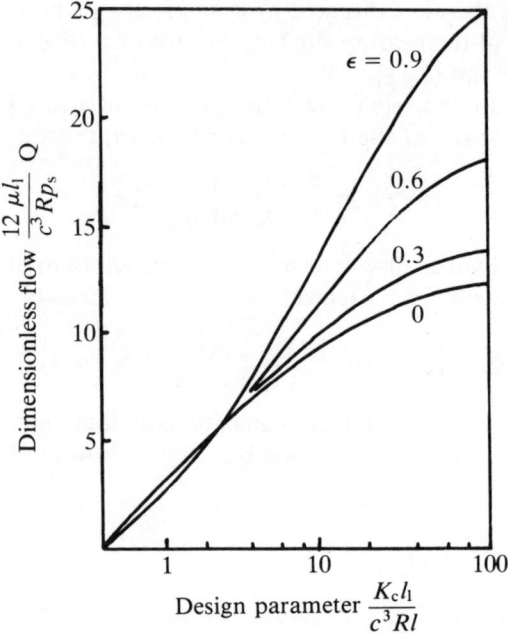

**Figure 2.42** Variation of bearing lubricant flow rate $Q$ with design parameter and eccentricity ratio for a capillary compensated hydrostatic journal bearing; $l_1$ is the axial pocket $l$ and length.

therefore be written to include the squeeze-film term as

$$\frac{\partial}{\partial \theta} h^3 \frac{\partial p}{\partial \theta} + \left(\frac{R}{L}\right)^2 \frac{\partial}{\partial y} h^3 \frac{\partial p}{\partial y} = 12\mu R^2 \frac{dh}{dt} \quad (2.125)$$

If the journal is assumed to execute small sinusoidal oscillations, in 'the direction of the original static load for example, the lubricant pressures at each point in the bearing may also be assumed to vary in a sinusoidal manner. The initial film clearance at any angle $\theta$ to the line of centres (see Figure 2.10) is given by

$$h_0 = c(1 + \varepsilon \cos \theta) \quad (2.126)$$

where $e$ is the journal eccentricity and $c$ the bearing radial clearance. Under dynamic load the clearance $h$ at any point varies as

$$h = h_0 + h_1 \exp(j\omega t)\cos \theta \quad (2.127)$$

whereupon the local pressure fluctuations take the form

$$p = p_0 + p_1 \exp(j\omega t) \quad (2.128)$$

where $p_0$ is the steady-state pressure and $p_1$ the dynamic component which is complex in general. Substituting Equations (2.126)–(2.128) into Equation

(2.125), and neglecting products of small terms, gives

$$\frac{\partial}{\partial \theta}(\overset{*}{h_0^3} + 3h_0^2 h_1 \cos\theta \exp(j\omega t)) \frac{\partial}{\partial \theta}(\overset{*}{p_0} + p_1 \exp(j\omega t))$$

$$+ \left(\frac{R}{L}\right)^2 \frac{\partial}{\partial y}(\overset{*}{h_0^3} + 3h_0^2 h_1 \cos\theta \exp(j\omega t)) \frac{\partial}{\partial y}(\overset{*}{p_0} + p_1 \exp(j\omega t))$$

$$= 12\mu R^2 \frac{\partial}{\partial t}(\overset{*}{h_0} + h_1 \exp(j\omega t)\cos\theta) \quad (2.129)$$

The products of terms marked * represent the bearing steady operating characteristics as given by Equation (2.116) where $\partial h_0/\partial t = 0$, and so may be deleted from the equation. The remaining terms describe the bearing dynamic characteristics and may be written separately, after dividing throughout by $\exp(j\omega t)$, as

$$\frac{\partial}{\partial \theta} h_0^3 \frac{\partial p_1}{\partial \theta} + \frac{\partial}{\partial \theta}\left(3h_0^2 h_1 \frac{\partial p_0}{\partial \theta} \cos\theta\right) + \left(\frac{R}{L}\right)^2 \frac{\partial}{\partial y} h_0^3 \frac{\partial p_1}{\partial y}$$

$$+ \left(\frac{R}{L}\right)^2 \frac{\partial}{\partial y} 3h_0^2 h_1 \frac{\partial p_0}{\partial y} \cos\theta = j\omega\, 12\mu R^2 h_1 \cos\theta \quad (2.130)$$

It is Equation (2.130) which must now be solved at each of the finite-difference mesh points to determine the lubricant pressure distribution associated with the dynamic load. In order to do this, the appropriate pocket pressure boundary conditions must first be determined.

Consideration of flow continuity for one bearing pocket $i$ reveals that the flow in through the restrictor must equal that out over the lands plus the increase in volume due to separation of the journal and bearing bush surfaces. That is,

$$K_c(p_s - p_i) = Q_i p_i + A \frac{dh}{dt} \quad (2.131)$$

or, for orifice compensation,

$$K_0(p_s - p_i)^{1/2} = Q_i p_i + A \frac{dh}{dt}$$

where $Q_i$ is the flow out of the bearing pocket per unit pocket pressure, $p_i$ is the pocket pressure, $A$ the pocket area and $dh/dt$ the rate of change of bearing clearance at the pocket location. If the clearance varies according to Equation (2.127), then for small-amplitude oscillations we may write

$$p_i = p_{i0} + p_{i1} \exp(j\omega t) \quad (2.132)$$

and

$$Q_i = Q_{i0} + Q_{i1} \exp(j\omega t) \quad (2.133)$$

where $Q_{io}$ is the steady-state flow out of the bearing pocket and $Q_{i1}$ is given by

$$Q_{i1} = \frac{\partial Q_{io}}{\partial h} h_1 \qquad (2.134)$$

which is evaluated from steady-state data. Substituting Equations (2.127) and (2.132)–(2.134) into Equation (2.131), and neglecting products of small terms, gives

$$p_{i1} = \frac{-(p_{io}Q_{i1} + j\omega A h_1 \cos \theta)}{(K_c + Q_{io})} \qquad (2.135)$$

or, for orifice compensation,

$$p_{i1} = \frac{-(p_{io}Q_{i1} + j\omega A h_1 \cos \theta)}{K_0(p_s - p_{io})^{-1/2} + Q_{io}} \qquad (2.136)$$

Equations (2.135) and (2.136) give the expressions for the pocket pressures which should be used as boundary conditions when solving Equation (2.130) numerically, using the original finite-difference mesh. Where the computer to be used cannot handle complex numbers, it will be necessary to separate 'in phase' and 'quadrature' components of the equations, solving each part of Equation (2.130) independently. The components of dynamic load which are 'in phase' and at 'quadrature' to the journal displacements, $W_r$ and $W_j$ respectively, may then be evaluated by integrating the resulting lubricant pressures over the bearing surface as before.

The bearing oil film may then be modelled as spring and damping coefficients given by $K$ and $C$ respectively where

$$K = \frac{W_r}{h_1} \qquad (2.137)$$

and

$$C = \frac{W_j}{\omega h_1} \qquad (2.138)$$

Variation of bearing stiffness and damping with design pressure ratio $P_p/P_s$ and steady eccentricity ratios are shown in Figures 2.43 and 2.44.

In general, when the bearing geometry is not symmetrical about the axis coincident with the direction of the journal eccentricity, bearing cross-coupling coefficients are required to model the bearing. Where appropriate, such coefficients could be determined as shown above for the direct stiffness and damping coefficients, except that the 'in-phase' and 'quadrature' loads $W_r$ and $W_j$ are replaced by those acting in a direction normal to the assumed journal motion. Cross-coupling coefficients obtained in this way have been described by Rowe and Chong (1986).

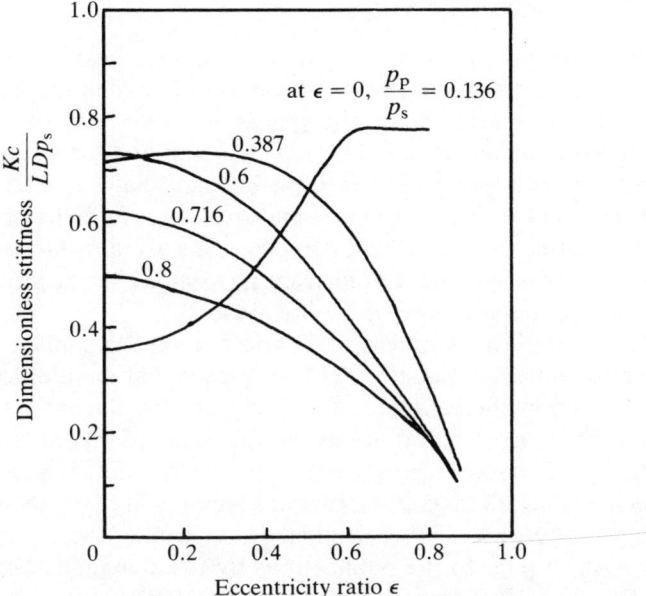

**Figure 2.43** Variation of bearing dimensionless stiffness $K$ with eccentricity ratio $\varepsilon$ and pressure ratio $p_p/p_s$.

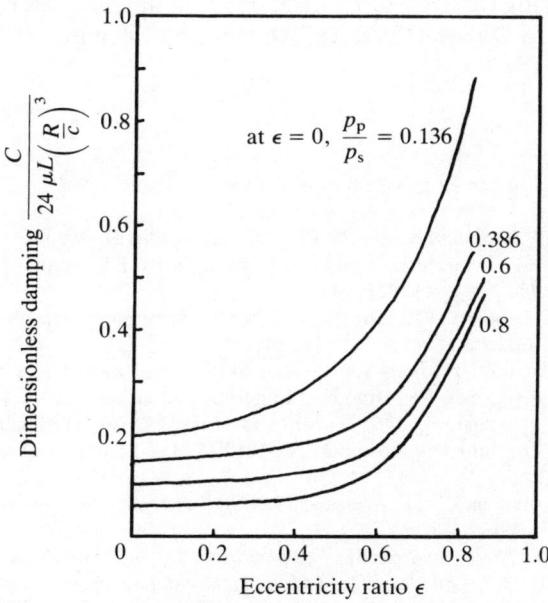

**Figure 2.44** Variation of bearing dimensionless damping with eccentricity ratio $\varepsilon$ and pressure ratio $P_p/P_s$.

## 2.6.4 Hybrid bearings

Hybrid bearings are a special class of hydrostatic bearing where a significant part of the bearing load-carrying capacity is generated by hydrodynamic action at the bearing lands. In appearance they may be similar to multi-recess hydrostatic bearings, or may be similar to hole or grooved entry externally pressurized gas bearings, but may have a relatively larger land area. Hybrid bearings have the same advantages as hydrodynamic bearings, but also offer increased load-carrying capacity and film stiffness without substantially increasing the lubricant flow rate. They also enable surface-to-surface contact at zero speed to be avoided.

The theoretical treatment of hybrid bearings is similar to that outlined above for pure hydrostatic bearings, except that an additional term has to be included in the equations for lubricant flow in order to allow for the effects of journal rotation, as in the case of hydrodynamic bearings. Lubricant pressure distribution, flow rates and bearing load-carrying capacities may all be calculated using numerical methods described earlier in this chapter. For a more detailed description of these methods the reader may wish to refer to the publications by O'Donoghue *et al.* (1970), Heller and Shapiro (1969), and Rohde and Ezzat (1976). Alternatively a 'lumped parameter' type of approximation may be employed which enables closed-form expressions to be written for lubricant flow rates over the bearing lands; these can be used to calculate the approximate pressure distribution and load carrying capacity of the bearing. Examples of this type of analysis can be found in Davies (1969) and Davies and Leonard (1970).

## References

Booker, J. F. 1965. A table of journal bearing integral. *Trans. ASME., J. Basic Eng.* June. **87D**, 533–535.

Cameron, A. 1981. *Basic lubrication theory*. Chichester: Ellis Horwood.

Davies, P. B. 1969. A general analysis of multi-recess hydrostatic journal bearings. *Proc. IMechE.* **184**, pt. 1, **43**, 827–838.

Davies, P. B. & R. Leonard 1970. The dynamic behaviour of multi-recess hydrostatic journal bearings. *Proc. IMechE.* **184**, pt.3L, 139–147.

ESDU 1966. Calculation methods for steadily loaded pressure fed hydrodynamic journal bearings. Engineering Science Data Unit, London, Item no. 66023.

Elrod, H. G. & S. B. Malanoski 1960. Theory and design data for continuous film, self-acting journal bearings of finite length. Report I-A 2049-13. The Franklin Institute, Philadelphia, Pa., November.

Eschmann, P., Hasbargen, L. & K. Weigand 1985. *Ball and roller bearings, theory, design, and application*. New York: Wiley.

Fuller, D. D. 1984. *Theory and practice of lubrication for engineers*. New York: Wiley.

Ghosh, M. K. & B. C. Majumdar 1978. Stiffness and damping characteristics of hydrostatic multirecess oil journal bearings. *Int. J. Mach. Tool Des. Res.* **18**, 139–151.

Goodwin, M. J., Hooke, C. J. & J. E. T. Penny 1983. Controlling the dynamic characteristics

# REFERENCES

of a hydrostatic bearing by using a pocket-connected accumulator. *Proc. IMechE.* **197C**, December, 225–258.

Grassam, N. S. & J. W. Powell 1964. *Gas lubricated bearings.* London: Butterworth.

Gumbel, L. 1921. Verleich der Ergebnisse der rechnerischen Behandlung des Lagerschmierungsproblem mit neueren Versuchsergebnissen. *Mbl. berl. bez.* September, Verein Deutscher Ingenieure, Dusseldorf.

Hahn, E. J. 1984. Equivalent stiffness and damping coefficients for squeeze film dampers. *IMechE Conf. on Vibrations in Rotating Machinery*, York. Paper C325/84. 507–514.

Harris, T. A. 1966. *Rolling bearing analysis.* New York: Wiley.

Heller, S. & W. Shapiro 1969. A numerical solution for the incompressible hybrid journal bearing with cavitation. *Trans. ASME., J. Lub. Tech.*, July, **91F**, 508–515.

Holmes, R. 1960. The vibration of a rigid shaft on short sleeve bearings. *J. Mech. Eng. Sci.* **2** (4), 337–341.

Holmes, R. & M. Dede. 1980. Dynamic pressure determinations in a squeeze-film damper. *MechE Conf. on Vibrations in Rotating Machinery*, York. Paper C260/80, 71–75.

Holmes, R. & M. Dogan 1982. Investigation of a rotor bearing assembly incorporating a squeeze film damper bearing. *J. Mech. Eng. Sci.* **24** (3), 129–137.

Lund, J. W. 1964. The hydrostatic gas journal bearing with journal rotation and vibration. *Trans. ASME, J. Basic Eng.* **86**, 328–336.

Lund, J. W., Arwas, E. B., Cheng, H. S., Ng, C. W. & C. H. Pan 1965. Rotor-bearing dynamics design technology, pt. III: design handbook for fluid film type bearings. Technical report AFAPL-TR-65-45, Aero Propulsion Lab, Wright-Patterson Air Force Base, Ohio.

Lund, J. W. 1966. Self excited stationary whirl orbits of a journal in a sleeve bearing. Ph.D. dissertation, Rensselaer Polytechnic Institute.

Majumdar, B. C. 1978. On the analytical solution of external pressurized porous gas journal bearings. *Trans. ASME., J. Lub. Tech.*, **100**, July, 442–444.

Mohan, S. & E. J. Hahn 1974. Design of squeeze film damper supports for rigid rotors. *Trans. ASME., J. Eng. Ind.*, August, **96B**, 976–982.

MTI 1972. *Gas bearing design manual*, D. F. Wilcock (ed.). Mechanical Technology Incorporated, Latham, NY.

Mullan, P. J. & H. H. Richardson 1964. Plane vibration of the inherently compensated gas journal bearing; analysis and comparison with experiment. *Trans. ASLE.*, **7**, 277–287.

Ocvirk, F. W. 1952. Short bearing approximation for full journal bearings. NASA. Techn. note 2808.

O'Donoghue, J. P., Rowe, W. B. & C. J. Hooke 1970. Computer anaysis of externally pressurized journal bearings. *Proc. IMechE.* **184**, pt. 3L, 48–53.

Palmgren, A. 1959. *Ball and roller bearing engineering*, 3rd edn., SKF Industries, Philadelphia, Pa.: Burbank.

Pinkus, O. & B. Sternlicht 1961. *Theory of hydrodynamic lubrication.* New York: McGraw-Hill.

Powell, J. W. 1970. *Design of aerostatic bearings* Brighton: Machinery Publishing.

Raimondi, A. S. 1961. A numerical solution for the gas lubricated full journal bearing of finite length. *Trans.* ASLE, April, 131–155.

Rao, J. S. & A. Mukherjee 1977. Stiffness and damping coefficients of tilted journal bearings. *Mechanism and Machine Theory.* **12**, 339ff.

Rao, N. S. 1977. Analysis of the stiffness and damping characteristics of an externally pressurized porous gas journal bearing. *Trans. ASME., J. Lub. Tech.*, April, **99**, 295–301.

Reynolds, O. 1886. On the theory of lubrication and its application to Mr. Beauchamp Tower's experiments, including an experimental determination of the viscosity of olive oil. *Phil. Trans. Roy. Soc. London* **177**, pt. 1, 157–234.

Rohde, S. M. & H. A. Ezzat 1976. On the dynamic behaviour of hybrid journal bearings. *Trans. ASME., J. Lub. Tech.*, **98F**, Jan., 90–94.

Rowe, W. B. & F. S. Chong 1986. Computation of dynamic force coefficients for hybrid (hydrostatic/hydrodynamic) journal bearings by the finite disturbance and perturbation techniques. *Tribology International* **19**, no. 5, Oct., 260–271.

Shires, G. L. & D. Pantall, 1963. The aerostatic jacking of a vented aerodynamic journal bearing. *Proc. IMechE.*, Lub. and Wear Grp. Conv., May, 87–96.

Stieber, W. 1933. *Das Schwimmlager*. Berlin: Krayn.

Swift, H. W. 1932. The stability of lubricating films in journal bearings. *Proc. Inst. Civ. Engrs.* **233**.

Szeri, A. Z. 1980. *Tribology – friction, lubrication, and wear*. New York: Hemisphere.

White, D. C. 1972. The dynamics of a rigid rotor supported on squeeze film bearings. *Proc. IMechE Conf. Vibrations in Rotating Systems*, London, February, 213–229.

White, M. F. 1979. Rolling element bearing vibration transfer characteristics: effect of stiffness. *J. Appl. Mech.*, **46**, September, 677–684.

# 3 Single-mass rotor dynamics

## 3.1 Introduction

In many rotating machines the rotor mass may be concentrated at one location on the shaft, and the shaft itself may be relatively light. In such situations relatively simple design calculations may be all that is required to predict the behaviour of the assembled system. In other situations the rotor mass may not be concentrated at one point, but may be approximated to being so in order to aid in the design of the machine. Even in situations where neither of these cases apply it is still possible to obtain a fundamental understanding of rotating machine dynamics in general by studying such systems, as described in this chapter. The chapter introduces the reader to the behaviour of rotating machines under the action of dynamic loading and shows how various design parameters influence machine characteristics, such as vibration amplitude and force transmitted to the foundations.

The first part of the chapter deals with the simplest system – that consisting of a flexible shaft running in rigid bearings; here the reader is introduced to the concept of machine critical speeds and their implications for the machine behaviour. Rotor classification is also discussed. The next two sections deal, in turn, with the effects of anisotropic bearing flexibility and of bearing damping and cross-coupling on the characteristics of a machine with a single-mass rigid rotor. This study is then extended, in the subsequent section, to include the effects of mounting the single-mass rotor on a light flexible shaft. The later parts of the chapter discuss the effects of foundation flexibility, aerodynamic effects and gyroscopic effects.

## 3.2 Flexible shaft in rigid bearings

In some situations the system under consideration may be idealized as one consisting of a single mass $m$ mounted on a light flexible shaft which runs in rigid bearings. Such a system is shown in Figure 3.1. The centre of gravity of the rotor $G$ is offset from the rotor geometric centre $C$ by a distance $e$. The instantaneous lateral and angular deflections of the shaft at the location of the rotor are $y$, in the vertical direction, and $\theta$ in the $y$–$z$ plane, as shown. The force and moment acting between the rotor and shaft are $F_y$ and $M_{yz}$; the rotor also has an unbalance force $me\omega^2$ acting in the direction

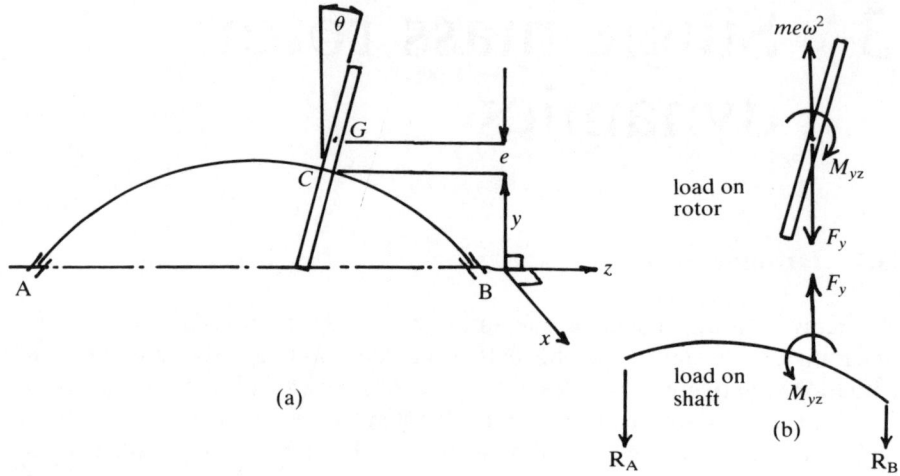

**Figure 3.1** (a) Single mass rotor mounted on a light flexible shaft running in rigid bearings, (b) Loads applied to rotor and shaft. See text for details.

shown. The relationship between the loading on the shaft, $F_y$ and $M_{yz}$, and the shaft deflection, $y$ and $\theta$, is given by simple beam deflection theory (Roark and Young 1984) and takes the form

$$\begin{Bmatrix} y \\ \theta \end{Bmatrix} = \begin{bmatrix} a^2 b^2/3EIl & -(3a^2l - 2a^3 - al^2)/3EIl \\ ab(b-a)/3EIl & -(3al - 3a^2 - l^2)3EIl \end{bmatrix} \begin{Bmatrix} F_y \\ M_{yz} \end{Bmatrix} \quad (3.1)$$

where $l$ is the shaft length AB, $E$ is Young's modulus of elasticity for the shaft material, $I$ is the shaft second moment of area, and where $a$ is the distance AC and $b$ the distance CB in Figure 3.1. Equation (3.1) can be written more simply as

$$\begin{Bmatrix} y \\ \theta \end{Bmatrix} = \begin{bmatrix} \alpha_{11} & \alpha_{12} \\ \alpha_{21} & \alpha_{22} \end{bmatrix} \begin{Bmatrix} F_y \\ m_{yz} \end{Bmatrix} \quad (3.2)$$

where the terms $\alpha_{11}$, $\alpha_{12}$, etc. are influence coefficients relating the shaft deformation to the applied loading. Equation (3.2) may then be rearranged to give

$$\begin{Bmatrix} F_y \\ M_{yz} \end{Bmatrix} = \begin{bmatrix} k_{11} & k_{12} \\ k_{21} & k_{22} \end{bmatrix} \begin{Bmatrix} y \\ \theta \end{Bmatrix} \quad (3.3)$$

which may be written more simply as

$$\{p\} = [k]\{d\} \quad (3.4)$$

where the matrix $[k]$ is the inverse of the matrix containing the elements $\alpha$ in Equation (3.2).

The equations of motion for the rotor mass, in the $y$ and $\theta$ directions, are written

$$me\omega^2 - F_y = m\ddot{y}, \qquad M_{yz} = I_d\ddot{\theta} \qquad (3.5)$$

where $me\omega^2$ is the imbalance force acting on the rotor, and $I_d$ is the rotor moment of inertia about a diameter. Noting that for sinusoidal vibrations $\ddot{y} = -\omega^2 y$ and $\ddot{\theta} = -\omega^2\theta$, expressions for $F_y$ and $M_{yz}$ may be substituted from Equations (3.3) into (3.5) which may then be written

$$\begin{Bmatrix} me\omega^2 \\ 0 \end{Bmatrix} = \begin{bmatrix} (k_{11} - m\omega^2) & k_{12} \\ k_{21} & (k_{22} - I_d\omega^2) \end{bmatrix} \begin{Bmatrix} y \\ \theta \end{Bmatrix} \qquad (3.6)$$

which may then be rearranged to give

$$\begin{Bmatrix} y \\ \theta \end{Bmatrix} = \begin{bmatrix} (k_{11} - m\omega^2) & k_{12} \\ k_{21} & (k_{22} - I_d\omega^2) \end{bmatrix}^{-1} \begin{Bmatrix} me\omega^2 \\ 0 \end{Bmatrix} \qquad (3.7)$$

or, more simply

$$\{d\} = [Z]\{f\} \qquad (3.8)$$

Equation (3.7) gives the variation of rotor deformation in the vertical plane with running speed. The motion of the rotor in the $x$–$z$ plane takes the same form, but is 90° out of phase with that in the $y$–$z$ plane because the forcing is 90° out of phase.

A plot of $y/e$ against frequency ratio $\omega/\omega_n$ for the case of a rotor mounted at shaft mid-span, a system first analysed by Jeffcott (1919), is shown in Figure 3.2. It can be seen that as speed increases from zero to $\omega_n$, vibration amplitude increases from zero to infinity. At speeds greater than $\omega_n$ vibration amplitude decreases, tending to a minimum value of $-e$ at very high speeds. The significance of the negative vibration amplitudes is that the shaft deflects in the opposite direction to the unbalance eccentricity $e$. It follows from this that at very high speeds where $y$ tends to $-e$, the rotor mass centre $G$ moves towards the line joining the bearing centres. Other analyses of Jeffcott rotor-type systems have also been carried out by Gunter (1966) and Rieger (1982).

The frequency $\omega_n$ is known as the system 'whirling speed' or 'critical speed'. It is noteworthy that whirling is not a vibration of the shaft in the sense that a point on the shaft surface is subjected to a fluctuating stress, it is simply that the shaft rotates in a bent configuration (in the shape of a banana). In the more general case, when the rotor is not mounted at the centre of the shaft, there are two critical speeds – one associated mainly with lateral motion of the rotor and one associated with angular motion.

In the above system the forces transmitted through the bearings are those which are related to the deflection of the shaft. Simple mechanics theory shows that the bearing reaction forces at A and B in Figure 3.1 are related to

# SINGLE-MASS ROTORS

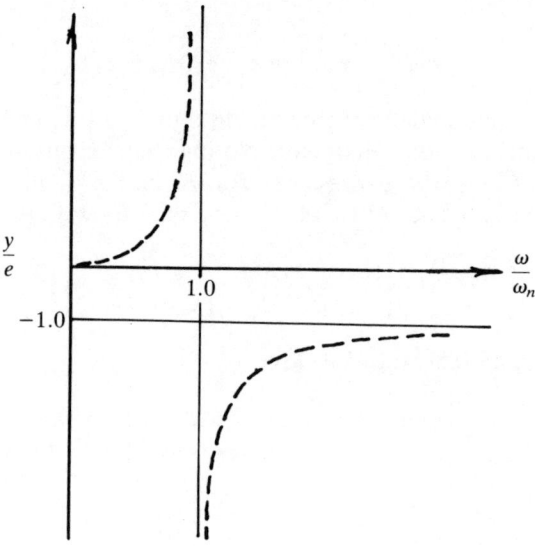

**Figure 3.2** Variation of dimensionless rotor vibration amplitude $y/e$ with frequency ratio $\omega/\omega_n$ for the system shown in Figure 3.1, rotor at mid-span.

the loading on the shaft, $F_y$ and $M_{yz}$, by the equations

$$\begin{Bmatrix} R_A \\ R_B \end{Bmatrix} = \begin{bmatrix} b/l & -1/l \\ a/l & 1/l \end{bmatrix} \begin{Bmatrix} F_y \\ M_{yz} \end{Bmatrix} \quad (3.9)$$

which may be written as

$$\{R\} = [I]\{p\} \quad (3.10)$$

Equation (3.10) may then be developed, using Equations (3.4) and (3.8), to give

$$\begin{aligned} \{R\} &= [I][k]\{d\} \\ &= [I][k][Z]\{f\} \\ &= [C]\{f\} \end{aligned} \quad (3.11)$$

which may be written more explicitly as

$$\begin{Bmatrix} R_A \\ R_B \end{Bmatrix} = \begin{bmatrix} c_{11} & c_{12} \\ c_{21} & c_{22} \end{bmatrix} \begin{Bmatrix} me\omega^2 \\ 0 \end{Bmatrix} \quad (3.12)$$

It can be shown from the above that the forces transmitted through the bearings are also a maximum at the system critical speed. These forces are of course, dynamic forces and are superimposed on any steady loads which may be present, due to gravity loading for example.

In real systems which are designed to operate above their critical speed the machine would normally be run through the critical speed very quickly so that the very large vibrations and forces associated with resonance do not have sufficient time to build up. The same may also be true during run-down when some form of braking may be employed. If the system is run at the critical speed and the vibrations are allowed to build up then either the shaft will fracture and a catastrophic failure will result, or there may be sufficient damping in the system (not allowed for in the above analysis) to simply limit the vibration and force amplitudes to some very high value.

In the system considered above the rotor vibration was accommodated by flexure of the shaft. In other types of machine the shaft itself is so rigid that it may be considered as part of the rotor, and most of the vibration is accommodated by flexure of the bearings. The mathematical treatment of such machines differs from that presented above and is developed below. The latter type of machine is also sometimes referred to as a 'rigid rotor' or 'rigid shaft' machine, as compared with the former which is sometimes referred to as a 'flexible rotor' or 'flexible shaft' machine. The term 'rigid rotor' may also include other types of machine when used in the context of balancing (ISO 1967), and is discussed more fully in Chapter 7.

EXAMPLE 3.1

A light horizontal shaft is supported at each end in bearings which behave as pinned supports (with a high radial stiffness and zero tilt stiffness). The shaft has a diameter of 15 mm and a length of 0.6 m, and carries a 9 kg rotor at mid-span. The rotor out-of-balance is 0.003 kg.m. Calculate the system critical speed associated with lateral motion of the rotor, and determine the operating speed range which will ensure that the shaft bending stress does not exceed 150 MN.m.$^{-2}$ Assume $E = 200$ GN.m$^{-2}$

SOLUTION

The influence coefficients in Equation (3.2) relating force and deflection at shaft mid-span (the rotor location) are given by standard beam theory as

$$\alpha_{11} = \frac{l^3}{48EI}, \qquad \alpha_{12} = \alpha_{21} = 0$$

whereupon

$$k_{11} = \frac{48EI}{l^3} = \frac{48 \times 200 \times 10^9 \times \pi \times 0.015^4}{0.6^3 \times 64} = 0.11 \text{ MN.m}^{-1}$$

where the symbols take their usual meaning. The rotor displacement $y$ may then be obtained from Equation (3.7) as

$$\begin{Bmatrix} y \\ \theta \end{Bmatrix} = \begin{bmatrix} k_{11} - m\omega^2 & 0 \\ 0 & k_{22} - I_d\omega^2 \end{bmatrix}^{-1} \begin{Bmatrix} me\omega^2 \\ 0 \end{Bmatrix}$$

that is

$$y = \frac{me\omega^2}{k_{11} - m\omega^2}$$

which is infinite when

$$\omega = \sqrt{\frac{k_{11}}{m}} = \sqrt{\frac{0.11 \times 10^6}{9}} = 111 \text{ rad.s}^{-1}$$

$$= 1057 \text{ rev.min}^{-1}$$

The maximum bending stress in the shaft due to gravity loading is at the shaft mid-span, and is given by

$$\sigma = \frac{My}{I} = \frac{9 \times 9.81 \times 0.6}{4} \times \frac{0.015}{2} \times \frac{64}{\pi \times 0.015^4} = 40 \text{ MN.m}^{-2}$$

where the symbols take their usual meaning. The 'static' deflection due to gravity loading is

$$\delta = mg \frac{l^3}{48EI} = \frac{9 \times 9.81 \times 0.6^3 \times 64}{48 \times 200 \times 10^9 \times \pi \times 0.015^4} = 0.8 \text{ mm}$$

If the allowable stress is 150 MN.m$^{-2}$, then the additional shaft deflection which can be permitted as a consequence of rotor vibration is

$$y = 0.8 \frac{(150 - 40)}{40} = 2.2 \text{ mm}$$

Then, because the rotor eccentricity is

$$e = \frac{0.003}{9} = 0.33 \text{ mm}$$

substitution into Equation (3.3) gives

$$\frac{y}{e} = \pm \frac{2.2}{0.33} = \frac{m\omega^2}{k_{11} - m\omega^2} = \frac{1}{k_{11}/m\omega^2 - 1} = \frac{1}{(\omega_n/\omega)^2 - 1}$$

which may be rearranged to give

$$\omega = \left[\frac{0.33}{\pm 2.2} + 1\right] 111$$

$$= 103.5 \text{ rad.s}^{-1} \text{ or } 120.4 \text{ rad.s}^{-1}$$

Therefore, the allowable running speed range is

$$\omega < 103.5 \text{ rad.s}^{-1} \text{ or } \omega > 120.4 \text{ rad.s}^{-1}$$

that is

$$\omega < 988 \text{ rev.min}^{-1} \text{ or } \omega > 1150 \text{ rev.min}^{-1}$$

## 3.3 Symmetrical rigid shaft in flexible anisotropic bearings

In systems where the bearings are far more flexible than the shaft it is the bearings which will have the greatest influence on the motion of the rotor. Such systems may be idealized as rigid rotor systems similar to that shown in Figure 3.3. In this system the shaft does not flex and the bearings are assumed to behave as linear springs having a stiffness $k_x$ in the horizontal direction and $k_y$ in the vertical direction. If the rotor has mass $m$ whose centre is offset from the rotor geometric centre by distances $e$ and $d$, and if $x$ and $y$ are the linear displacements of the rotor in the horizontal and vertical directions respectively, and $\phi$ and $\theta$ are the angular displacements of the rotor in the $x$–$z$ and $y$–$z$ planes respectively, then the equations of motion for the rotor may be written as

$$\left.\begin{array}{l} me\omega^2 \cos \omega t - k_x x = m\ddot{x} \\ me\omega^2 \sin \omega t - k_y y = m\ddot{y} \\ me\omega^2 d \cos \omega t - k_x l\phi = I_d \ddot{\phi} \\ me\omega^2 d \sin \omega t - k_y l\theta = I_d \ddot{\theta} \end{array}\right\} \quad (3.13)$$

Remembering that for sinusoidal vibrations $\ddot{x} = -\omega^2 x$, $\ddot{y} = -\omega^2 y$,

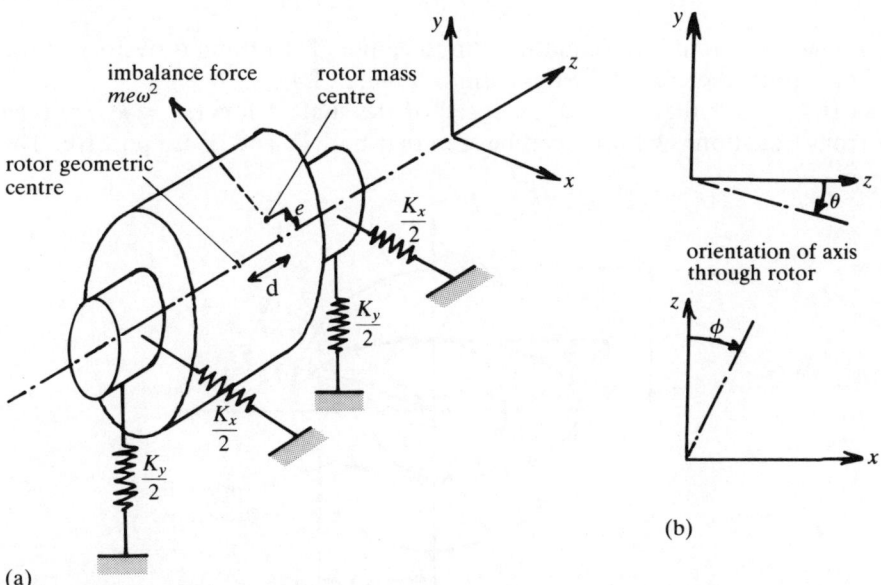

**Figure 3.3** (a) Rigid rotor running in flexible isotropic bearings, (b) Convention for positive angular motion of rotor relative to cartesian axes.

## SINGLE-MASS ROTORS

$\ddot{\phi} = -\omega^2\phi$ and $\ddot{\theta} = -\omega^2\theta$, Equations (3.13) may be rearranged to give

$$\left.\begin{array}{l} x = \dfrac{me\omega^2}{k_x - m\omega^2} \cos \omega t = X \cos \omega t \\[2ex] y = \dfrac{me\omega^2}{k_y - m\omega^2} \sin \omega t = Y \sin \omega t \end{array}\right\} \quad (3.14a)$$

$$\left.\begin{array}{l} \phi = \dfrac{me\omega^2 d}{k_x l - \omega^2 I_d} \cos \omega t = \Phi \cos \omega t \\[2ex] \theta = \dfrac{me\omega^2 d}{k_y l - \omega^2 I_d} \sin \omega t = \Theta \sin \omega t \end{array}\right\} \quad (3.14b)$$

From consideration of Equations (3.14) it is apparent that this system has four critical speeds, two for lateral vibration in the horizontal and vertical directions and two for angular vibration in the $x$–$z$ and $y$–$z$ planes.

Squaring both sides of Equations (3.14a) and then adding them together gives

$$\frac{x^2}{X^2} + \frac{y^2}{Y^2} = 1 \quad (3.15a)$$

and similarly from (3.14b) we obtain

$$\frac{\phi^2}{\Phi^2} + \frac{\theta^2}{\Theta^2} = 1 \quad (3.15b)$$

Equation (3.15a) is the equation for an ellipse. The orbital trajectory of the rotor must therefore take the form in Figure (3.4).

If $k_x$ is less than $k_y$ and the speed of the shaft is low ($\omega^2 < k_x/m$) then from Equations (3.14a) it can be seen that both $X$ and $Y$ are positive. The

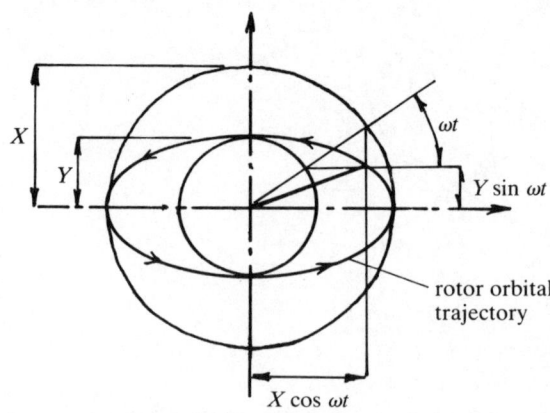

**Figure 3.4** Rotor whirl orbits for the system shown in Figure 3.3.

rotor therefore whirls in the same direction as the rotation of the shaft. If however the speed of rotation is between those corresponding to the critical speeds in the horizontal and vertical directions ($k_x/m < \omega^2 < k_y/m$), then $X$ becomes negative indicating that the rotor horizontal displacement is always in the opposite direction to that it adopted when operated at low speeds. It follows from this that the rotor must whirl in the opposite direction to that in which the shaft is rotating. At higher shaft speeds ($\omega^2 > k_y/m$) both $X$ and $Y$ are negative and consideration of Figure 3.4, with both displacements in the opposite direction to that shown, reveals that the rotor whirls once more in the same direction as its rotation.

Equation (3.15b), relating to the angular motion of the rotor, is also the equation of an ellipse; this means that there is an elliptical orbital trajectory of the rotor ends due to angular motion of the rotor. This rotor motion is caused by the imbalance couple $me\omega^2 d$ acting on the rotor, and it is superimposed on the lateral motion of the rotor described in the preceding paragraph. A reversal of the direction of the orbit associated with this motion also occurs, between the two critical speeds associated with angular motion of the rotor.

The amplitude of the force transmitted to the bearings is now different in the horizontal and vertical directions, as well as at each end of the rotor. The force transmitted is that which causes the bearings to deform and is given by the product of spring stiffness and rotor deflection at the bearing. The force amplitudes are accordingly

$$F_x = k_x(x \pm l\theta)/2$$
$$F_y = k_y(y \pm l\phi)/2 \quad (3.16)$$

in the horizontal and vertical directions respectively. The sign in the bracketed terms depends upon which end of the rotor is being considered, that is it depends upon whether the angular motion of the rotor causes the rotor end to deflect in the same or opposite direction to the lateral deflection of the rotor. Once more it is clear that these bearing forces must take on maximum values when the system is operated at the critical speeds, where $X$, $Y$, $\phi$ and $\theta$ are maximum.

## 3.4 Symmetrical rigid shaft in flexible anisotropic bearings with damping and cross-coupling

In some machines the bearings have associated damping properties as well as spring stiffness properties, for example in the case of oil-film lubricated bearings. Furthermore in the case of hydrodynamic bearings the shaft motion in the horizontal direction is coupled with that in the vertical direction. In most applications the properties of such bearings are described

# SINGLE-MASS ROTORS

in terms of the eight linearized bearing stiffness and damping coefficients as introduced in Chapter 2. This section describes how the response of a rigid rotor supported in such bearings may be calculated.

Consider the system shown in Figure 3.5 where the rotor is subjected to an imbalance force $me\omega^2$ located some distance $d$ from the rotor centre of gravity. The out of balance forces in the horizontal and vertical directions may then be written as

$$F_x = me\omega^2 \sin \omega t$$
$$F_y = me\omega^2 \cos \omega t \qquad (3.17a)$$

and the moments about the rotor centre of gravity caused by these forces are

$$M_{xz} = me\omega^2 d \sin \omega t$$
$$M_{yz} = me\omega^2 d \cos \omega t \qquad (3.17b)$$

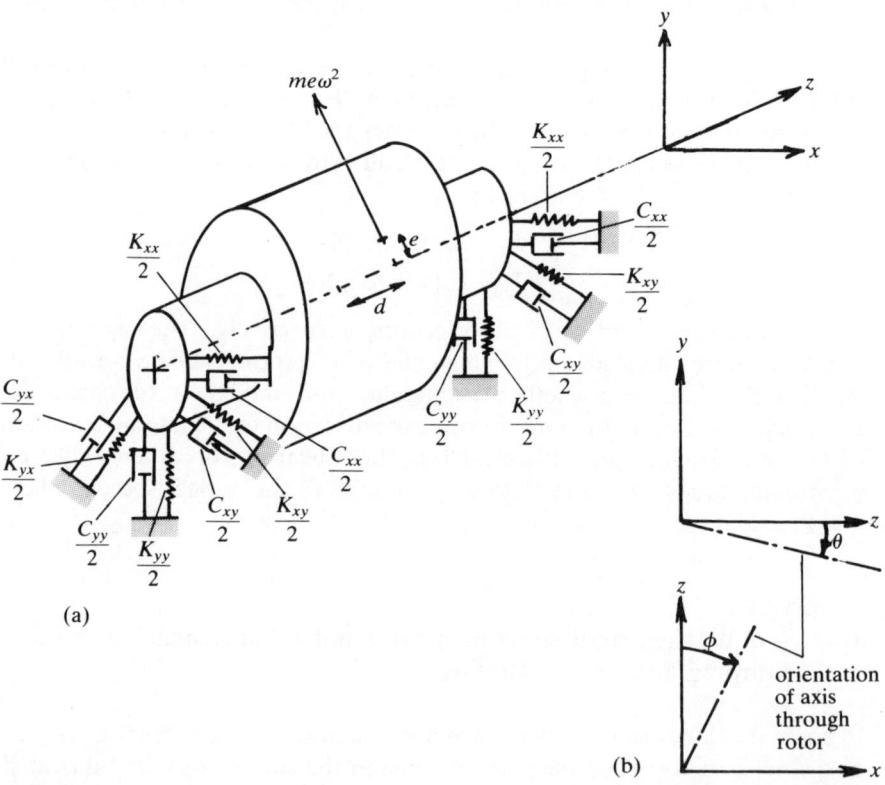

**Figure 3.5** (a) Rigid rotor running in flexible anisotropic bearings with damping and cross-coupling. See text for explanation; (b) Convention for positive angular motion of rotor relative to cartesian axes.

## DAMPING AND CROSS-COUPLING

in the $x$–$z$ and $y$–$z$ planes respectively. The subsequent displacements of the rotor will lag behind the respective forcing because of the effects of damping and cross-coupling. The displacements of the rotor centre of gravity can be written in the form

$$\left.\begin{aligned} x = X \sin(\omega t - \alpha) &= X(\sin \omega t \cos \alpha - \cos \omega t \sin \alpha) \\ &= X_1 \sin \omega t + X_2 \cos \omega t \\ y = Y \cos(\omega t - \beta) &= Y(\cos \omega t \cos \beta + \sin \omega t \sin \beta) \\ &= Y_1 \cos \omega t + Y_2 \sin \omega t \end{aligned}\right\} \quad (3.18a)$$

whilst the angular displacements of the rotor about its centre of gravity may be written as

$$\left.\begin{aligned} \phi = \Phi \sin(\omega t - \gamma) &= \Phi(\sin \omega t \cos \gamma - \cos \omega t \sin \gamma) \\ &= \Phi_1 \sin \omega t + \Phi_2 \cos \omega t \\ \theta = \Theta \cos(\omega t - \delta) &= \Theta(\cos \omega t \cos \delta + \sin \omega t \sin \delta) \\ &= \Theta_1 \cos \omega t + \Theta_2 \sin \omega t \end{aligned}\right\} \quad (3.18b)$$

from which it follows that the corresponding rotor linear and angular velocities are given by

$$\left.\begin{aligned} \dot{x} &= \omega X_1 \cos \omega t - \omega X_2 \sin \omega t \\ \dot{y} &= -\omega Y_1 \sin \omega t + \omega Y_2 \cos \omega t \\ \dot{\phi} &= \omega \Phi_1 \cos \omega t - \omega \Phi_2 \sin \omega t \\ \dot{\theta} &= -\omega \Phi_1 \sin \omega t + \omega \Theta_2 \cos \omega t \end{aligned}\right\} \quad (3.19)$$

and the corresponding accelerations are given by

$$\left.\begin{aligned} \ddot{x} &= -\omega^2 X_1 \sin \omega t - \omega^2 X_2 \cos \omega t \\ \ddot{y} &= -\omega^2 Y_1 \cos \omega t - \omega^2 Y_2 \sin \omega t \\ \ddot{\phi} &= -\omega^2 \Phi_1 \sin \omega t - \omega^2 \Phi_2 \cos \omega t \\ \ddot{\theta} &= -\omega^2 \Theta_1 \cos \omega t - \omega^2 \Theta_2 \sin \omega t \end{aligned}\right\} \quad (3.20)$$

The equations of motion of the rotor are given by

$$\left.\begin{aligned} F_x - K_{xx}x - K_{xy}y - C_{xx}\dot{x} - C_{xy}\dot{y} &= m\ddot{x} \\ F_y - K_{yx}x - K_{yy}y - C_{yx}\dot{x} - C_{yy}\dot{y} &= m\ddot{y} \\ M_{xz} - K_{xx}l\phi - K_{xy}l\theta - C_{xx}l\dot{\phi} - C_{xy}l\dot{\theta} &= I_d\ddot{\phi} \\ M_{yz} - K_{yx}l\phi - K_{yy}l\theta - C_{yx}l\dot{\phi} - C_{yy}l\dot{\theta} &= I_d\ddot{\theta} \end{aligned}\right\} \quad (3.21)$$

in the horizontal and vertical directions, and about the $y$ and $x$ axes respectively, where $K_{xx}$, $K_{xy}$, etc. are the eight linearized bearing stiffness and damping coefficients for both bearings (twice those for one bearing, assuming a symmetrical system).

Substituting Equations (3.17)–(3.20) into Equations (3.21), and comparing coefficients of sin $\omega t$ and cos $\omega t$ on each side of each equation, results in the following eight equations:

$$\left.\begin{array}{l} me\omega^2 = -m\omega^2 X_1 + K_{xx}X_1 + K_{xy}Y_2 - C_{xx}\omega X_2 - C_{xy}\omega Y_1 \\ 0 = -m\omega^2 X_2 + K_{xx}X_2 + K_{xy}Y_1 + C_{xx}\omega X_1 + C_{xy}\omega Y_2 \\ me\omega^2 = -m\omega^2 Y_1 + K_{yx}X_2 + K_{yy}Y_1 + C_{yx}\omega X_1 + C_{yy}\omega Y_2 \\ 0 = -m\omega^2 Y_2 + K_{yx}X_1 + K_{yy}Y_2 - C_{yx}\omega X_2 - C_{yy}\omega Y_1 \end{array}\right\} \quad (3.22a)$$

$$\left.\begin{array}{l} me\omega^2 d = -\omega^2 I_d \Phi_1 + K_{xx}l^2\Phi_1 + K_{xy}l^2\Theta_2 - \omega C_{xx}l^2\Phi_2 - \omega C_{xy}l^2\Theta_1 \\ 0 = -\omega^2 I_d \Phi_2 + K_{xx}l^2\Phi_2 + K_{xy}l^2\Theta_1 + \omega C_{xx}l^2\Phi_1 + \omega C_{xy}l^2\Theta_2 \\ me\omega^2 d = -\omega^2 I_d \Theta_1 + K_{yx}l^2\Phi_2 + K_{yy}l^2\Theta_1 + \omega C_{yx}l^2\Phi_1 + \omega C_{yy}l^2\Theta_2 \\ 0 = -\omega^2 I_d \Theta_2 + K_{yx}l^2\Phi_1 + K_{yy}l^2\Theta_2 - \omega C_{yx}l^2\Phi_2 - \omega C_{yy}l^2\Theta_1 \end{array}\right\} \quad (3.22b)$$

Equation (3.22a) can be expressed in matrix form as

$$\begin{Bmatrix} me\omega^2 \\ 0 \\ me\omega^2 \\ 0 \end{Bmatrix} = \begin{bmatrix} K_{xx} - m\omega^2 & -\omega C_{xx} & -\omega C_{xy} & K_{xy} \\ \omega C_{xx} & K_{xx} - m\omega^2 & K_{xy} & \omega C_{xy} \\ \omega C_{yx} & K_{yx} & K_{yy} - m\omega^2 & \omega C_{yy} \\ K_{yx} & -\omega C_{yx} & -\omega C_{yy} & K_{yy} - m\omega^2 \end{bmatrix} \begin{Bmatrix} X_1 \\ X_2 \\ Y_1 \\ Y_2 \end{Bmatrix}$$

(3.23a)

or more simply as

$$\{F_l\} = [Z_l]\{u_l\} \quad (3.24a)$$

where the suffix '*l*' indicates reference to linear displacements and forces. Similarly Equation (3.22b) can be written in matrix form as

$$\begin{Bmatrix} me\omega^2 d \\ 0 \\ me\omega^2 d \\ 0 \end{Bmatrix} = \begin{bmatrix} K_{xx}l^2 - \omega^2 I_d & -\omega C_{xx}l^2 & -\omega C_{xy}l^2 & K_{xy}l^2 \\ \omega C_{xx}l^2 & K_{xx}l^2 - \omega^2 I_d & K_{xy}l^2 & \omega C_{xy}l^2 \\ \omega C_{yx}l^2 & K_{yx}l^2 & K_{yy}l^2 - \omega^2 I_d & \omega C_{yy}l^2 \\ K_{yx}l^2 & -\omega C_{yx}l^2 & -\omega C_{yy}l^2 & K_{yy}l^2 - \omega^2 I_d \end{bmatrix} \begin{Bmatrix} \Phi_1 \\ \Phi_2 \\ \Theta_1 \\ \Theta_2 \end{Bmatrix}$$

(3.23b)

which may be written more simply as

$$\{F_a\} = [Z_a]\{u_a\} \quad (3.24b)$$

where the suffix '*a*' indicates reference to angular displacements and forcing. Equations (3.24a) and (3.24b) may be combined to give the matrix equation

$$\begin{Bmatrix} \{F_l\} \\ \{F_a\} \end{Bmatrix} = \begin{bmatrix} [Z_l] & 0 \\ 0 & [Z_a] \end{bmatrix} \begin{Bmatrix} \{u_l\} \\ \{u_a\} \end{Bmatrix}$$

which can also be written as

$$\{F\} = [Z]\{u\} \tag{3.25}$$

Both sides of Equation (3.25) can now be pre-multiplied by the inverse of the $[Z]$ matrix to enable the displacement matrix to be defined as

$$\{u\} = [Z]^{-1}\{F\} \tag{3.26}$$

so that the amplitudes $X_1 X_2 Y_1 Y_2$ and $\Phi_1 \Phi_2 \Theta_1 \Theta_2$ are now known. The displacement amplitudes of the rotor are then given by

$$\left. \begin{array}{ll} X = (X_1^2 + X_2^2)^{1/2}, & Y = (Y_1^2 + Y_2^2)^{1/2} \\ \Phi = (\Phi_1^2 + \Phi_2^2)^{1/2}, & \Theta = (\Theta_1^2 + \Theta_2^2)^{1/2} \end{array} \right\} \tag{3.27}$$

and the corresponding phase lag angles are given by

$$\left. \begin{array}{ll} \alpha = \tan^{-1}\left[\dfrac{-X_2}{X_1}\right], & \beta = \tan^{-1}\left[\dfrac{Y_2}{Y_1}\right] \\ \gamma = \tan^{-1}\left[\dfrac{-\Phi_2}{\Phi_1}\right], & \delta = \tan^{-1}\left[\dfrac{\Theta_2}{\Theta_1}\right] \end{array} \right\} \tag{3.28}$$

The resulting shaft whirl orbit can be plotted using Equations (3.18) and will be found to take the form in Figure 3.6. The form of the orbit is still elliptical but the major and minor axes no longer line up with the bearing $x$ and $y$ axes. The force vectors given by Equations (3.17) are also shown on the diagram and can be seen to precede the displacement vectors.

The forces which are transmitted to the bearings are those which deform the bearing lubricant film, and do not include rotor inertia terms. In general the bearing force will lag behind the imbalance force such that the bearing horizontal and vertical force components, at end A of the machine, can be

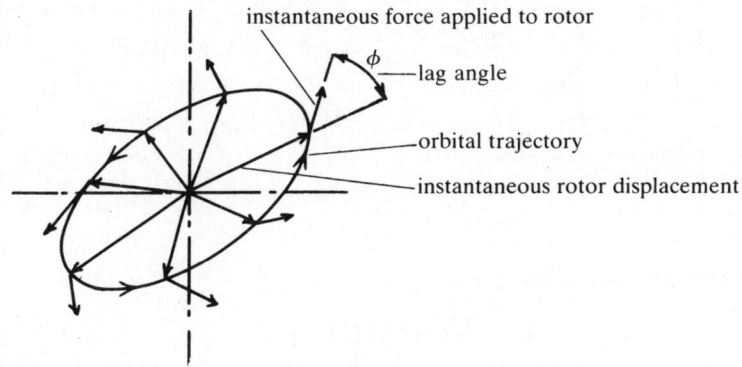

**Figure 3.6** Rotor whirl orbits for the system shown in Figure 3.5, showing force and displacement vectors at different time intervals.

represented as

$$_A f_{bx} = {_A F_{bx}} \sin(\omega t - \varepsilon) = {_A F_{bx1}}(\sin \omega t \cos \varepsilon - \cos \omega t \sin \varepsilon)$$
$$= {_A F_{bx1}} \sin \omega t + {_A F_{bx2}} \cos \omega t$$
$$_A f_{by} = {_A F_{by}} \cos(\omega t - \zeta) = {_A F_{by}}(\cos \omega t \cos \zeta + \sin \omega t \sin \zeta)$$
$$= {_A F_{by1}} \cos \omega t + {_A F_{by2}} \sin \omega t$$
(3.29a)

and at end B of the machine as

$$_B f_{bx} = {_B F_{bx}} \sin(\omega t - \eta) = {_B F_{bx}}(\sin \omega t \cos \eta - \cos \omega t \sin \eta)$$
$$= {_B F_{bx1}} \sin \omega t + {_B F_{bx2}} \cos \omega t$$
$$_B f_{by} = {_B F_{by}} \cos(\omega t - \lambda) = {_B F_{by}}(\cos \omega t \cos \lambda + \sin \omega t \sin \lambda)$$
$$= {_B F_{by1}} \cos \omega t + {_B F_{by2}} \sin \omega t$$
(3.29b)

These forces are related to the shaft displacement by the eight bearing stiffness and damping coefficients for the individual bearings at ends A and B by

$$_A f_{bx} = {_A K_{xx}}x + {_A K_{xy}}y + {_A C_{xx}}\dot{x} + {_A C_{xy}}\dot{y} + {_A K_{xx}}l\phi + {_A K_{xy}}l\theta + {_A C_{xx}}l\dot{\phi} + {_A C_{xy}}l\dot{\theta}$$
$$_A f_{by} = {_A K_{yx}}x + {_A K_{yy}}y + {_A C_{yx}}\dot{x} + {_A C_{yy}}\dot{y} + {_A K_{yx}}l\phi + {_A K_{yy}}l\theta + {_A C_{yx}}l\dot{\phi} + {_A C_{yy}}l\dot{\theta}$$
$$_B f_{bx} = {_B K_{xx}}x + {_B K_{xy}}y + {_B C_{xx}}\dot{x} + {_B C_{xy}}\dot{y} - {_B K_{xx}}l\phi - {_B K_{xy}}l\theta - {_B C_{xx}}l\dot{\phi} - {_B C_{xy}}l\dot{\theta}$$
$$_B f_{by} = {_B K_{yx}}x + {_B K_{yy}}y + {_B C_{yx}}\dot{x} + {_B C_{yy}}\dot{y} - {_B K_{yx}}l\phi - {_B K_{yy}}l\theta - {_B C_{yx}}l\dot{\phi} - {_B C_{yy}}l\dot{\theta}$$

(3.30)

Substituting Equations (3.18), (3.19) and (3.29) into Equation (3.30) and comparing coefficients of $\sin \omega t$ and $\cos \omega t$ on each side of both equations as before leads to the following matrix equation:

$$\begin{Bmatrix} _A F_{bx1} \\ _A F_{bx2} \\ _A F_{by1} \\ _A F_{by1} \\ _B F_{bx1} \\ _B F_{bx2} \\ _B F_{by1} \\ _B F_{by2} \end{Bmatrix} = \begin{bmatrix} \begin{bmatrix} K_{xx} & -\omega C_{xx} & -\omega C_{xy} & K_{xy} & K_{xx}l & -\omega C_{xx}l & -\omega C_{xy}l & K_{xy}l \\ \omega C_{xx} & K_{xx} & K_{xy} & \omega C_{xy} & \omega C_{xx}l & K_{xx}l & K_{xy}l & \omega C_{xy}l \\ \omega C_{yx} & K_{yx} & K_{yy} & \omega C_{yy} & \omega C_{yx}l & K_{yx}l & K_{yy}l & \omega C_{yy}l \\ K_{yx} & -\omega C_{yx} & -\omega C_{yy} & K_{yy} & K_{yx} & -\omega C_{yx}l & -\omega C_{yy}l & K_{yy} \end{bmatrix}_A \\ \begin{bmatrix} K_{xx} & -\omega C_{xx} & -\omega C_{xy} & K_{xy} & -K_{xx}l & \omega C_{xx}l & \omega C_{xy}l & -K_{xy}l \\ \omega C_{xx} & K_{xx} & K_{xy} & \omega C_{xy} & -\omega C_{xx}l & -K_{xx}l & -K_{xy}l & -\omega C_{xy}l \\ \omega C_{yx} & K_{yx} & K_{yy} & \omega C_{yy} & -\omega C_{yx}l & -K_{yx}l & -K_{yy}l & -\omega C_{yy}l \\ K_{yx} & -\omega C_{yx} & -\omega C_{yy} & K_{yy} & -K_{yx} & \omega C_{yx}l & \omega C_{yy}l & -K_{yy} \end{bmatrix}_B \end{bmatrix} \begin{Bmatrix} X_1 \\ X_2 \\ Y_1 \\ Y_2 \\ \Phi_1 \\ \Phi_2 \\ \Theta_1 \\ \Theta_2 \end{Bmatrix}$$

(3.31)

which can be written more simply as

$$\{F_b\} = [I]\{u\}$$
(3.32)

and which can be evaluated to yield the 'in-phase' and 'quadrature' bearing force amplitudes. The amplitudes of the forces transmitted to the bearings are then given by

$$_A F_{bx} = (_A F_{bx1}^2 + _A F_{bx2}^2)^{1/2}, \quad _A F_{by} = (_A F_{by1}^2 + _A F_{by2}^2)^{1/2}$$
$$_B F_{bx} = (_B F_{bx1}^2 + _B F_{bx2}^2)^{1/2}, \quad _B F_{by} = (_B F_{by1}^2 + _B F_{by2}^2)^{1/2}$$
(3.33)

whilst the corresponding phase lag angles are given by

$$\varepsilon = \tan^{-1}\left[\frac{-_A F_{bx2}}{_A F_{bx1}}\right], \quad \zeta = \tan^{-1}\left[\frac{_A F_{by2}}{_A F_{by1}}\right]$$
$$\eta = \tan^{-1}\left[\frac{-_B F_{bx2}}{_B F_{bx1}}\right], \quad \lambda = \tan^{-1}\left[\frac{_B F_{by2}}{_B F_{by1}}\right]$$
(3.34)

## 3.5 Asymmetrical flexible shaft in flexible anisotropic bearings with damping and cross-coupling

In systems which have both a flexible shaft and flexible bearings, as shown in Figure 3.7, the analysis must allow for different instantaneous displace-

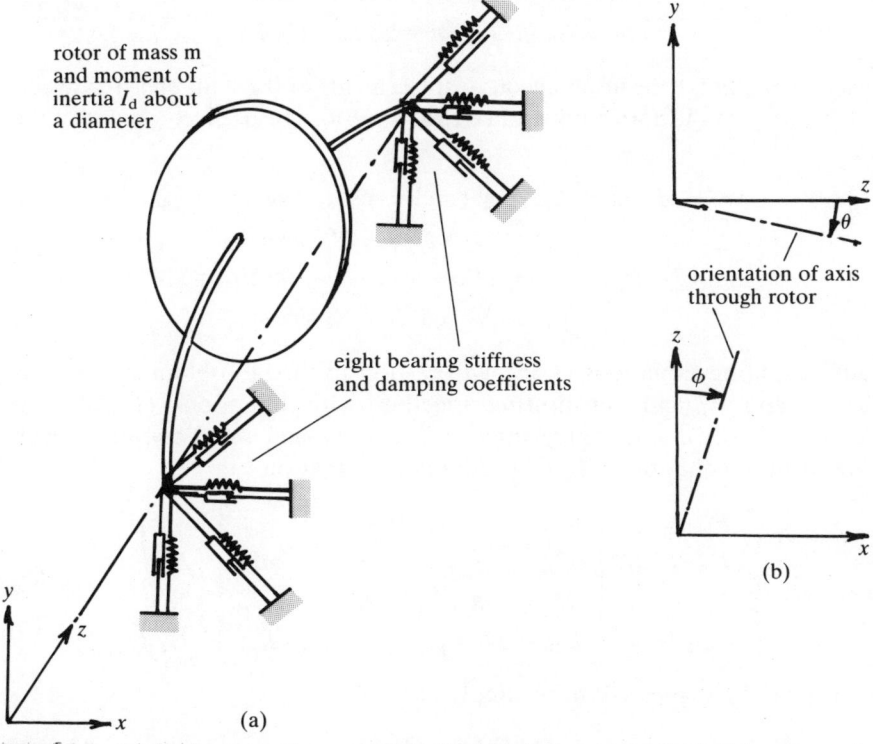

**Figure 3.7** (a) Single mass rotor on a light flexible shaft, running in flexible anisotropic bearings (represented as in Figure 3.5) with damping and cross-coupling; (b) Convention for positive angular motion of shaft relative to cartesian axes.

ments of the shaft at the rotor and at the bearings. In general the system will behave in a similar manner to that described in Section 3.4 except that the flexibility of the shaft will increase the overall flexibility of the support system as seen by the rotor. The analytical treatment described below, for this type of system, is one which is similar in nature to that used in Section 3.4 with the exception that an equivalent set of system stiffness and damping coefficients is first evaluated, which allows for the flexibility of the shaft in addition to that of the bearings, and is used in place of the bearing coefficients in Section 3.4. This approach has also been described by Goodwin *et al.* (1984) and Ogrodnik *et al.* (1985); it also enables the designer to include in the analysis the effects of aerodynamic cross-coupling, should the likely magnitude of this be known.

The total deflection of the rotor shown in Figure 3.7 is the vector sum of the deflection of the rotor relative to the shaft ends plus that of the shaft ends in the bearings. The deflection of the shaft ends in the bearings is related to the force transmitted through the bearings by the bearing stiffness and damping coefficients as shown in Equations (3.35):

$$f_{bx} = K_{xx}m + K_{xy}n + C_{xx}\dot{m} + C_{xy}\dot{n}$$
$$f_{by} = K_{yx}m + K_{yy}n + C_{yx}\dot{m} + C_{yy}\dot{n} \qquad (3.35)$$

where $m$ and $n$ are the instantaneous displacements of the shaft ends relative to the bearings in the horizontal and vertical directions respectively, and take the form

$$m = M \sin(\omega t - \mu) = M(\sin \omega t \cos \mu - \cos \omega t \sin \mu)$$
$$= M_1 \sin \omega t + M_2 \cos \omega t$$
$$n = N \cos(\omega t - \nu) = N(\cos \omega t \cos \nu + \sin \omega t \sin \nu) \qquad (3.36)$$
$$= N_1 \cos \omega t + N_2 \sin \omega t$$

Differentiating Equations (3.36) with respect to time to obtain expressions for $\dot{m}$ and $\dot{n}$, and substituting together with Equations (3.29a) into Equations (3.35), and comparing coefficients of $\sin \omega t$ and $\cos \omega t$ on each side of both equations results in the matrix equation

$$\begin{Bmatrix} {}_A F_{bx1} \\ {}_A F_{bx2} \\ {}_A F_{by1} \\ {}_A F_{by2} \end{Bmatrix} = \begin{bmatrix} K_{xx} & -\omega C_{xx} & -\omega C_{xy} & K_{xy} \\ \omega C_{xx} & K_{xx} & K_{xy} & \omega C_{xy} \\ \omega C_{yx} & K_{yx} & K_{yy} & \omega C_{yy} \\ K_{yx} & -\omega C_{yx} & -\omega C_{yy} & K_{yy} \end{bmatrix} \begin{Bmatrix} M_1 \\ M_2 \\ N_1 \\ N_2 \end{Bmatrix} \qquad (3.37)$$

which can be expressed more simply as

$$\{{}_A F_b\} = [{}_A I]\{{}_A v\} \qquad (3.38a)$$

Similarly, another equation describing the forces transmitted through the bearing at the other end of the machine may be written as

$$\{_BF_b\} = [_BI]\{_Bv\} \tag{3.38b}$$

This may be combined with Equation (3.38a) to give

$$\begin{Bmatrix}\{_AF_b\}\\ \{_BF_b\}\end{Bmatrix} = \begin{bmatrix}[_AI] & 0 \\ 0 & [_BI]\end{bmatrix}\begin{Bmatrix}_Av\\ _Bv\end{Bmatrix} \tag{3.39}$$

which can be abbreviated to

$$\{F_b\} = [I]\{v\} \tag{3.40}$$

The magnitude of the reaction forces transmitted by the bearings can also be evaluated in terms of the forces applied to the shaft by the rotor. Considering the shaft to behave as a simply supported beam carrying a point force and moment at the location of the rotor shown in Figure 3.7, the vertical reaction forces at the shaft ends are

$$\begin{aligned}_Af_{by} &= (1 - d/l)F_y + (1/l)M_{yz} \\ _Bf_{by} &= (d/l)F_y - (1/l)M_{yz}\end{aligned} \tag{3.41a}$$

and similarly the forces in the horizontal direction may be written as

$$\begin{aligned}_Af_{bx} &= (1 - d/l)F_x + (1/l)M_{xz} \\ _Bf_{bx} &= (d/l)F_x - (1/l)M_{xz}\end{aligned} \tag{3.41b}$$

Substituting Equations (3.29) into Equations (3.41), and recognizing that the forces and moments applied to the shaft by the rotor vary sinusoidally and may be written as

$$\left.\begin{aligned}F_y &= F_{y1}\sin\omega t + F_{y2}\cos\omega t \\ M_{yz} &= M_{yz1}\sin\omega t + M_{yz2}\cos\omega t \\ F_x &= F_{x1}\cos\omega t + F_{x2}\sin\omega t \\ M_{xz} &= M_{xz1}\cos\omega t + M_{xz2}\sin\omega t\end{aligned}\right\} \tag{3.42}$$

and substituting these expressions into Equations (3.41), and then comparing coefficients of $\cos\omega t$ and $\sin\omega t$ leads to the matrix equation

$$\begin{Bmatrix}_AF_{bx1}\\ _AF_{bx2}\\ _AF_{by1}\\ _AF_{by2}\\ _BF_{bx1}\\ _BF_{bx2}\\ _BF_{by1}\\ _BF_{by2}\end{Bmatrix} = \begin{bmatrix}0 & 0 & 1-d/l & 1/l & 0 & 0 & 0 & 0 \\ 0 & 0 & 0 & 0 & 0 & 0 & 1-d/l & 1/l \\ 1-d/l & 1/l & 0 & 0 & 0 & 0 & 0 & 0 \\ 0 & 0 & 0 & 0 & 1-d/l & 1/l & 0 & 0 \\ 0 & 0 & d/l & -1/l & 0 & 0 & 0 & 0 \\ 0 & 0 & 0 & 0 & 0 & 0 & d/l & -1/l \\ d/l & -1/l & 0 & 0 & 0 & 0 & 0 & 0 \\ 0 & 0 & 0 & 0 & d/l & -1/l & 0 & 0\end{bmatrix}\begin{Bmatrix}F_{y1}\\ M_{yz1}\\ F_{x1}\\ M_{xz1}\\ F_{y2}\\ M_{yz2}\\ F_{x2}\\ M_{xz2}\end{Bmatrix} \tag{3.43}$$

Equation (3.43) may be written more simply as

$$\{F_b\} = [A]\{F_s\} \tag{3.44}$$

where the subscripts b and s indicate loads applied to the bearing and shaft respectively. It is now possible to obtain expressions for the deflection at the bearings in terms of the forces and moments applied to the shaft. This can be done by pre-multiplying Equation (3.40) by $[I]$ and substituting for $\{F_b\}$ from Equation (3.44) to give

$$\{v\} = [I]^{-1}[A]\{F_s\} \tag{3.45}$$

Having obtained the deflection of the bearings caused by the loading on the shaft we are now able to evaluate the deflection at the location of the rotor associated with this movement of the shaft ends. Considering the shaft to be rigid for the moment, and assuming the deflection of the shaft ends in the horizontal direction at ends A and B to be $_Am$ and $_Bm$, then the horizontal deflection of the shaft at the rotor must be

$$\begin{aligned} x &= {}_Am + ({}_Bm - {}_Am)d/l \\ &= (1 - d/l){}_Am + (d/l){}_Bm \end{aligned} \tag{3.46}$$

and the slope of the shaft in the $x$–$z$ plane will be

$$\phi = ({}_Am - {}_Bm)/l \tag{3.47}$$

Similarly, for motion in the $y$ direction and $y$–$z$ plane

$$y = (1 - d/l){}_An + (d/l){}_Bn \tag{3.48}$$

$$\theta = ({}_An - {}_Bn)/l \tag{3.49}$$

Recognizing that these shaft displacements vary sinusoidally such that $x$, $\phi$, $y$ and $\theta$ may be written in the forms of Equations (3.18), and substituting these, together with expressions for $m$, $n$, $\dot{m}$ and $\dot{n}$ from Equations (3.36), into Equations (3.46)–(3.49) gives, on comparing coefficients of $\sin \omega t$ and $\cos \omega t$ on each side of the resulting four equations, the following matrix equation:

$$\begin{Bmatrix} Y_1 \\ \Theta_1 \\ X_1 \\ \Phi_1 \\ Y_2 \\ \Theta_2 \\ X_2 \\ \Phi_2 \end{Bmatrix} = \begin{bmatrix} 0 & 0 & 1-d/l & 0 & 0 & 0 & d/l & 0 \\ 0 & 0 & 1/l & 0 & 0 & 0 & -1/l & 0 \\ 1-d/l & 0 & 0 & 0 & d/l & 0 & 0 & 0 \\ 1/l & 0 & 0 & 0 & -1/l & 0 & 0 & 0 \\ 0 & 0 & 0 & 1-d/l & 0 & 0 & 0 & d/l \\ 0 & 0 & 0 & 1/l & 0 & 0 & 0 & -1/l \\ 0 & 1-d/l & 0 & 0 & 0 & d/l & 0 & 0 \\ 0 & 1/l & 0 & 0 & 0 & -1/l & 0 & 0 \end{bmatrix} \begin{Bmatrix} {}_AM_1 \\ {}_AM_2 \\ {}_AN_1 \\ {}_AN_2 \\ {}_BM_1 \\ {}_BM_2 \\ {}_BN_1 \\ {}_BN_2 \end{Bmatrix}$$

$$\tag{3.50}$$

which may be abbreviated to

$$\{u_b\} = [B]\{v\} \tag{3.51}$$

Substituting for $\{v\}$ from Equation (3.45) gives

$$\{u_b\} = [B][I]^{-1}[A]\{F_s\} = [C]\{F_s\} \tag{3.52}$$

Equation (3.52) now gives the deflection of the rotor that is caused only by movement of the shaft ends in the flexible bearings.

In order to obtain the net rotor deflection under a given load we must add the deflection due to deformation of the shaft to that which has been calculated in Equation (3.52) above. The deflection associated with flexure of the shaft alone has already been calculated in Section 3.2 and is given by Equation (3.2). Equation (3.2) may also be written for displacements in the $x$ direction and in the $x$–$z$ plane as

$$\begin{Bmatrix} x \\ \phi \end{Bmatrix} = \begin{bmatrix} \alpha_{11} & \alpha_{12} \\ \alpha_{21} & \alpha_{22} \end{bmatrix} \begin{Bmatrix} F_x \\ M_{xz} \end{Bmatrix} \tag{3.53}$$

and both this equation and Equation (3.2) may be expanded to allow for in-phase and quadrature components of load and deflection by substituting for $x$, $\phi$, $y$ and $\theta$ from Equations (3.18), and for $F_x$, $M_{xz}$, $F_y$ and $M_{yz}$ from Equations (3.42). Once more, coefficients of $\sin \omega t$ and $\cos \omega t$ on each side of the resulting equations may be compared to give

$$\begin{Bmatrix} Y_1 \\ \Theta_1 \\ X_1 \\ \Phi_1 \\ Y_2 \\ \Theta_2 \\ X_2 \\ \Phi_2 \end{Bmatrix} = \begin{bmatrix} \alpha_{11} & \alpha_{12} & 0 & 0 & 0 & 0 & 0 & 0 \\ \alpha_{21} & \alpha_{22} & 0 & 0 & 0 & 0 & 0 & 0 \\ 0 & 0 & \alpha_{11} & \alpha_{12} & 0 & 0 & 0 & 0 \\ 0 & 0 & \alpha_{21} & \alpha_{22} & 0 & 0 & 0 & 0 \\ 0 & 0 & 0 & 0 & \alpha_{11} & \alpha_{12} & 0 & 0 \\ 0 & 0 & 0 & 0 & \alpha_{21} & \alpha_{22} & 0 & 0 \\ 0 & 0 & 0 & 0 & 0 & 0 & \alpha_{11} & \alpha_{12} \\ 0 & 0 & 0 & 0 & 0 & 0 & \alpha_{21} & \alpha_{22} \end{bmatrix} \begin{Bmatrix} F_{y1} \\ M_{yz1} \\ F_{x1} \\ M_{xz1} \\ F_{y2} \\ M_{yz2} \\ F_x^2 \\ M_{xz}^2 \end{Bmatrix} \tag{3.54}$$

which may be written as

$$\{u_s\} = [\alpha]\{F_s\} \tag{3.55}$$

The net deflection of the rotor, that caused by deflection of the bearings plus that attributable to flexure of the shaft is then given by

$$\{u\} = \{u_b\} + \{u_s\}$$
$$= [[C] + [\alpha]]\{F_s\} \tag{3.56}$$
$$= [D]\{F_s\}$$

The matrix $[D]$ above describes the displacement of the shaft at the rotor

## SINGLE-MASS ROTORS

under the action of sinusoidal forces and moments applied there. It is very similar to the influence coefficient matrix in Equation (3.2).

The analysis of the system may now proceed in a manner similar to that described in Section 3.2 for the case of a flexible shaft running in rigid bearings. Pre-multiplying both sides of Equation (3.56) by the inverse of the matrix $[D]$ results in the equation

$$\{F_s\} = [D]^{-1}\{u\} = [E]\{u\} \qquad (3.57)$$

which may be compared with Equation (3.4) in Section 3.2. The equation of motion for the rotor, Equation (3.5), may also be expanded to allow for motion in the $x$ and $\phi$ senses and written as

$$\begin{aligned} me\omega^2 \sin \omega t - F_y &= m\ddot{y} \\ -M_{yz} &= I_d\ddot{\theta} \\ me\omega^2 \cos \omega t - F_x &= m\ddot{x} \\ -M_{xz} &= I_d\ddot{\phi} \end{aligned} \qquad (3.58)$$

Substituting for $F_y$, $M_{yz}$, $F_x$ and $M_{xz}$ from Equations (3.42) and for $x$, $\phi$, $y$ and $\theta$ from Equations (3.18), and comparing coefficients of $\sin \omega t$ and $\cos \omega t$ on each side of each equation once more, leads to the equation

$$\begin{Bmatrix} me\omega^2 \\ 0 \\ me\omega^2 \\ 0 \\ 0 \\ 0 \\ 0 \\ 0 \end{Bmatrix} = \begin{Bmatrix} F_{y1} \\ M_{y21} \\ F_{x1} \\ M_{xz1} \\ F_{y2} \\ M_{yz2} \\ F_{x2} \\ M_{xz2} \end{Bmatrix} + \begin{bmatrix} -\omega^2 m & 0 & 0 & 0 & 0 & 0 & 0 & 0 \\ 0 & -\omega^2 I_d & 0 & 0 & 0 & 0 & 0 & 0 \\ 0 & 0 & -\omega^2 Im & 0 & 0 & 0 & 0 & 0 \\ 0 & 0 & 0 & -\omega^2 I_d & 0 & 0 & 0 & 0 \\ 0 & 0 & 0 & 0 & -\omega^2 m & 0 & 0 & 0 \\ 0 & 0 & 0 & 0 & 0 & -\omega^2 I_d & 0 & 0 \\ 0 & 0 & 0 & 0 & 0 & 0 & -\omega^2 m & 0 \\ 0 & 0 & 0 & 0 & 0 & 0 & 0 & -\omega^2 I_d \end{bmatrix} \begin{Bmatrix} Y_1 \\ \theta_1 \\ X_1 \\ \phi_1 \\ Y_2 \\ \theta_2 \\ X_2 \\ \phi_2 \end{Bmatrix}$$

(3.59)

which can be written as

$$\{f\} = \{F_s\} + [G]\{u\} \qquad (3.60)$$

## ASYMMETRY

Substituting for $\{F_s\}$ from Equation (3.57) gives

$$\{f\} = [[E] + [G]]\{u\} = [H]\{u\} \qquad (3.61)$$

which is similar to Equation (3.6) in Section 3.2. Equation (3.61) can then be rearranged to give the response of the rotor as

$$\{u\} = [H]^{-1}\{f\} \qquad (3.62)$$

Once the response of the rotor has been obtained it is possible to calculate the forces transmitted through the bearings by substituting for $\{u\}$ back into Equation (3.57) to find the loading applied to the shaft, $\{F_s\}$; the value of $\{F_s\}$ so obtained may in turn be substituted into Equation (3.45) to establish the displacements at each bearing $\{v\}$, and these can finally be substituted back into Equation (3.40) to calculate the bearing forces. The respective phase lag angles of all of these parameters can be evaluated by calculating the arctangent of the ratio of the in-phase/quadrature components, as has been described for the other cases in the previous sections.

For the reader requiring further information, investigations of systems comprising a single rotor mounted on flexible shafts which run in flexible damped bearings have also been reported by Lund and Sternlicht (1962), Gunter (1970), Kirk and Gunter (1972) and Barrett *et al.* (1978).

EXAMPLE 3.2

A light shaft of diameter 0.1 m and length 2.54 m is made of material whose modulus of elasticity is 208 GN.m$^{-2}$. It carries a rotor whose mass is 2000 kg and whose moment of inertia about a diameter is 1000 kg.m$^2$. The shaft runs in bearings which have an effective stiffness of 5 MN.m$^{-1}$ in the vertical direction and 10 MN.m$^{-1}$ in the horizontal direction, and negligible cross-coupling stiffness. Estimate the system critical speeds.

SOLUTION

The bearing dynamic stiffness matrix $[I]$ (see Equation 3.40) is given by

$$[I] = \begin{bmatrix} 10 & 0 & 0 & 0 & 0 & 0 & 0 & 0 \\ 0 & 10 & 0 & 0 & 0 & 0 & 0 & 0 \\ 0 & 0 & 5 & 0 & 0 & 0 & 0 & 0 \\ 0 & 0 & 0 & 5 & 0 & 0 & 0 & 0 \\ 0 & 0 & 0 & 0 & 10 & 0 & 0 & 0 \\ 0 & 0 & 0 & 0 & 0 & 10 & 0 & 0 \\ 0 & 0 & 0 & 0 & 0 & 0 & 5 & 0 \\ 0 & 0 & 0 & 0 & 0 & 0 & 0 & 5 \end{bmatrix} \times 10^6$$

and the matrix $A$ describing how the loading on the shaft is distributed to the bearings is found, using Equation (3.43) with $a = l/2$, to be

$$[A] = \begin{bmatrix} 0 & 0 & 0.5 & 1/l & 0 & 0 & 0 & 0 \\ 0 & 0 & 0 & 0 & 0 & 0 & 0.5 & 1/l \\ 0.5 & 1/l & 0 & 0 & 0 & 0 & 0 & 0 \\ 0 & 0 & 0 & 0 & 0.5 & 1/l & 0 & 0 \\ 0 & 0 & 0.5 & -1/l & 0 & 0 & 0 & 0 \\ 0 & 0 & 0 & 0 & 0 & 0 & 0.5 & -1/l \\ 0.5 & -1/l & 0 & 0 & 0 & 0 & 0 & 0 \\ 0 & 0 & 0 & 0 & 0.5 & -1/l & 0 & 0 \end{bmatrix}$$

The relationship between the shaft displacement at the rotor and the shaft displacement at the bearings is given by Equation (3.50), where for the system under consideration the matrix $[B]$ is

$$[B] = \begin{bmatrix} 0 & 0 & 0.5 & 0 & 0 & 0 & 0.5 & 0 \\ 0 & 0 & 1/l & 0 & 0 & 0 & -1/l & 0 \\ 0.5 & 0 & 0 & 0 & 0.5 & 0 & 0 & 0 \\ 1/l & 0 & 0 & 0 & -1/l & 0 & 0 & 0 \\ 0 & 0 & 0 & 0.5 & 0 & 0 & 0 & 0.5 \\ 0 & 0 & 0 & 1/l & 0 & 0 & 0 & -1/l \\ 0 & 0.5 & 0 & 0 & 0 & 0.5 & 0 & 0 \\ 0 & 1/l & 0 & 0 & 0 & -1/l & 0 & 0 \end{bmatrix}$$

Since, in this example the matrix $[I]$ is a diagonal matrix its inverse is also a diagonal matrix whose elements are the reciprocal of the corresponding elements in the matrix $[I]$. The matrix $[C]$ in Equation (3.52) may then be obtained, where

$$[C] = [B][I]^{-1}[A]$$
$$= \begin{bmatrix} 0.1 & 0 & 0 & 0 & 0 & 0 & 0 & 0 \\ 0 & 0.4/l & 0 & 0 & 0 & 0 & 0 & 0 \\ 0 & 0 & 0.05 & 0 & 0 & 0 & 0 & 0 \\ 0 & 0 & 0 & 0.2/l & 0 & 0 & 0 & 0 \\ 0 & 0 & 0 & 0 & 0.1 & 0 & 0 & 0 \\ 0 & 0 & 0 & 0 & 0 & 0.4/l & 0 & 0 \\ 0 & 0 & 0 & 0 & 0 & 0 & 0.05 & 0 \\ 0 & 0 & 0 & 0 & 0 & 0 & 0 & 0.2/l \end{bmatrix} \times 10^{-6}$$

For the system details given, the influence coefficients in the matrix $[\alpha]$ of Equation (3.54), describing the deformation of the shaft relative to the shaft ends, may be calculated using the expressions given in Equations (3.1) as

$\alpha_{11} = a^2b^2/3EI = 1.27^2 \times 1.27^2 \times 64/(3 \times 208 \times 10^9 \times \pi \times 0.1^4) = 0.33 \times 10^{-6}$

$\alpha_{22} = -(3al - 3a^2 - l^2)/3EIl$

$= -(3 \times 1.27 \times 2.54 - 3 \times 1.27^2 - 2.54^2) \times 64/(3 \times 208 \times 10^9 \times 0.1^4)$

$= 0.207 \times 10^{-6}$

$\alpha_{12} = \alpha_{21} = 0$

The matrix $[D]$ in Equation (3.56) may then be evaluated, relating the shaft deflection at the rotor to the loading on the shaft, as

$[D] = [C] + [\alpha]$

$$= \begin{bmatrix} 0.433 & 0 & 0 & 0 & 0 & 0 & 0 & 0 \\ 0 & 0.269 & 0 & 0 & 0 & 0 & 0 & 0 \\ 0 & 0 & 0.383 & 0 & 0 & 0 & 0 & 0 \\ 0 & 0 & 0 & 0.238 & 0 & 0 & 0 & 0 \\ 0 & 0 & 0 & 0 & 0.433 & 0 & 0 & 0 \\ 0 & 0 & 0 & 0 & 0 & 0.269 & 0 & 0 \\ 0 & 0 & 0 & 0 & 0 & 0 & 0.383 & 0 \\ 0 & 0 & 0 & 0 & 0 & 0 & 0 & 0.238 \end{bmatrix} \times 10^{-6}$$

Inverting the above matrix to obtain the matrix $[E]$ in Equation (3.57), and allowing for the inertia terms by adding matrix $[G]$ of Equation (3.59) results in matrix $[H]$ which, for this example, is also a diagonal matrix. The system response is infinite at the critical speeds where the determinant of $[H]$ is equal to zero. This will be when any one of the diagonal terms is zero, that is when

$2.31 \times 10^6 - 2000\omega^2 = 0$, or $2.61 \times 10^6 - 2000\omega^2 = 0$,

or $3.72 \times 10^6 - 1000\omega^2 = 0$, or $4.20 \times 10^6 - 1000\omega^2 = 0$

(The other four diagonal terms are the same as those above in this example). This means that the critical speeds are at $\omega = 34, 36, 61$ and $65$ rad.s.$^{-1}$

## 3.6 Effects of flexible foundations

In some machines the bearings themselves may be mounted on flexible foundations which will in turn influence the motion of the rotor mass. In such machines the net displacement of the rotor is given by the vector sum of that relative to the shaft ends, plus that of the shaft ends relative to the bearings, plus that of the bearings relative to space (the latter being due to the bearings moving on their flexible foundations relative to space). A diagram of a system which might behave in such a manner is shown in

## SINGLE-MASS ROTORS

Figure 3.8. The theoretical analysis of the rotor, shaft, and bearing response, and that of the force transmissibility of such a system, may be carried out in a manner similar to that described in the previous section when dealing with a flexible rotor in flexible bearings. In this case, however, it is necessary to first calculate the values of an equivalent set of bearing stiffness and damping coefficients which not only allow for the flexibility of the bearing but also allow for that of the foundation. These equivalent bearing coefficients may then be used in place of the bearing coefficients used in the analyses discussed previously.

The relationship between the force transmitted through the bearings and the displacement of the shaft ends in the bearings is governed by the bearing stiffness and damping coefficients and is described by Equation (3.40) in the previous section as

$$\{F_b\} = [I]\{v\} \tag{3.40}$$

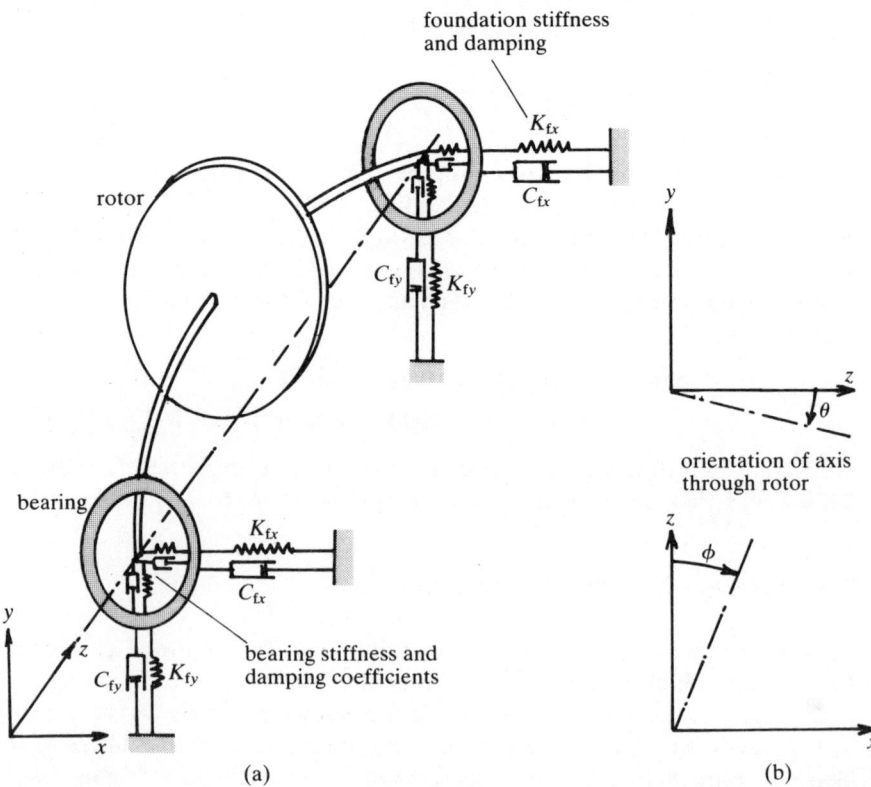

**Figure 3.8** (a) Single mass rotor on a light flexible shaft, running in flexible anisotropic bearings (cross-coupling coefficients not shown), which are mounted on flexible foundations. The rotor has mass $m$ and moment of inertia $I_d$ about a diameter; (b) Convention for positive angular motion of shaft relative to cartesian axes.

## FLEXIBLE FOUNDATIONS

A similar expression may be developed which describes the relationship between the displacements of the bearings on their foundations and the force transmitted through the bearings. If consideration is first given to the effects of applying a force $f_{bx}$ to the bearing in the horizontal direction (assuming, for the moment, that the rest of the system is not assembled in the bearings) as shown in Figure 3.9, then the bearing will respond according to the equation

$$f_{bx} - K_{fx}a - C_{fx}\dot{a} = m'\ddot{a} \tag{3.63}$$

where $a$ is the horizontal displacement of the bearing on the flexible foundations, $m'$ is the mass of one bearing, and $K_{fx}$ and $C_{fx}$ are the horizontal stiffness and damping of the foundation, respectively, under one bearing. Similarly the response of the bearing in the vertical direction to a force $f_{by}$ is given by

$$f_{by} - K_{fy}b - C_{fy}\dot{b} = m'\ddot{b} \tag{3.64}$$

where $b$ is the bearing vertical displacement and $K_{fy}$ and $C_{fy}$ are the vertical stiffness and damping of the foundation, respectively, under one bearing. The displacements of the bearing will take the form

$$\begin{aligned}
a &= A\sin(\omega t - \sigma) = A(\sin \omega t \cos \sigma - \cos \omega t \sin \sigma) \\
&= A_1 \sin \omega t + A_2 \cos \omega t \\
b &= B\cos(\omega t - \tau) = B(\cos \omega t \cos \tau + \sin \omega t \sin \tau) \\
&= B_1 \cos \omega t + B_2 \sin \omega t
\end{aligned} \tag{3.65}$$

Differentiating Equations (3.65) with respect to time obtain expressions for $\dot{a}$, $\ddot{a}$, $\dot{b}$ and $\ddot{b}$, and substituting, together with Equations (3.29a) into

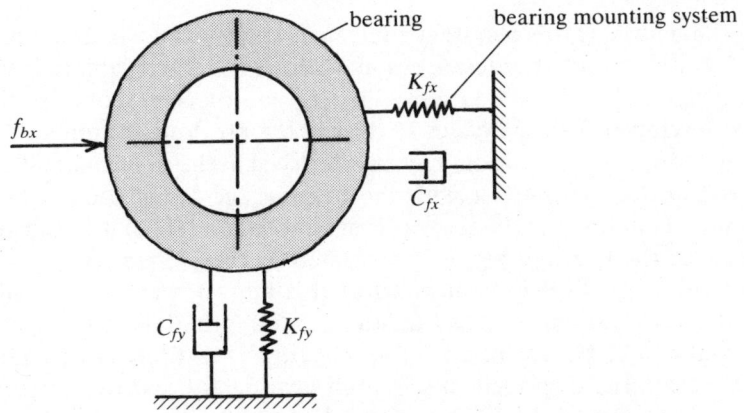

**Figure 3.9** A bearing mounted on flexible foundations, under the action of an externally applied load.

Equations (3.63) and (3.64), then comparing coefficients of sin $\omega t$ and cos $\omega t$ gives the following matrix equation:

$$\begin{Bmatrix} F_{bx1} \\ F_{bx2} \\ F_{by1} \\ F_{by2} \end{Bmatrix} = \begin{bmatrix} K_{fx} - m'\omega^2 & -\omega C_{fx} & 0 & 0 \\ \omega C_{fx} & K_{fx} - m'\omega^2 & 0 & 0 \\ 0 & 0 & K_{fy} - m'\omega^2 & \omega C_{fy} \\ 0 & 0 & -\omega C_{fy} & K_{fy} - m'\omega^2 \end{bmatrix} \begin{Bmatrix} A_1 \\ A_2 \\ B_1 \\ B_2 \end{Bmatrix} \qquad (3.66)$$

which may be written more simply as

$$\{_A F_b\} = [D]\{e\} \qquad (3.67)$$

This may be expanded to allow for both bearings as

$$\begin{Bmatrix} _A F_b \\ _B F_b \end{Bmatrix} = \begin{bmatrix} [_A D] & 0 \\ 0 & [_B D] \end{bmatrix} \begin{Bmatrix} _A e \\ _B e \end{Bmatrix} \qquad (3.68)$$

which can be abbreviated as

$$\{F_b\} = [D]\{e\} \qquad (3.69)$$

The total displacement of the shaft ends under the action of an applied force $\{F_b\}$ is given by the vector sum of the individual displacements $\{v\}$ and $\{e\}$ as

$$\begin{aligned} \{w\} &= \{v\} + \{e\} \\ &= [[I]^{-1} + [D]^{-1}]\{F_b\} \\ &= [I']^{-1}\{F_b\} \end{aligned} \qquad (3.70)$$

which may be restated in the form

$$\{F_b\} = [I']\{w\} \qquad (3.71)$$

where the matrix $[I']$ is a system equivalent stiffness matrix describing the overall shaft support characteristics and allows for the flexibilities of both the bearings and their foundations. The $[I']$ matrix corresponds to the $[I]$ matrix developed for the system studied in the previous section.

A study of the rotor motion may now proceed in the same manner as that indicated in Section 3.4 but with the bearing and foundation equivalent stiffness and damping coefficients as evaluated in the $[I']$ matrix substituted for those of the bearings alone, represented by the $[I]$ matrix.

Once the rotor displacement matrix $\{u\}$ is known, then it is possible to substitute back, as before, into Equations (3.57) to find the loading applied to the shaft $\{F_s\}$. The loading $\{F_s\}$ may again be substituted into equations 3.45 to obtain the displacements at each bearing $\{v\}$ and this in turn can again be substituted back into Equation (3.40) to obtain the forces transmitted to the bearings $\{F_b\}$.

The force transmitted to the foundations is, of course, not the same as that transmitted through the bearings (because of the influence of the bearing inertia forces). Its value is given by the force–displacement relationship for the foundations which takes the form

$$\begin{Bmatrix} F_{fx1} \\ F_{fx2} \\ F_{fy1} \\ F_{fy2} \end{Bmatrix} = \begin{bmatrix} K_{fx} & -\omega C_{fx} & 0 & 0 \\ \omega C_{fx} & K_{fx} & 0 & 0 \\ 0 & 0 & K_{fy} & C_{fy} \\ 0 & 0 & -\omega C_{fy} & K_{fy} \end{bmatrix} \begin{Bmatrix} A_1 \\ A_2 \\ B_1 \\ B_2 \end{Bmatrix} \quad (3.72)$$

The amplitudes of the forces transmitted to the foundations may then be evaluated from

$$F_{fx} = (F_{fx1}^2 + F_{fx2}^2)^{1/2}, \qquad F_{fy} = (F_{fy1}^2 + F_{fy2}^2)^{1/2} \quad (3.73)$$

in the horizontal and vertical directions respectively. The corresponding phase lag angles are then given by

$$\sigma = \tan^{-1}\left[\frac{-F_{fx2}}{F_{fx1}}\right], \qquad \tau = \tan^{-1}\left[\frac{F_{fy2}}{F_{fy1}}\right] \quad (3.74)$$

## 3.7 Gyroscopic effects

In some machines the rotor mass is mounted on the shaft such that, at the location of the rotor, deflection of the shaft also tends to result in a change in slope of the shaft. Typically this might be the case when the rotor is not mounted at the centre of a symmetrical shaft, or when the rotor is overhung. When this is the case the rotor will tend to whirl in a conical path and centrifugal forces will act on the rotor mass as shown in Figure 3.10(a). The effect of these centrifugal forces is to tend to straighten the shaft, making it behave more as though it was constrained not to change slope at the rotor location. Because of this the deflection of the shaft will tend to not be so great and its stiffness will effectively increase; as a consequence there will be a corresponding increase in the system natural frequency.

The system shown in Figure 3.10(b) spins with angular velocity $\omega$ about the spin axis $OZ$. At the instant in time shown the rotor is also precessing with angular velocity $\dot{\phi}$ about a vertical axis $OY$. Figure 3.10(c) shows angular momentum vectors drawn (parallel with the axis of rotation to which they refer, the direction of rotation being clockwise when viewed from the arrow tail) representing the system angular momentum at some time $t$ and at a small interval of time later. It can be seen that the change in angular momentum over the time interval considered is $I_p\omega\dot{\phi}$, where $I_p$ is the polar moment of inertia of the rotor, and so the gyroscopic moment which must be applied to produce this change, equal to the rate of change of

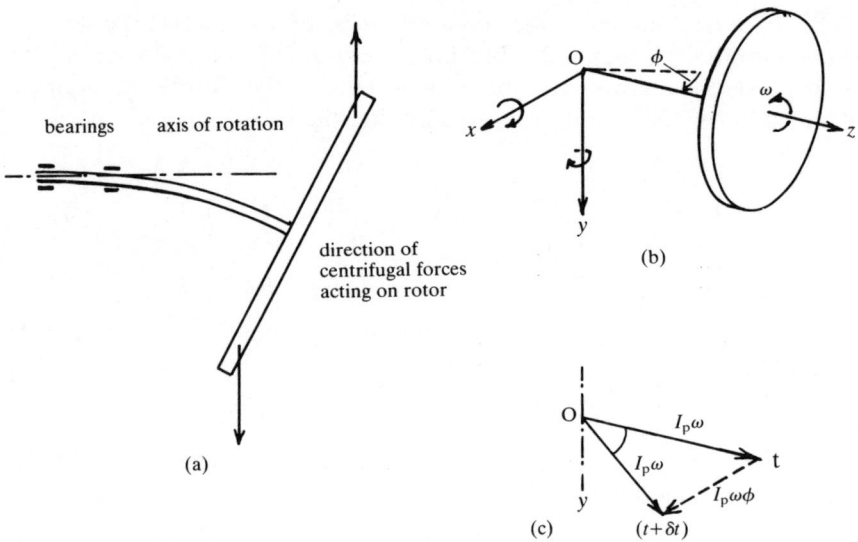

**Figure 3.10** Gyroscopic effects associated with a spinning rotor, (a) an overhung rotor mounted on a light flexible shaft; (b) general angular motion of a rotor; (c) angular momentum vectors for the system shown in (b). $I_p\omega$ are the vectors representing angular momentum about spin axis $Ox/Oz$, $I_p\omega\phi$ is the change in momentum during time $\delta t$.

angular momentum, is given by $I_p\omega\dot{\phi}$. In Figure 3.10(b) the gyroscopic couple applied to the rotor (or strictly, that which reacts the gyroscopic couple applied by the rotor to the support structure), $M_g$ say, must act about the axis $Ox$, since this axis corresponds with the orientation of the vector $I_p\omega\phi$ in Figure 3.10(c). The angular sense of this couple is clockwise about $Ox$ when viewed from $O$. The net moment about the axis $Ox$ is equal to the product of rotor moment of inertia about the $Ox$ axis and the angular acceleration about the $Ox$ axis, that is

$$M_x - M_g = I_d\ddot{\theta} \tag{3.75}$$

where $M_x$ is a moment which is applied to the rotor by the shaft and $I_d$ is the rotor moment of inertia about a diameter. Alternatively this may be written as

$$M_x = I_p\omega\dot{\phi} + I_d\ddot{\theta} \tag{3.76}$$

A similar expression may also be developed describing rotor motion in the $xz$ plane as

$$M_y = -I_p\omega\dot{\theta} + I_d\ddot{\phi} \tag{3.77}$$

The angular displacements of the rotor will be periodic and will take the form

$$\phi = \phi_2 \cos \omega t + \phi_1 \sin \omega t, \quad \theta = \theta_1 \cos \omega t + \theta_2 \sin \omega t \tag{3.78}$$

in general, but for an isotropic system where the rotor support characteristics are the same in both directions then the amplitudes of $\phi$ and $\theta$ will be equal and Equations (3.78) will take the form

$$\theta = \psi \cos \omega t, \qquad \phi = \psi \sin \omega t \tag{3.79}$$

where the motion in one direction lags behind that in the perpendicular direction by one quarter of a cycle. Differentiating Equations (3.79) and substituting into Equation (3.76) (or 3.77) gives the magnitude of the net moment applied to the shaft, which is equal and opposite to that which the shaft applies to the rotor, as

$$M = -(I_p - I_d)\omega^2 \psi \tag{3.80}$$

For the isotropic system the deflection and slope of the shaft will take the form

$$z = C_1 F + C_2 M, \qquad \psi = C_3 F + C_4 M \tag{3.81}$$

where $C_1$, $C_2$ etc. are coefficients relating forces and moments, applied to the shaft at the location of the rotor, to the shaft deformation. For anisotropic systems a similar set of equations may be written for shaft deformation in the normal direction. In the case of an overhung rotor mounted on rigid bearings Equations (3.81) take the form

$$z = \frac{l^3 F}{3EI} + \frac{l^2 M}{2EI}, \qquad \psi = \frac{l^2 F}{2EI} + \frac{l M}{EI} \tag{3.82}$$

The centrifugal force $F$ is equal to $m z \omega^2$ where $m$ is the rotor mass, while the applied moment $M$ is given by Equation (3.80). Also, for a rotor with a small length/diameter ratio, $I_p = 2I_d$. When these substitutions are made, Equations (3.82) become

$$(1 - m\omega^2 l^3/3EI)z + (I_d \omega^2 l^2/2EI)\psi = 0$$
$$-(m\omega^2 l^2/2EI)z + (1 + I_d \omega^2 l^2/EI)\psi = 0 \tag{3.83}$$

These equations are satisfied for any $z$ and $\psi$ when

$$(1 - m\omega^2 l^3/3EI)(1 + I_d \omega^2 l/EI) + (m\omega^2 l^2/2EI)(I_d \omega^2 l^2/2EI) = 0 \tag{3.84}$$

Substituting $\omega_p = (3EI/ml^3)^{1/2}$, the system natural frequency for a point mass rotor, and $\alpha = 3I_d/ml^2$, equation (3.84) becomes

$$\omega^4 + 4\omega_p^2(1/\alpha - 1)\omega^2 - 4\omega_p^4/\alpha = 0 \tag{3.85}$$

This relationship is shown graphically in Figure 3.11 where it can be seen that for rotors with large $\alpha$ parameter values the system natural frequency is almost double that for the point-mass rotor system.

Gyroscopic effects tend to be particularly significant in situations where there is either an overhung rotor (so that any forcing causes a considerable

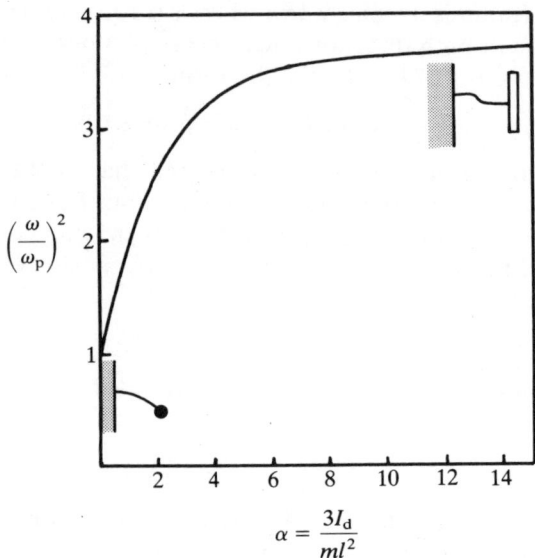

**Figure 3.11** Effect of gyroscopic couple on the resonant frequency of the system shown in Figure 3.10.

slope of the shaft at the rotor) or where the rotor runs at very high speeds. In instances where the rotor itself forms most of the shaft, as opposed to the case where the rotor is disc-like and is mounted on a relatively light shaft, gyroscopic effects are less significant. The analyses in the preceding sections can be extended to allow for gyroscopic effects by modifying the right-hand sides of the equations of motion of the rotor, Equations (3.5), (3.13), (3.17) and (3.58), for motion in the $\theta$ and $\phi$ directions, to include the gyroscopic couple terms as shown in Equations (3.76) and (3.77) above; the analysis then proceeds as before.

Further discussion of gyroscopic effects in rotating machinery is given in Thomson (1978), Den Hartog (1985), Green (1948) and Alba (1973).

## 3.8 Aerodynamic effects

In some types of rotating machinery the local static pressure on the surface of the rotor changes considerably along the length of the machine, for example in the case of turbines and compressors. Ideally, at any particular station along the length of the shaft, the local pressure of the surrounding medium does not change with position around the rotor circumference and so there are no net lateral forces on the shaft affecting its motion. In some cases, however, the lateral pressure balance described above may be upset as a consequence of some initial movement of the rotor away from its normal near-concentric (with the machine stator) running position. Under these circumstances an additional force, created by the pressure of the surrounding medium (usually a gas), acts on the rotor and affects its motion. Such 'aerodynamic' forces are usually related either to the operation of labyrinth seals between the shaft and stator, or to the operation of a turbine stage.

Aerodynamic forces related to seals are usually set up as a result of misalignment between the shaft and housing, as shown in Figure 3.12. In this instance the deflection of the shaft causes a decrease in the high-pressure side clearance of the seal plate, whilst that on the low-pressure side is increased, on the side of the machine towards which the rotor has become displaced. As a consequence of this there will be a decrease in pressure in the seal itself, on that side of the shaft, which will have the effect of reducing the restoring force on the shaft. The restoring force continues to decrease throughout the entire vibration half-cycle during which the high-pressure side clearance of the seal plate is reduced below average, such that the shaft restoring force in effect leads shaft deflection by up to $90°$. In this situation the aerodynamic force is a stabilizing factor. If the pressure gradient across the seals is reversed, however, there is a phase change of up to $180°$ in the aerodynamic force, which can then tend to destabilize the rotor motion.

In some cases it is possible for seals to provide a significant force on the rotating shaft even in the absence of misalignment. Because seals around a rotating shaft invariably involve very small clearances, they can sometimes behave as fluid film bearings in their own right, and give rise to additional spring and damping coefficients associated with their operation. In most instances this will not be the case, however, since the seal radial clearance will generally be much greater than that at the bearings, and so the hydrodynamic pressures generated will be lower than those at the bearings. Whenever there is cause for uncertainty the seal 'bearing' forces should be calculated at the design stage and used in subsequent system response calculations.

In the case of aerodynamic forces related to turbine stage operation, consider the turbine blade stage shown in Figure 3.13, where the blade ring

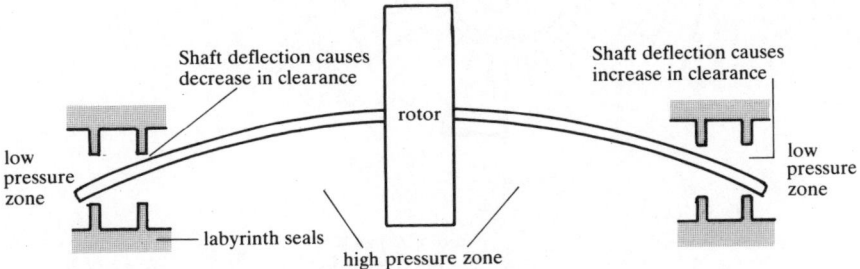

**Figure 3.12** Misalignment of a shaft causing aerodynamic effects at the seals.

is momentarily displaced towards the stator away from its equilibrium position. Because the tangential gas forces applied to the blade ring are related to the blade tip clearance, there is a variation in the magnitude of the tangential blade force around the circumference of the turbine blade ring as shown in the diagram. Clearly, in the case shown there is a net tangential force on the rotor which leads the shaft deflection by $90°$, and is therefore destabilizing, causing the rotor to whirl in the direction of rotation.

In both of the situations discussed above a component of force is caused to act on the rotor in a direction which is perpendicular to the instantaneous direction of rotor displacement, and whose mangitude is related to the magnitude of the rotor displacement. The phenomenon therefore has the effect of coupling force in, say, the horizontal direction to displacement in the vertical direction, and vice versa. The effect is similar to that related to the cross-coupling coefficients used to describe journal bearing operating characteristics. For this reason aerodynamic cross-coupling is usually represented in mathematical analyses by additional cross-coupling terms, added to the overall equivalent stiffness and damping coefficients discussed in earlier sections of this chapter.

Further discussion on aerodynamic effects in rotating machinery, together with sample evaluations of cross-coupling coefficients, can be found in Alford (1965).

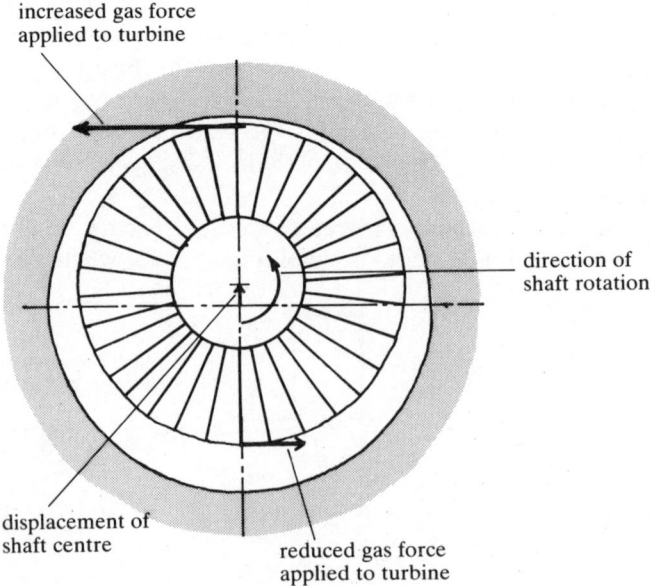

**Figure 3.13** Aerodynamic effects related to turbine blade stages.

# References

Aiba, S. 1973. The effect of gyroscopic moment and distributed mass on the vibration of a rotating shaft with a rotor. *Bull. JSME* **16**, 1550–1561.

Alford, J. S. 1965. Protecting turbomachinery from self-excited rotor whirl. *Trans ASME J. Eng. Power*, October, 333–344.

Barrett, L. E., Gunter, E. J. & P. E. Allaire 1978. Optimum bearing and support damping for unbalance response and stability of turbomachinery. *Trans. ASME., J. En. Power*, **100**, January, 89–94.

Den Hartog, J. P. 1985. *Mechanical vibrations*, 4th edn. New York: McGraw-Hill.

Goodwin, M. J., Penny, J. E. T. & C. J. Hooke 1984. Variable impedance bearings for large rotating machinery. *IMechE Conf. on Vibrations in Rotating Machinery*. York. Dec. Paper C288/84. 535–542.

Green, R. B. 1948. Gyroscopic effects on the critical speeds of flexible rotors. *Trans. ASME, J. Appl. Mech.* **15**, 369–376.

Gunter, E. J. 1966. Dynamic stability of rotor-bearing systems. NASA report SP-113, 29.

Gunter, E. J. 1970. Influence of flexibility mounted rolling elements bearings on rotor response part 1 – linear analysis. *Trans. ASME., J. Lub. Tech.* **92**, January, 59–75.

ISO, 1967. International Standards Organization, Geneva. Draft document No. TC108/WG 6.

Jeffcott, H. H. 1919. The lateral vibration of loaded shafts in the neighbourhood of a whirling speed. The effect of want of balance. *Phil. Mag.*, series 6, **57**, 304ff.

Kirk, R. G. & E. J. Gunter 1972. Effect of support flexibility and damping on the dynamic response of a single mass flexible rotor in elastic bearings. NASA, Houston. Report No. Cr-2083.

Lund, J. W. & B. Sternlicht 1962. Rotor-bearing dynamics with emphasis on attenuation. *Trans. ASME., J. Basic Eng.* Ser. D., **84**, 4, 491–502.

Ogrodnik, P. J., M. J. Goodwin & J. E. T. Penny 1985. The influence of design parameters on oil whirl in rotor-bearing systems. *Symp. Instability in Rotating Machinery* (sponsored by Bently Rotor Dynamics Corporation). Carson City, Nevada.

Rieger, N. F. 1982. Vibrations of rotating machinery, Pt. I: Rotor bearing dynamics. The Vibration Institute, Clarendon Hills, Ilinois, 1.6.

Roark, R. J. & W. C. Young 1984. *Formulas for stress and strain*, 5th edn. Tokyo: McGraw-Hill Kogakusha.

Thomson, W. T. 1978. *Vibration theory and applications*. London: George Allen & Unwin.

# 4 Systems with many degrees of freedom

## 4.1 Introduction

In many machines the rotor mass is concentrated at a single location on the shaft. In other cases, the rotor mass may be concentrated at a number of locations, or alternatively distributed along the shaft length. Furthermore, the machine may not be symmetrical about the shaft centre. In these cases the simplified theoretical analyses described in the preceding chapter cannot be used to obtain accurate predictions of machine dynamic behaviour, and an alternative approach is required. Because the behaviour of such machines is described in terms of the movements of each of the components of the rotor mass, rather than only one movement of one mass, such systems have several degrees of freedom. We use the abbreviation MDOF to denote these multi-degree-of-freedom systems.

This chapter describes a number of techniques which can be used to predict the behaviour of MDOF systems. The first four sections outline alternative commonly used numerical methods which can be used to calculate both machine critical speeds and forced response; the subsequent section shows how the effects of flexible bearings and foundations can be incorporated in these analyses. Magnetic effects, phenomena particularly important in electrical machines, are discussed in the following section which describes how they can be allowed for when calculating machine critical speeds and response. The last two sections of the chapter present simple methods of predicting machine natural frequencies by hand calculation, suitable for use when a rapid, albeit approximate, solution is required. It should be noted that this chapter is not exhaustive, and that other techniques are available for rotor dynamics analysis which are not discussed here owing to spatial limitations. Examples include finite element methods (Nelson and McVaugh 1976; Rouch and Kao 1980) and modal analysis methods (Lund 1974a, Gunter *et al.* 1978, Nicholas 1986).

## 4.2 Method of influence coefficients

The method of influence coefficients is a mathematical technique which can be used to calculate the natural frequencies and the forced response of rotating machines. The method can be used to predict machine operating characteristics by hand for two-mass systems, but computerized versions are more useful, particularly if the system under consideration has more than two degrees of freedom, when the mathematics becomes long and tedious. The method is described in several publications (see for example Thomson 1978) and is outlined below using the example of a three-mass rotor mounted on a flexible shaft, running in rigid bearings; it can, of course, be easily extended to allow for more degrees of freedom and for damped flexible supports.

Consider the three-mass rotor shown in Figure 4.1, where the masses $m_1$, $m_2$ and $m_3$ are mounted on a light flexible shaft. Consider the application of steady forces to the masses, of magnitudes $f_1$, $f_2$ and $f_3$ respectively, so that the corresponding shaft deflections are $x_1$, $x_2$ and $x_3$. If a force $f$ is applied only at $m_1$ then the deflection of $m_1$ will be proportional to the applied force and will be given by

$$x_1 = a_{11} f \tag{4.1}$$

If instead the force $f$ was applied at the location of mass $m_2$, the deflection of $m_1$ would again be proportional to the magnitude of the force and would be given by

$$x_1 = a_{12} f \tag{4.2}$$

Similarly, if the force were applied at $m_3$, the deflection of $m_1$ would take the form

$$x_1 = a_{13} f \tag{4.3}$$

If forces $f_1$, $f_2$ and $f_3$ were applied at the locations of all of the masses simultaneously, then the total deflection of $m_1$ would be

$$x_1 = a_{11} f_1 + a_{12} f_2 + a_{13} f_3 \tag{4.4a}$$

and the corresponding deflections of $m_2$ and $m_3$ would be

$$x_2 = a_{21} f_1 + a_{22} f_2 + a_{23} f_3 \tag{4.4b}$$

$$x_3 = a_{31} f_1 + a_{32} f_2 + a_{33} f_3 \tag{4.4c}$$

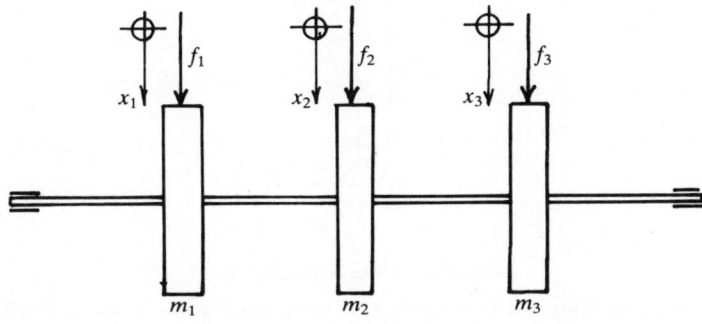

**Figure 4.1** A system with three rotor masses on a light flexible shaft. See text for explanation.

Equations (4.4) can be written in matrix form as

$$\begin{Bmatrix} x_1 \\ x_2 \\ x_3 \end{Bmatrix} = \begin{bmatrix} a_{11} & a_{12} & a_{13} \\ a_{21} & a_{22} & a_{23} \\ a_{31} & a_{32} & a_{32} \end{bmatrix} \begin{Bmatrix} f_1 \\ f_2 \\ f_3 \end{Bmatrix} \quad (4.5a)$$

and the coefficients $a_{11}$, $a_{12}$, etc. are known as 'influence coefficents'.

The technique can also be extended to account for angular movements $\phi$ of the rotor, and for the application of point moments $M$ at various locations along the shaft as part of the loading on the system. Equation (4.5a) would then take the form

$$\begin{Bmatrix} x_1 \\ x_2 \\ x_3 \\ \phi_1 \\ \phi_2 \\ \phi_2 \end{Bmatrix} = \begin{bmatrix} a_{11} & a_{12} & a_{13} & a_{14} & a_{15} & a_{16} \\ a_{21} & a_{22} & a_{23} & a_{24} & a_{25} & a_{26} \\ a_{31} & a_{32} & a_{33} & a_{34} & a_{35} & a_{36} \\ a_{41} & a_{42} & a_{43} & a_{44} & a_{45} & a_{46} \\ a_{51} & a_{52} & a_{53} & a_{54} & a_{55} & a_{56} \\ a_{61} & a_{62} & a_{63} & a_{64} & a_{65} & a_{66} \end{bmatrix} \begin{Bmatrix} f_1 \\ f_2 \\ f_3 \\ M_1 \\ M^2 \\ M_3 \end{Bmatrix} \quad (4.5b)$$

The analysis so far has referred only to static loads applied to the shaft. When the mass displacements are changing rapidly with time, the applied force has to overcome the mass inertia as well as to deform the shaft. If the force and moment applied externally to the mass $m_1$ are $f_1'$ and $M_1'$ and those transmitted to the shaft are $f_1$ and $M_1$ (equal and opposite to the reaction force of the shaft on the mass) as shown in Figure 4.2, then the relationships between $f_1'$ and $f_1$ and between $M_1'$ and $M_1'$ are given by

$$f_1' - f_1 = m_1 \ddot{x}_1, \qquad M_1' - M_1 = I_1 \ddot{\phi}_1 \quad (4.6a)$$

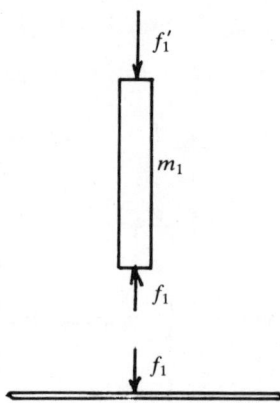

**Figure 4.2** Reaction forces acting between a rotor mass and the shaft. See text for explanation.

INFLUENCE COEFFICIENTS

where $I$ is the rotor moment of inertia about a diameter; similarly at the other rotor mass stations

$$f_2' - f_2 = m_2\ddot{x}_2, \qquad M_2' - M_2 = I_2\ddot{\phi}_2 \qquad (4.6b)$$

$$f_3' - f_3 = m_3\ddot{x}_3, \qquad M_3' - M_3 = I_3\ddot{\phi}_3 \qquad (4.6c)$$

Substituting for $f_1$, $f_2$, $f_3$, $M_1$, $M_2$ and $M_3$ from Equations (4.6), and remembering that $\ddot{x} = -\omega^2 x$ and $\ddot{\phi} = -\omega^2 \phi$, gives

$$\begin{Bmatrix} x_1 \\ x_2 \\ x_3 \\ \phi_1 \\ \phi_2 \\ \phi_3 \end{Bmatrix} = \begin{bmatrix} a_{11} & a_{12} & a_{13} & a_{14} & a_{15} & a_{16} \\ a_{21} & a_{22} & a_{23} & a_{24} & a_{25} & a_{26} \\ a_{31} & a_{32} & a_{33} & a_{34} & a_{35} & a_{36} \\ a_{41} & a_{42} & a_{43} & a_{44} & a_{45} & a_{46} \\ a_{51} & a_{52} & a_{53} & a_{54} & a_{55} & a_{56} \\ a_{61} & a_{62} & a_{63} & a_{64} & a_{65} & a_{66} \end{bmatrix} \begin{Bmatrix} f_1' \\ f_2' \\ f_3' \\ M_1' \\ M_2' \\ M_3' \end{Bmatrix}$$

$$+ \omega^2 \begin{bmatrix} a_{11}m_1 & a_{12}m_2 & a_{13}m_3 & a_{14}I_1 & a_{15}I_2 & a_{16}I_3 \\ a_{21}m_1 & a_{22}m_2 & a_{23}m_3 & a_{24}I_1 & a_{25}I_2 & a_{26}I_3 \\ a_{31}m_1 & a_{32}m_2 & a_{33}m_3 & a_{34}I_1 & a_{35}I_2 & a_{36}I_3 \\ a_{41}m_1 & a_{42}m_2 & a_{43}m_3 & a_{44}I_1 & a_{45}I_2 & a_{46}I_3 \\ a_{51}m_1 & a_{52}m_2 & a_{53}m_3 & a_{54}I_1 & a_{55}I_2 & a_{56}I_2 \\ a_{61}m_1 & a_{62}m_2 & a_{63}m_3 & a_{64}I_1 & a_{65}I_2 & a_{66}I_3 \end{bmatrix} \begin{Bmatrix} x_1 \\ x_2 \\ x_3 \\ \phi_1 \\ \phi_2 \\ \phi_3 \end{Bmatrix} \quad (4.7)$$

which may be rearranged to give

$$\begin{bmatrix} a_{11}m_1 - 1/\omega^2 & a_{12}m_2 & a_{13}m_3 & a_{14}I_1 & a_{15}I_2 & a_{16}I_3 \\ a_{21}m_1 & a_{22}m_2 - 1/\omega^2 & a_{23}m_3 & a_{24}I_1 & a_{25}I_2 & a_{26}I_3 \\ a_{31}m_1 & a_{32}m_2 & a_{33}m_3 - 1/\omega^2 & a_{34}I_1 & a_{35}I_2 & a_{36}I_3 \\ a_{41}m_1 & a_{42}m_2 & a_{43}m_3 & a_{44}I_1 - 1/\omega^2 & a_{45}I_2 & a_{46}I_3 \\ a_{51}m_1 & a_{52}m_2 & a_{53}m_3 & a_{54}I_1 & a_{55}I_2 - 1/\omega^2 & a_{56}I_3 \\ a_{61}m_1 & a_{62}m_2 & a_{63}m_3 & a_{64}I_1 & a_{65}I_2 & a_{66}I_3 - 1/\omega^2 \end{bmatrix} \begin{Bmatrix} x_1 \\ x_2 \\ x_3 \\ \phi_1 \\ \phi_2 \\ \phi_3 \end{Bmatrix}$$

$$= \frac{-1}{\omega^2} \begin{bmatrix} a_{11} & a_{12} & a_{13} & a_{14} & a_{15} & a_{16} \\ a_{21} & a_{22} & a_{23} & a_{24} & a_{25} & a_{26} \\ a_{31} & a_{32} & a_{33} & a_{34} & a_{35} & a_{36} \\ a_{41} & a_{42} & a_{43} & a_{44} & a_{45} & a_{46} \\ a_{51} & a_{52} & a_{53} & a_{54} & a_{55} & a_{56} \\ a_{61} & a_{62} & a_{63} & a_{64} & a_{65} & a_{66} \end{bmatrix} \begin{Bmatrix} f_1' \\ f_2' \\ f_3' \\ M_1' \\ M_2' \\ M_3' \end{Bmatrix} \quad (4.8)$$

or alternatively

$$[A']\{X\} = [A]\{F'\} \qquad (4.9)$$

The only unknowns in Equation (4.8) are the mass displacements $x$ and $\phi$, since values for the applied loads (for example imbalance forces) can be assumed, for the purposes of response and critical speed calculations, and the shaft influence coefficients can be evaluated from beam theory using, for example, area moments. The equation may be solved by pre-multiplying both sides by the inverse of the matrix $A'$ to give

$$\{X\} = [A']^{-1}[A]\{F'\} = [R]\{F'\} \tag{4.10}$$

In general the applied forces and mass displacements will not all be in phase with one another, and damping forces may also act upon the shaft, so the more general form of Equation (4.10) will be that which includes both in-phase and quadrature terms for the $x$ and $f'$ terms, that is

$$\begin{Bmatrix} X_r \\ X_j \end{Bmatrix} = \begin{bmatrix} R & 0 \\ 0 & R \end{bmatrix} \begin{Bmatrix} F'_r \\ F'_j \end{Bmatrix} \tag{4.11}$$

where the suffixes $r$ and $j$ denote in-phase and quadrature terms respectively.

In the case of free vibrations the right-hand side of Equation (4.9) reduces to zero to give

$$[A']\{X\} = 0 \tag{4.12}$$

which is only satisfied when the determinant of $[A']$ is zero (except for the trivial case of $\{X\}$ equal to zero). Since the elements of the matrix $[A']$ contain terms involving $\omega^2$ it is possible then to use an appropriate mathematical routine to compute the values of $\omega$ which satisfy this condition. These will be the roots of the equation

$$|A'| = 0 \tag{4.13}$$

and will be the system natural frequencies.

EXAMPLE 4.1

A light shaft of length 3 m has a modulus of $EI = 2$ MN.m$^2$ and is supported at each end in self-aligning bearings which have a very high radial stiffness. The shaft carries a concentrated rotor mass $A$ of 80 kg at shaft mid-span, and a second rotor $B$ of mass 100 kg at 0.75 m from one bearing. Calculate the system natural frequencies associated with lateral motion of the rotor masses. Assume that the values for rotor moments of inertia about their diameters are very small.

SOLUTION

For a simply supported shaft of length $l$ carrying a load $W$ at a distance $a$ from one end (and a distance $b$ from the other end), as shown in Figure 4.3, the deflection at a distance $x$ from the first end is given from basic

# TRANSFER MATRIX

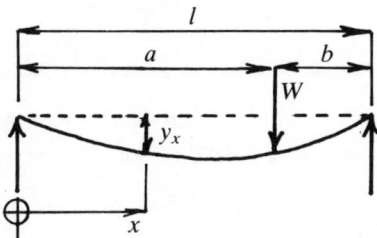

**Figure 4.3** Light shaft carrying a load at some general position. See text (Example 4.1) for explanation.

mechanics theory as

$$y_x = \frac{Wbx(l^2 - x^2 - b^2)}{6EIl} \quad (x \leq l - b)$$

Using the above equation, the influence coefficients may be evaluated for the shaft in question as

$$a_{AA} = 0.281 \times 10^{-6} \text{ m.N}^{-1}$$
$$a_{AB} = 0.193 \times 10^{-6} \text{ m.N}^{-1}$$
$$a_{BA} = 0.193\ 10^{-6} \times \text{m.N}^{-1}$$
$$a_{BB} = 0.158\ 10^{-6} \times \text{m.N}^{-1}$$

For free vibrations Equation (4.9) then takes the form

$$[A']\{x\} = 0$$

which, except for the trivial case of $\{x\} = 0$, is satisfied only when

$$|A'| = 0$$

that is when

$$(a_{AA}m_A - 1/\omega^2)(a_{BB}m_B - 1/\omega^2) - a_{BA}m_A a_{AB}m_B = 0$$

Substituting in the above values of the system parameters, and rearranging gives

$$57.2 \times 10^{-12}\omega^4 - 38.3 \times 10^{-6}\omega^2 + 1 = 0$$

which may be solved to give the two system natural frequencies as

$$\omega_{n1} = 165 \text{ rad.s}^{-1}, \quad \omega_{n2} = 801 \text{ rad.s}^{-1}$$

## 4.3 Transfer matrix method

The transfer matrix method is an alternative numerical method suitable for predicting the dynamic characteristics of rotating machinery. It has an

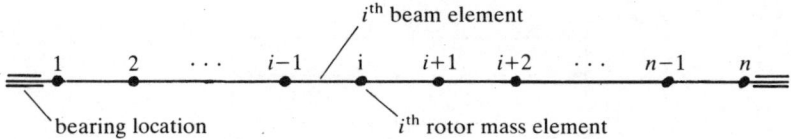

Figure 4.4  Modelling a real rotor with discrete elements.

advantage over the method of influence coefficients in that the size of the matrices being handled does not increase with the number of degrees of freedom, and so there is less demand on computer memory. This, together with its relative simplicity, is probably the reason for its popularity. In contrast to the influence coefficient method, instead of treating the shaft as a beam in its entirety in order to examine the deflections, the shaft is divided up into a number of imaginary smaller beam elements and attention is focused on the behaviour of each of these elements in order to determine the overall system behaviour. The rotor mass is similarly divided up into a number of smaller imaginary masses concentrated at the junctions of the beam elements so that a rotor and shaft may be modelled as shown in Figure 4.4. The transfer matrix method was developed independently by Myklestad (1944) and Prohl (1945); more recent commentary on its application in rotor dynamics has been published by Kikuchi (1970); Lund and Orcutt (1967), and Lund (1974b); it is also discussed by Pestel and Leckie (1963), Rao (1983), and Vance (1988).

Considering, first, the beam shown in Figure 4.5 as though it were a cantilever, the displacement and slope at the free end are related to the applied moment $M$ and shear force $Q$, and to the displacement and slope at

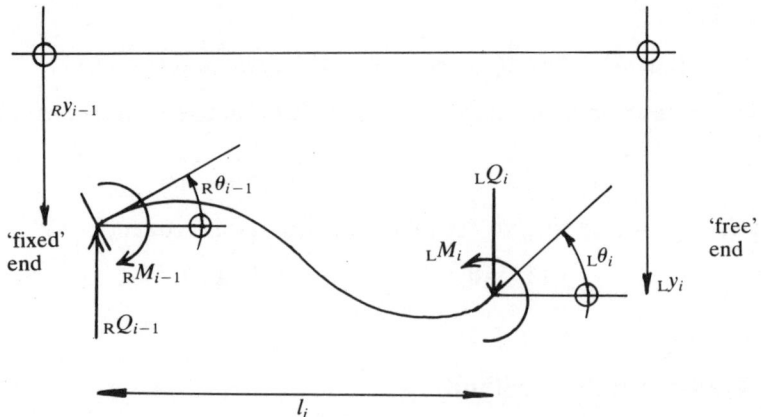

Figure 4.5  Forces and deflections relating to an elemental length of shaft. See text for explanation.

the fixed end, by the expressions

$$_L y_i = {_R}y_{i-1} - {_R}\theta_{i-1}l - \frac{{_L}M_i l^2}{2EI} + \frac{{_L}Q_i l^3}{3EI}$$

$$_L \theta_i = {_R}\theta_i + \frac{{_L}M_i l}{EI} - \frac{{_L}Q_i l^2}{2EI}$$

(4.14)

while the shear force and bending moment at the fixed end are related to those at the free end by

$$_R Q_{i-1} = {_L}Q_i, \qquad {_R}M_{i-1} = {_L}M_i + {_L}Q_i l \qquad (4.15)$$

After first substituting for $_L Q_i$ and $_L M_i$ from Equations (4.15) into Equations (4.14), both sets of equations may be expressed in matrix form, for the beam element at station $i$ in Figure 4.4, as

$$\left\{\begin{matrix} -y \\ \theta \\ M \\ Q \end{matrix}\right\}_{L,i} = \begin{bmatrix} 1 & 1 & l^2/2EI & l^3/6EI \\ 0 & 1 & l/EI & l^2/2EI \\ 0 & 0 & 1 & 1 \\ 0 & 0 & 0 & 1 \end{bmatrix}_i \left\{\begin{matrix} -y \\ \theta \\ M \\ Q \end{matrix}\right\}_{R, i-1} \qquad (4.16)$$

or more simply as

$$_L \{S\}_i = [F]_i {_R}\{S\}_{i-1} \qquad (4.17)$$

where $[F]$ is known as the field transfer matrix, and the matrix $\{S\}$ is the state vector. A similar set of equations may be written for motion in the horizontal direction, and in general the state vectors will have both in-phase and quadrature components relative to some phase reference, so that Equation (4.17) may be expanded to give the more general form as

$$\left\{\begin{matrix} \{S\}_{hr} \\ \{S\}_{hj} \\ \{S\}_{vr} \\ \{S\}_{vj} \\ 1 \end{matrix}\right\}_{L,i} = \begin{bmatrix} [F] & 0 & 0 & 0 & 0 \\ 0 & [F] & 0 & 0 & 0 \\ 0 & 0 & [F] & 0 & 0 \\ 0 & 0 & 0 & [F] & 0 \\ 0 & 0 & 0 & 0 & 1 \end{bmatrix}_i \left\{\begin{matrix} \{S\}_{hr} \\ \{S\}_{hj} \\ \{S\}_{vr} \\ \{S\}_{vj} \\ 1 \end{matrix}\right\}_{R, i-1} \qquad (4.18)$$

where the subscripts $h$ and $v$ indicate reference to the horizontal and vertical directions, and subscripts $r$ and $j$ indicate in-phase and quadrature terms respectively. The last line has been added to facilitate inclusion of imbalance in the analysis, as will be made clear later. Equation (4.18) may alternatively be expressed as

$$_L \{S'\}_i = [F']_i {_R}\{S'\}_{i-1} \qquad (4.19)$$

where $\{S'\}$ and $[F']$ are the modified state vector and field matrix.

If we now consider the mass at station $i$ shown in Figure 4.6, the relationships between the loading acting on the mass and the mass

## MDOF SYSTEMS

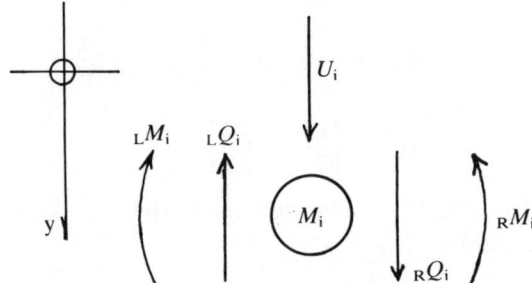

**Figure 4.6** Forces and deflections relating to a concentrated mass at the junction between two elemental lengths of shaft. $u_i$ is the imbalance force at location $i$. See text for explanation.

displacement are given by the equations of motion for the mass which are

$$_RQ_i - {_LQ_i} + u_i = m\ddot{y}_i = -\omega^2 m y_i$$
$$_RM_i - {_LM_i} = I\ddot{\theta}_i = -\omega^2 I \theta_i \qquad (4.20)$$

Also it is clear that the displacements and slopes on each side of the mass are equal, and so we may write the matrix equation

$$_R\begin{Bmatrix} -y \\ \theta \\ M \\ Q \end{Bmatrix}_i = \begin{bmatrix} 1 & 0 & 0 & 0 \\ 0 & 1 & 0 & 0 \\ 0 & -\omega^2 I_d & 1 & 0 \\ m\omega^2 & 0 & 0 & 1 \end{bmatrix}_i {_L}\begin{Bmatrix} -y \\ \theta \\ M \\ Q \end{Bmatrix}_i + \begin{Bmatrix} 0 \\ 0 \\ 0 \\ -u \end{Bmatrix}_i \qquad (4.21)$$

which may be written more simply as

$$_R\{S\}_i = [P]_i {_L}\{S\}_i + \{U\}_i \qquad (4.22)$$

where $[P]$ is known as the point matrix, and $\{U\}$ is an imbalance matrix. The above equations can, of course, be written also for motion in the horizontal direction, and in general the imbalance $u$ will have both in-phase and quadrature components (unless it happens to be that which forms the phase datum). Equation (4.22) may therefore be expanded to the more general form

$$_R\begin{Bmatrix} \{S\}_{hr} \\ \{S\}_{hj} \\ \{S\}_{vr} \\ \{S\}_{vj} \\ 1 \end{Bmatrix}_i = \begin{bmatrix} [P] & 0 & 0 & 0 & \{U\}_{hr} \\ 0 & [P] & 0 & 0 & \{U\}_{hj} \\ 0 & 0 & [P] & 0 & \{U\}_{vr} \\ 0 & 0 & 0 & [P] & \{U\}_{vj} \\ 0 & 0 & 0 & 0 & 1 \end{bmatrix}_i {_L}\begin{Bmatrix} \{S\}_{hr} \\ \{S\}_{hj} \\ \{S\}_{vr} \\ \{S\}_{vj} \\ 1 \end{Bmatrix}_i \qquad (4.23)$$

which can be expressed more simply as

$$_R\{S'\}_i = [P']_i {_L}\{S'\}_i \qquad (4.24)$$

where $[P']$ is a modified point matrix. If gyroscopic effects are allowed for, the second of Equations (4.20) is modified as described in Section 3.7 with the consequence that the following elements of the $[P']$ matrix in Equation (4.24) are changed to $-P_{3,14} = P_{7,10} = P_{11,6} = -P_{15,2} = I_p\omega^2$, where $I_p$ is the rotor polar moment of inertia, and where the positive displacements in the horizontal plane are $x$ and $\phi$ as indicated in Figure 3.8(b).

Equations (4.19) and (4.24) may be combined for a particular station to give

$$_R\{S'\}_i = [P]_i {_L}\{S'\}_i$$
$$= [P]_i[F']_i {_R}\{S'\}_{i-1} \qquad (4.25)$$
$$= [U]_i {_R}\{S'\}_{i-1}$$

where $[U]_i$ is the transfer matrix for station $i$. It is noteworthy that if a continuously distributed mass is to be modelled, the corresponding element transfer matrix can be determined by rewriting Equation (4.55) in the form of Equation (4.25). The transfer matrix for all $n$ stations in the system may be obtained in this manner, and may be combined to give

$$_R\{S'\}_1 = [U]_1 {_R}\{S'\}_0$$
$$_R\{S'\}_2 = [U]_2 {_R}\{S'\}_1 = [U]_2[U]_1 {_R}\{S'\}_0$$
$$_R\{S'\}_3 = [U]_3 {_R}\{S'\}_2 = [U]_3[U]_2[U]_1 {_R}\{S'\}_0$$
$$\cdots$$
$$_R\{S'\}_{n-1} = [U]_{n-1}[U]_{n-2}\ldots[U]_3[U]_2[U]_1 {_R}\{S'\}_0$$
$$= [T] {_R}\{S'\}_0 \qquad (4.26).$$

where $[T]$ is an overall transfer matrix for the complete system, and is still only $17 \times 17$ in size.

To determine the system characteristics it is first necessary to define the system boundary conditions which describe the system supports. For example, if the shaft were simply supported at each end then both displacements and moments would be zero at the supports. Assigning zero values to these parameters in Equation (4.26), and writing down only those lines which have zero, now, on the left-hand side, whilst neglecting those terms on the right-hand side which are zero or are multiplied by zero, leads to a modified form of Equation (4.26):

$$\begin{bmatrix} t_{1,2} & t_{1,4} & t_{1,6} & t_{1,8} & \cdots \\ t_{3,2} & t_{3,4} & t_{3,6} & \cdots \\ t_{5,2} & t_{5,4} & \cdots & \text{etc.} \\ t_{7,2} & \cdots \\ \cdots \\ t_{15,2} & & & & t_{15,16} \end{bmatrix} \begin{Bmatrix} \theta_{hr} \\ Q_{hr} \\ \theta_{nj} \\ Q_{nj} \\ \theta_{vr} \\ Q_{vr} \\ \theta_{vj} \\ Q_{vj} \end{Bmatrix} = - \begin{Bmatrix} t_{1,17} \\ t_{3,17} \\ \cdots \\ \cdots \\ \cdots \\ t_{15,17} \end{Bmatrix} \qquad (4.27)$$

which may be written more simply as

$$[T']\{S''\} = \{T''\} \qquad (4.28)$$

Equation (4.28) may now be solved to determine the unknown values of shear force and slope at station 0, whereupon these values may be substituted back into the matrix $_R\{S'\}_0$ whose elements are now all known.

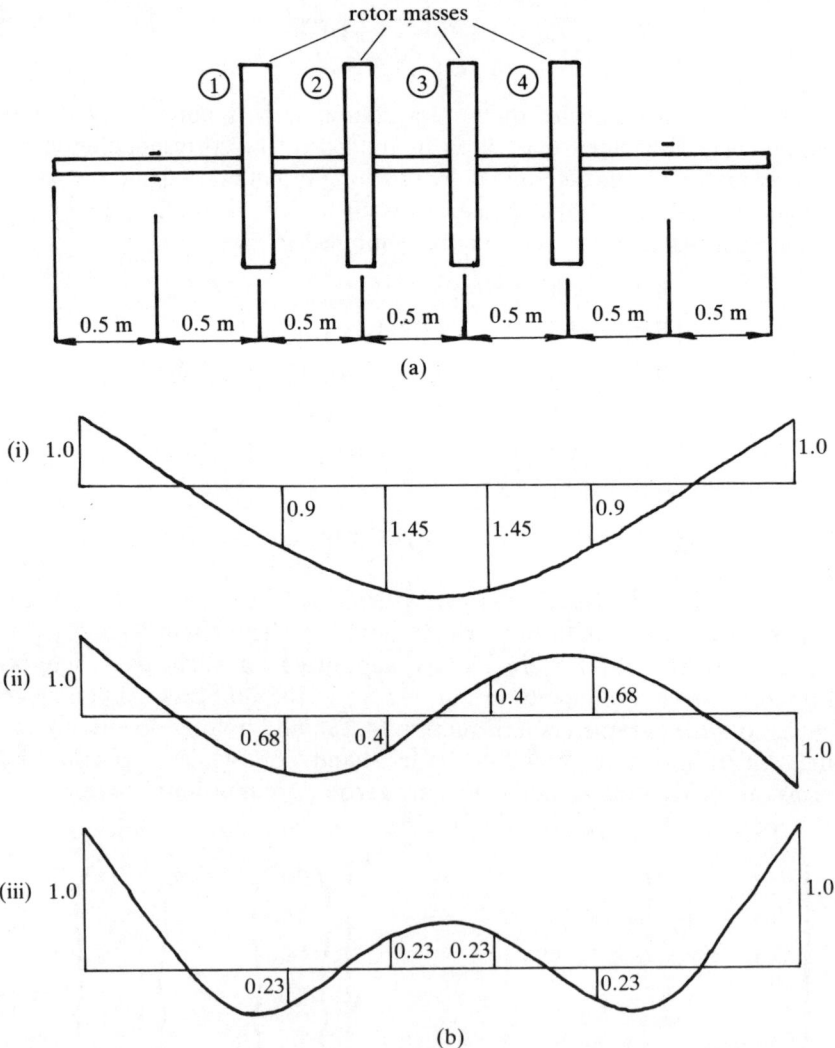

**Figure 4.7** A four-rotor system, running in pinned bearings, (a) Details of shaft design. The mass of each rotor is 60.3 kg, shaft diameter 0.1 m, $E = 208$ GN.m$^{-2}$, shaft mass 61.6 kg.m$^{-1}$; (b) critical speeds and mode shapes (Figures indicate relative amplitudes), (i) 1150 rev.min$^{-1}$, (ii) 4250 rev.min$^{-1}$ and (iii) 8150 rev.min$^{-1}$.

Equations (4.25) may then be used to determine the system response at all other stations. Having obtained the in-phase and quadrature components of all parameters, the net parameter amplitudes may then be calculated by vector summation.

If only the system critical speeds are required, and not the forced response, then the right-hand side of Equation (4.28) reduces to zero and the system natural frequencies are then those which satisfy the equation

$$[T']\{S''\} = 0 \qquad (4.29)$$

that is

$$|T'| = 0 \qquad (4.30)$$

The frequencies can be found using an iterative method, or by some other convenient mathematical routine.

EXAMPLE 4.2

Application of the transfer matrix method to determine the natural frequencies and mode shapes of the system shown in Figure 4.7(a) yielded the results indicated in Figure 4.7(b). The shaft modulus of elasticity was taken to be $E = 208$ GN.m$^{-2}$, and the mass density of the shaft material as $\rho = 7830$ kg.m$^{-3}$. The bearings were assumed to be pinned supports.

## 4.4 Mechanical impedance (and receptance) methods

Impedance and receptance methods enable the engineer to evaluate the behaviour of a complete system from considerations of the behaviour of individual components of the system. They are particularly convenient to use when, for example, the characteristics of the shaft and those of the bearings are determined from independent experimental or theoretical investigations carried out over a range of frequencies. A simple addition of the shaft impedance to the bearing impedance then gives the impedance of the complete system, which may be used to find system critical speeds and forced response. It is also possible to identify system natural frequencies as being those where the shaft impedance 'matches' the bearing impedance. The general principles are first discussed below with reference to a simple spring–mass system, before an outline of their applications to rotor dynamics is given.

Mechanical impedance, $Z$, is defined as the force required to produce unit displacement; it is generally a complex quantity because of the phase lag of displacement behind force. (*Note:* The modern definition of impedance is the ratio force/velocity, not force/displacement, the earlier definition arising from a fallacious consideration of electrical effects. In the context of this section the older definition is more appropriate and will be adhered to.) In the case of the spring-mass system shown in Figure 4.8(a) the impedance of the spring alone is then given by

$$Z_s = k \tag{4.31}$$

Also, since the mass acting in isolation responds according to Newton's second law,

$$f = m\ddot{x} = -\omega^2 m x \tag{4.32}$$

the impedance of the mass alone is

$$Z_m = f/x = -\omega^2 m \tag{4.33}$$

In Figure 4.8(a) the spring and mass are effectively connected in parallel as far as the force is concerned (see the equivalent system in Figure 4.8b). For subsystems connected in parallel the net system impedance is given by the sum of the individual subsystem impedances (for example the equivalent stiffness of two springs connected in parallel is given by the sum of the individual spring stiffnesses), so that for the system in Figures 4.8 the impedance at the forcing point is

$$Z = Z_s + Z_m = k - \omega^2 m \tag{4.34}$$

At the natural frequency the system impedance will be zero (since any force produces an infinite amplitude, the exception being when the forcing point is at a node whereupon the impedance tends towards infinity) so that

$$k - \omega^2 m = 0 \tag{4.35}$$

that is

$$\omega^2 = k/m \tag{4.36}$$

It is noteworthy that for subsystems connected in series, the net system 'receptance' is given by the sum of the individual subsystem receptances

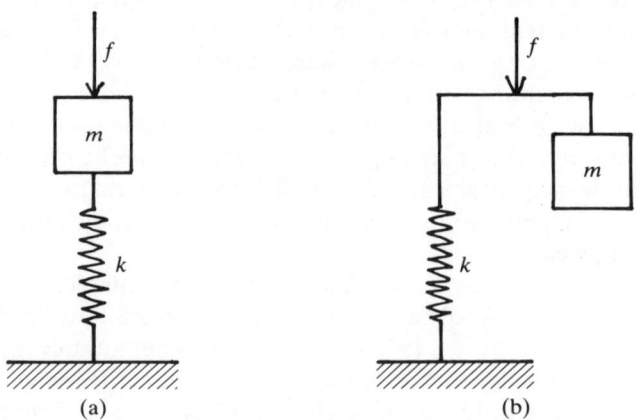

**Figure 4.8** (a) A simple spring and mass; (b) equivalent system for analysis purposes.

(receptance being the inverse of impedance). Some applications of this rule are discussed in Section 5.6.

The approach described above may be applied to machines whose shafts carry many rotor inertias when the shaft free–free impedance has been determined independently of those of the bearings, pedestals, and foundations. (Shaft free–free impedance means the impedance when the shaft is not constrained at either support point; this can be determined experimentally by suspending the shaft so that it is supported only in the vertical direction, then determining the impedance from the horizontal response to a known horizontal forcing). The individual component impedances may then be combined according to the rules described above to determine the impedance of the complete system.

The impedance of a light flexible shaft carrying a number of rotor masses as shown in Figure 4.9 may be determined theoretically as follows. The shaft is considered to be forced, in the first instance, at locations A and B. The forcing causes reaction forces and moments $F_1, M_1, F_2, M_2, \ldots$ etc. to be set up as a consequence of the rotor mass inertias and these deform the shaft according to the relationship

$$\begin{Bmatrix} {}_l\delta_1 \\ {}_l\delta_2 \\ \vdots \\ {}_l\delta_n \\ {}_a\delta_1 \\ {}_a\delta_2 \\ \vdots \\ {}_a\delta_n \end{Bmatrix} = \begin{bmatrix} a_{11} & a_{12} & a_{13} & a_{14} & \cdots & a_{a,2n} \\ a_{21} & a_{22} & a_{23} & a_{24} & \cdots & a_{2,2n} \\ a_{31} & \cdots & \text{etc.} & & & \\ a_{n1} & a_{n2} & a_{n3} & a_{n4} & \cdots & a_{n,2n} \\ a_{n+1,1} & a_{n+1,2} & a_{n+1,3} & a_{n+1,4} & \cdots & a_{n+1,2n} \\ a_{n+2,1} & a_{n+2,2} & a_{n+2,3} & a_{n+2,4} & \cdots & a_{n+2,2n} \\ a_{n+3,1} & \cdots & \text{etc.} & & & \\ a_{2n,1} & a_{2n,2} & a_{2n,3} & a_{2n,4} & \cdots & a_{2n,2n} \end{bmatrix} \begin{Bmatrix} F_1 \\ F_2 \\ \vdots \\ F_n \\ M_1 \\ M_2 \\ \vdots \\ M_n \end{Bmatrix} \quad (4.37)$$

where the values of $a_{ij}$ are determined from beam theory, and where the subscripts $l$ and $a$ indicate linear and angular shaft deformations respectively. In general the loading applied to any rotor mass $i$ to cause its acceleration is

$$_mF_i = m_i({}_l\ddot{\delta} + \ddot{x})$$
$$= -\omega^2 m_i({}_l\delta + x) \quad (4.38a)$$

and

$$_mM_i = I_i({}_a\ddot{\delta} + \ddot{\phi})$$
$$= -\omega^2 I_i({}_a\delta + \phi) \quad (4.38b)$$

where $x$ and $\phi$ are shaft displacements caused by movement at A and B and are not caused by the inertia forces. The reaction loading on the shaft is equal and opposite and is given by

$$_sF_i = \omega^2 m_i({}_l\delta + x), \quad _sM_i = \omega^2 I_i({}_a\delta + \phi) \quad (4.39)$$

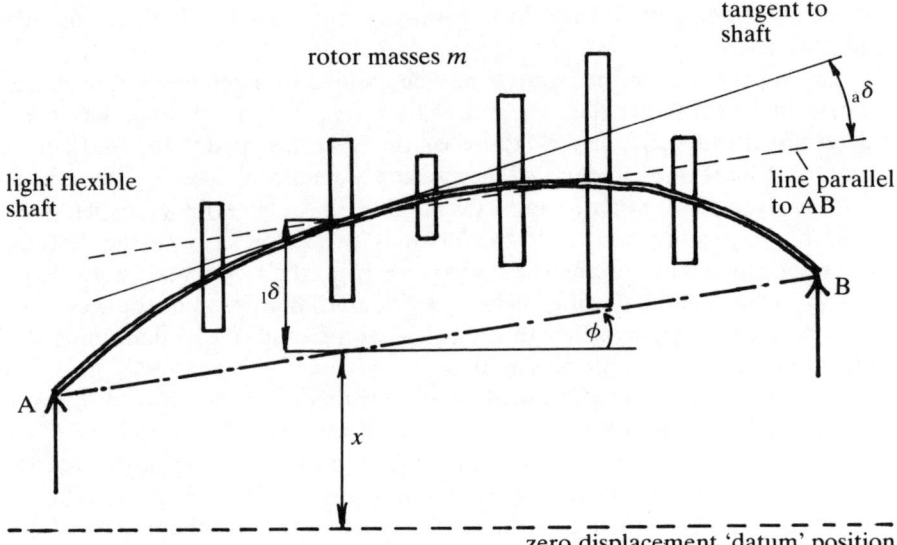

**Figure 4.9** A light shaft carrying many concentrated masses.

Substituting Equations (4.39), written for each rotor, into Equations (4.37)

$$\begin{Bmatrix} {}_1\delta_1 \\ {}_1\delta_2 \\ \vdots \\ {}_1\delta_n \\ {}_a\delta_1 \\ {}_a\delta_2 \\ \vdots \\ {}_a\delta_n \end{Bmatrix} = \begin{bmatrix} a_{11} & a_{12} & a_{13} & \cdots & a_{1,2n} \\ a_{21} & a_{22} & a_{23} & \cdots & a_{2,2n} \\ a_{31} & \cdots & & \text{etc.} & \\ a_{n,1} & a_{n2} & a_{n3} & \cdots & a_{n,2n} \\ a_{n+1,1} & a_{n+1,2} & a_{n+1,3} & \cdots & a_{n+1,2n} \\ a_{n+2,1} & a_{n+2,2} & a_{n+2,3} & \cdots & a_{n+2,2n} \\ a_{n+3,1} & \cdots & & \text{etc.} & \\ a_{2n,1} & a_{an,2} & a_{2n,3} & \cdots & a_{2n,2n} \end{bmatrix} \begin{Bmatrix} m_1({}_1\delta_1 + x_1) \\ m_2({}_1\delta_2 + x_2) \\ \cdots \\ m_n({}_1\delta_n + x_n) \\ I_1({}_a\delta_1 + \phi_1) \\ I_2({}_a\delta_2 + \phi_2) \\ \cdots \\ I_n({}_a\delta_n + \phi_n) \end{Bmatrix}$$

(4.40)

which may be rearranged to give

$$\begin{bmatrix} 1 - \omega^2 m_1 a_{11} & -\omega^2 m_2 a_{12} & \cdots & -\omega^2 I_n a_{1,2n} \\ -\omega^2 m_1 a_{21} & 1 - \omega^2 m_2 a_{22} & \cdots & -\omega^2 I_n a_{2,2n} \\ & \cdots & & \vdots \\ -\omega^2 m_1 a_{2n,1} & -\omega^2 m_2 a_{2n,2} & \cdots & 1 - \omega^2 I_n a_{2n,2n} \end{bmatrix} \begin{Bmatrix} {}_1\delta_1 \\ {}_1\delta_2 \\ \vdots \\ {}_a\delta_n \end{Bmatrix}$$

$$= \begin{bmatrix} a_{11}m_1 & a_{12}m_2 & \cdots & a_{1,2n}I_n \\ a_{21}m_1 & a_{22}m_2 & \cdots & a_{2,2n}I_n \\ & \cdots & & \\ a_{2n,1}m_1 & a_{2n,2}m_2 & & a_{2n,2n}I_n \end{bmatrix} \begin{Bmatrix} x_1 \\ x_2 \\ \cdots \\ \phi_n \end{Bmatrix} \quad (4.41)$$

## MECHANICAL IMPEDANCE

which may be written more simply as
$$[A]\{\delta\} = [R]\{x\} \quad (4.42)$$
and rearranged to give
$$\{\delta\} = [A]^{-1}[R]\{x\} = [C]\{x\} \quad (4.43)$$

In general the application of inertia loads $F_i$ and $M_i$ to the shaft at some point causes proportional reaction forces $F_A$ and $F_B$ at points A and B, where

$$\begin{Bmatrix} F_A \\ F_B \end{Bmatrix} = \begin{bmatrix} b_{A1} & b_{A2} \\ b_{B1} & b_{B2} \end{bmatrix} \begin{Bmatrix} F_i \\ M_i \end{Bmatrix} \quad (4.44)$$

The generalized form of Equation (4.44), allowing for all inertia forces, is

$$\begin{Bmatrix} F_A \\ F_B \end{Bmatrix} = \begin{bmatrix} b_{A1} & b_{A2} & b_{A3} & \dots & b_{A,2n} \\ b_{B1} & b_{B2} & b_{B2} & \dots & b_{B,2n} \end{bmatrix} \begin{Bmatrix} F_1 \\ F_2 \\ \dots \\ F_n \\ M_1 \\ M_2 \\ \dots \\ M_n \end{Bmatrix} \quad (4.45)$$

where the elements $b_{ij}$ may be determined from equilibrium considerations. Substituting for the rotor inertia forces from Equations (4.39) into Equations (4.45) gives

$$\begin{Bmatrix} F_A \\ F_B \end{Bmatrix} = \begin{bmatrix} b_{A1} & b_{A2} & b_{A3} & \dots & b_{A,2n} \\ b_{B1} & b_{B2} & b_{B3} & \dots & b_{B,2n} \end{bmatrix} \begin{Bmatrix} m_1({}_I\delta_1 + x_1) \\ m_2({}_I\delta_2 + x_2) \\ \dots \\ m_n({}_I\delta_n + x_n) \\ I_1({}_a\delta_1 + \phi_1) \\ I_2({}_a\delta_2 + \phi_2) \\ \dots \\ I_n({}_a\delta_n + \phi_n) \end{Bmatrix} \quad (4.46)$$

Substituting for $\delta$ from Equations (4.43) into Equations (4.46) then gives

$$\begin{Bmatrix} F_A \\ F_B \end{Bmatrix} = \begin{bmatrix} b_{A1} & b_{A2} & b_{A3} & \dots & b_{A,2n} \\ b_{B1} & b_{B2} & b_{B3} & \dots & b_{B,2n} \end{bmatrix}$$
$$\times \begin{bmatrix} m_1(1+c_{11}) & m_1 c_{12} & m_1 c_{13} & \dots & m_1 c_{1,2n} \\ m_2 c_{21} & m_2 1 + c_{22} & m_2 c_{23} & \dots & m_2 c_{2,2n} \\ & & \dots & & \\ I_n c_{2n,1} & I_n c_{2n,2} & I_n c_{2n,3} & \dots & I_n(1+c_{2n,2n}) \end{bmatrix} \begin{Bmatrix} x_1 \\ x_2 \\ \dots \\ \phi_n \end{Bmatrix}$$
$$= [D]\{x\} \quad (4.47)$$

But the displacements $x$, $\phi_1$, $x_2$, $\phi_2$, ..., etc. are related to the displacements at the forcing points, $x_A$ and $x_B$ by a relationship of the form

$$\begin{Bmatrix} x_1 \\ x_2 \\ \vdots \\ x_n \\ \phi_1 \\ \phi_2 \\ \vdots \\ \phi_n \end{Bmatrix} = \begin{bmatrix} g_{11} & g_{12} \\ g_{21} & g_{22} \\ \vdots & \\ g_{n1} & g_{n2} \\ g_{n+1,1} & g_{n+1,2} \\ g_{n+2,1} & g_{n+2,2} \\ \vdots & \\ g_{2n,1} & g_{2n,2} \end{bmatrix} \begin{Bmatrix} x_A \\ x_B \end{Bmatrix} \qquad (4.48)$$

which may be written as

$$\{x\} = [G] \begin{Bmatrix} x_A \\ x_B \end{Bmatrix} \qquad (4.49)$$

The elements of the matrix $[G]$ may, of course, be determined from simple considerations of geometry. Substituting Equations (4.49) into Equations (4.47) gives

$$\begin{Bmatrix} F_A \\ F_B \end{Bmatrix} = [D][G] \begin{Bmatrix} x_A \\ x_B \end{Bmatrix}$$

$$= [Z] \begin{Bmatrix} x_A \\ x_B \end{Bmatrix}$$

$$= \begin{bmatrix} z_{AA} & z_{AB} \\ z_{BA} & z_{BB} \end{bmatrix} \begin{Bmatrix} x_A \\ x_B \end{Bmatrix} \qquad (4.50)$$

where $[Z]$ is the impedance matrix for the shaft and rotor assembly, relating forcing and displacements at points A and B. The more general form of Equation (4.50) which allows also for motion in the $y$ direction is

$$\begin{Bmatrix} F_{xA} \\ F_{xB} \\ \hdashline F_{yA} \\ F_{yB} \end{Bmatrix} = \left[ \begin{array}{cc:cc} z_{AA} & z_{AB} & & \\ z_{BA} & z_{BB} & \multicolumn{2}{c}{\raisebox{1ex}{0}} \\ \hdashline & & z_{AA} & z_{AB} \\ \multicolumn{2}{c:}{\raisebox{1ex}{0}} & z_{BA} & z_{BB} \end{array} \right] \begin{Bmatrix} x_A \\ x_B \\ \hdashline y_A \\ y_B \end{Bmatrix} \qquad (4.51)$$

The response at any other location along the shaft can be found by pre-multiplying both sides of Equation (4.51) by the inverse of the impedance matrix to yield the displacements at $A$ and $B$, and substituting these values into Equations (4.48). If $F_B$ is chosen to be zero, point A can be chosen as any location along the shaft and the corresponding impedances so evaluated. Equation (4.51) can then be expanded to allow for any number

## MECHANICAL IMPEDANCE

of forcing points; for example, when there is a third forcing point, C, the equation takes form

$$\begin{Bmatrix} F_{xA} \\ F_{xB} \\ F_{xC} \\ \hline F_{yA} \\ F_{yB} \\ F_{yC} \end{Bmatrix} = \begin{bmatrix} z_{AA} & z_{AB} & z_{AC} & & & \\ z_{BA} & z_{BB} & z_{BC} & & 0 & \\ z_{CA} & z_{CB} & z_{CC} & & & \\ \hline & & & z_{AA} & z_{AB} & z_{AC} \\ & 0 & & z_{BA} & z_{BB} & z_{BC} \\ & & & z_{CA} & z_{CB} & z_{CC} \end{bmatrix} \begin{Bmatrix} x_A \\ x_B \\ x_C \\ \hline y_A \\ y_B \\ y_C \end{Bmatrix} \quad (4.52)$$

If $F_C$ is a force caused by imbalance, and $F_A$ and $F_B$ are reaction forces at pinned bearing supports, the forced response of the system can be determined by assigning $x_A$, $x_B$, $y_A$ and $y_B$ to zero values. The bearing reaction forces may then be determined from Equation (4.52) which then gives

$$\begin{Bmatrix} F_{xA} \\ F_{xB} \\ F_{yA} \\ F_{yB} \end{Bmatrix} = \begin{bmatrix} z_{AC} & 0 \\ z_{BC} & 0 \\ 0 & z_{AC} \\ 0 & z_{BC} \end{bmatrix} \begin{Bmatrix} F_{xC}/z_{CC} \\ F_{yC}/z_{CC} \end{Bmatrix} \quad (4.53)$$

These reaction forces may be substituted back into Equation (4.52), which may be written more generally, to give the response at any other location, as

$$\begin{Bmatrix} x_A \\ x_B \\ x_C \\ \hline y_A \\ y_B \\ y_C \end{Bmatrix} = \begin{bmatrix} z_{AA} & z_{AB} & z_{AC} & & & \\ z_{BA} & z_{BB} & z_{BC} & & 0 & \\ z_{CA} & z_{CB} & z_{CC} & & & \\ \hline & & & z_{AA} & z_{AB} & z_{AC} \\ & 0 & & z_{BA} & z_{BB} & z_{BC} \\ & & & z_{CA} & z_{CB} & z_{CC} \end{bmatrix}^{-1} \begin{Bmatrix} F_{xA} \\ F_{xB} \\ F_{xC} \\ \hline F_{yA} \\ F_{yB} \\ F_{yC} \end{Bmatrix} \quad (4.54)$$

Series expressions for the elements of the impedance matrix, for a free–free shaft with uniformly distributed mass and flexibility, have been evaluated by Bishop and Johnson (1960); similar expressions have also been published by Gladwell and Bishop (1959) for shafts with non-uniformly distributed mass and flexibility.

EXAMPLE 4.3

A concentrated rotor mass of 100 kg and negligible moment of inertia about its diameter is carried at the centre of a 2 m long shaft of modulus

## MDOF SYSTEMS

$EI = 1.17$ MN.m$^2$ as shown in Figure 4.10. Calculate the impedance of the shaft and rotor as measured at the bearing locations at each end of the shaft, and determine the system pin–pin natural frequency (i.e. that based on an assumption of the shaft being carried by pin hinge supports at both ends).

SOLUTION

The shaft influence coefficient as measured at the location of the mass is

$$a_{11} = \frac{1^3}{48EI} = \frac{2^3}{48 \times 1.17 \times 10^6} = 0.142 \ \mu\text{m.N}^{-1}$$

There are no other mass stations to consider, and since the moment of inertia of the rotor about its diameter is negligible there will be no reaction moments between the rotor and shaft. This influence coefficient $a_{11}$ is therefore the only one of concern. Equation (4.5), then, takes the form

$$(1 - m_1\omega^2 a_{11})_1\delta_1 = (\omega^2 a_{11} m_1) x_1$$

whereupon

$$\delta_1 = \frac{x}{(1/\omega^2 a_{11} m_1) - 1}$$

Since the mass is located at the centre of the shaft, the bearing reaction to any mass inertia force will be the same at each bearing, and so Equation (4.43) becomes

$$\begin{Bmatrix} F_A \\ F_B \end{Bmatrix} = \begin{bmatrix} 0.5 \\ 0.5 \end{bmatrix} F_1$$

$$= \begin{bmatrix} 0.5 \\ 0.5 \end{bmatrix} \omega^2 m_1 (_1\delta_1 + x_1)$$

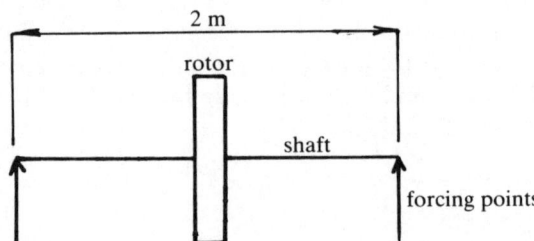

**Figure 4.10** A single mass rotor ($m = 100$ kg) mounted on a light flexible shaft of modulus ($EI = 1.17$ MN.m$^2$). See Example 4.3.

Substituting for ${}_1\delta_1$ from above and rearranging gives

$$\begin{Bmatrix} F_A \\ F_B \end{Bmatrix} = \begin{bmatrix} 0.5 \\ 0.5 \end{bmatrix} \left(\frac{\omega^2 m_1}{(1 - \omega^2 a_{11} m_1)}\right) x_1$$

$$= \begin{bmatrix} 0.5 \\ 0.5 \end{bmatrix} \left(\frac{\omega^2 m_1}{(1 - \omega^2 a_{11} m_1)}\right) (0.5 x_A + 0.5 x_B)$$

$$= \left(\frac{\omega^2 m_1}{(1 - \omega^2 a_{11} m_1)}\right) \begin{bmatrix} 0.25 & 0.25 \\ 0.25 & 0.25 \end{bmatrix} \begin{Bmatrix} x_A \\ x_B \end{Bmatrix}$$

$$= \begin{bmatrix} z_{AA} & z_{AB} \\ z_{BA} & z_{BB} \end{bmatrix} \begin{Bmatrix} x_A \\ x_B \end{Bmatrix}$$

Since the system in question is symmetrical, $F_A = F_B = F$ and $x_A = x_B = x$ for symmetrical vibration modes. The above equation may therefore be further simplified to give the force at each bearing as

$$F = \left(\frac{0.5 \omega^2 m_1}{1 - \omega^2 m_1 a_{11}}\right) x$$

$$= z_{sym} x$$

If the bearings behave as pinned supports then the bearing locations on the shaft will be node points at the system critical speeds, and so the impedance at the bearings will be infinite. Therefore

$$1 - \omega^2 m_1 a_{11} = 0$$

and so

$$\omega = (m_1 a_{11})^{1/2}$$
$$= (10^2 \times 0.142 \times 10^{-6})^{1/2}$$
$$= 265 \text{ rad.s}^{-1}$$

(It is also noteworthy that for excitation of asymmetric vibration modes $x_A = -x_B = x$, whereupon $z_{asym} = 0$ at all frequencies).

## 4.5 Dynamic stiffness matrix method

The dynamic stiffness matrix method is another means of analysis available for use with MDOF systems. It is similar to the transfer matrix method discussed in Section 4.3 in that it involves a division of the shaft into a number of smaller elements for the purposes of analysis; however, the equations which describe the system are formulated in a different manner and the response at all stations in the system is determined simultaneously. The method has the advantage that once the system matrix has been assembled it can be used to calculate system stability thresholds (Rieger et al. 1976) as well as response and critical speeds; the major disadvantage associated with its use is that it requires the storing and manipulation of large matrices, and so is more demanding of computer power than is the transfer matrix method, for example.

In the analysis described below each of the smaller elements of the whole shaft are assumed to possess uniformly distributed mass, with the option of an additional concentrated mass at one end of the element as indicated in Figures 4.11. Consider each element of shaft to be acted upon by forces and moments, applied by neighbouring elements, as shown in the diagram. If the beam element alone, without the concentrated mass, is first considered (Figure 4.11b), the relationship between the applied forces and moments and the resulting deflections and slopes is given by Euler's beam theory as

$$\begin{Bmatrix} M_A \\ Q_A \\ M_B \\ Q_B \end{Bmatrix} = \begin{bmatrix} k_{11} & k_{12} & k_{13} & k_{14} \\ k_{21} & k_{22} & k_{23} & k_{24} \\ k_{31} & k_{32} & k_{33} & k_{34} \\ k_{41} & k_{42} & k_{43} & k_{44} \end{bmatrix} \begin{Bmatrix} y_A \\ \theta_A \\ y_B \\ \theta_B \end{Bmatrix} \quad (4.55)$$

where

$k_{11} = k_{33} = \lambda k_{22} = \lambda k_{44} = EI\lambda^2(\sin \lambda l \sinh \lambda l)/(\cos \lambda l \cosh \lambda l - 1)$

$k_{13} = k_{31} = -\lambda k_{42} = -\lambda k_{24} - EI\lambda^2(\cos \lambda l - \cosh \lambda l)/(\cos \lambda l \cosh \lambda l - 1)$

$k_{12} = -k_{34} = -EI\lambda(\cos \lambda l \sinh \lambda l - \sin \lambda l \cosh \lambda l)/(\cos \lambda l \cosh \lambda l - 1)$

$k_{14} = -k_{32} = -EI\lambda(\sin \lambda l + \sinh \lambda l)/(\cos \lambda l \cosh \lambda l - 1)$

$k_{21} = -k_{43} = EI\lambda^3(\cos \lambda l \sinh \lambda l + \sin \lambda l \cosh \lambda l)/(\cos \lambda l \cosh \lambda l - 1)$

$k_{23} = -k_{41} = EI\lambda^3(\sin \lambda l + \sinh \lambda l)/(\cos \lambda l \cosh \lambda l - 1)$

and

$\lambda^4 = \rho a \omega^2 / EI$

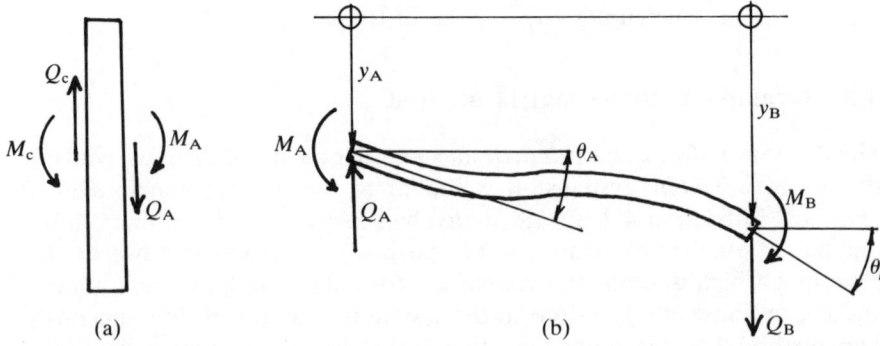

**Figure 4.11** Components of one form of shaft element that can be used with the dynamic stiffness matrix method, (a) concentrated mass section of element; (b) beam section of element. See text for details.

and where $\rho$ is the material mass density, $a$ is the cross-sectional area, $\omega$ is the vibration frequency, $E$ is Young's modulus of elasticity and $I$ is the section second moment of area. (Alternatively a light beam model may be used by re-arranging Equations (4.16) into the same form as Equation 4.55).

In general the applied moments and shear forces take the form

$$\begin{aligned} M_A &= {}_cM_A \cos \omega t + {}_sM_A \sin \omega t \\ M_B &= {}_cM_B \cos \omega t + {}_sM_B \sin \omega t \\ Q_A &= {}_cQ_A \cos \omega t + {}_sQ_A \sin \omega t \\ Q_B &= {}_cQ_B \cos \omega t + {}_sQ_B \sin \omega t \end{aligned} \quad (4.56)$$

and the deflections and slopes may similarly be written as

$$\begin{aligned} y_A &= {}_cY_A \cos \omega t + {}_sY_A \sin \omega t \\ y_B &= {}_cY_B \cos \omega t + {}_sY_B \sin \omega t \\ \theta_A &= {}_c\theta_A \cos \omega t + {}_s\theta_A \sin \omega t \\ \theta_B &= {}_c\theta_B \cos \omega t + {}_s\theta_B \sin \omega t \end{aligned} \quad (4.57)$$

Substitution of Equations (4.56) and (4.57) into Equations (4.55), and comparing coefficients of $\sin \omega t$ and $\cos \omega t$ gives

$$\begin{Bmatrix} {}_cM_A \\ {}_cQ_A \\ {}_sM_A \\ {}_sQ_A \\ {}_cM_B \\ {}_cQ_B \\ {}_sM_B \\ {}_sQ_B \end{Bmatrix} = \begin{bmatrix} k_{11} & k_{12} & 0 & 0 & k_{13} & k_{14} & 0 & 0 \\ k_{21} & k_{22} & 0 & 0 & k_{23} & k_{24} & 0 & 0 \\ 0 & 0 & k_{11} & k_{12} & 0 & 0 & k_{13} & k_{14} \\ 0 & 0 & k_{21} & k_{22} & 0 & 0 & k_{23} & k_{24} \\ k_{31} & k_{32} & 0 & 0 & k_{33} & k_{34} & 0 & 0 \\ k_{41} & k_{42} & 0 & 0 & k_{43} & k_{44} & 0 & 0 \\ 0 & 0 & k_{31} & k_{32} & 0 & 0 & k_{33} & k_{34} \\ 0 & 0 & k_{41} & k_{42} & 0 & 0 & k_{43} & k_{44} \end{bmatrix} \begin{Bmatrix} {}_cy_A \\ {}_c\theta_A \\ {}_sy_A \\ {}_s\theta_A \\ {}_cy_B \\ {}_c\theta_B \\ {}_sy_B \\ {}_s\theta_B \end{Bmatrix} \quad (4.58)$$

which can be written more simply as

$$\begin{Bmatrix} \{F_A\} \\ \{F_B\} \end{Bmatrix} = \begin{bmatrix} [u_{11}] & [u_{12}] \\ [u_{21}] & [u_{22}] \end{bmatrix} \begin{Bmatrix} \{d_A\} \\ \{d_B\} \end{Bmatrix} \quad (4.59)$$

Equation (4.59) may be expanded to allow for shear forces, moments, slopes and displacements in the horizontal direction, and written then as

$$\begin{Bmatrix} \{F_{Av}\} \\ \{F_{Ah}\} \\ \{F_{Bv}\} \\ \{F_{Bh}\} \end{Bmatrix} \begin{bmatrix} [u_{11}] & 0 & [u_{12}] & 0 \\ 0 & [u_{11}] & 0 & [u_{12}] \\ [u_{21}] & 0 & [u_{22}] & 0 \\ 0 & [u_{21}] & 0 & [u_{22}] \end{bmatrix} \begin{Bmatrix} \{d_{Av}\} \\ \{d_{Ah}\} \\ \{d_{Bv}\} \\ \{d_{Bh}\} \end{Bmatrix} \quad (4.60)$$

If we now consider the forces acting on the concentrated mass at the end of the element (Figure 4.11a), the equations of motion for the mass are

$$\begin{aligned}
Q_{Av} + U_v - Q_{Cv} - B_v &= m\ddot{y} \\
Q_{Ah} + U_h - Q_{Ch} - B_h &= m\ddot{x} \\
M_{Cv} - M_{Av} - I_p\omega\dot{\theta}_{Ah} &= I_d\ddot{\theta}_{Av} \\
M_{Ch} - M_{Ah} + I_p\omega\dot{\theta}_{Av} &= I_d\ddot{\theta}_{Ah}
\end{aligned} \qquad (4.61)$$

where $B$ is a bearing reaction force (equal to zero if the station considered is not at a bearing location), and $U$ is a known imbalance load ($U$ and $B$ are not shown in the diagram); $m$ is the magnitude of the concentrated mass, and $I_p$ and $I_d$ the polar and transverse moments of inertia of the concentration mass (the quantity $I_p$ can be set equal to zero if gyroscopic effects are to be ignored). Noting that the slopes and displacements on each side of the concentrated mass are the same and take the form given in Equations (4.57), and writing the bearing reaction forces also in the form

$$\begin{aligned}
B_v &= {}_cB_v \cos \omega t + {}_sB_v \sin \omega t \\
B_h &= {}_cB_h \cos \omega t + {}_sB_h \sin \omega t
\end{aligned} \qquad (4.62)$$

enables substitution of Equation (4.51) and (4.62) into Equations (4.61), and comparison of coefficients of $\cos \omega t$ and $\sin \omega t$ on each side of each equation gives

$$\begin{aligned}
{}_cQ_{Av} &= {}_cB_v - m\omega^2{}_cY + {}_cQ_{Cv} - {}_cU_v \\
{}_cQ_{Ah} &= {}_cB_h - m\omega^2{}_cX + {}_cQ_{Ch} - {}_cU_h \\
{}_sQ_{Av} &= {}_sB_v - m\omega^2{}_sY + {}_sQ_{Cv} - {}_sU_v \\
{}_sQ_{Ah} &= {}_sB_h - m\omega^2{}_sX + {}_sQ_{Ch} - {}_sU_h \\
{}_cM_{Av} &= {}_cM_{Cv} - I_p\omega^2{}_s\theta_{Ah} + I_d\omega_c^2\theta_{AV} \\
{}_cM_{Ah} &= {}_cM_{Ch} + I_p\omega^2{}_s\theta_{Av} + I_d\omega_c^2\theta_{Ah} \\
{}_sM_{Av} &= {}_sM_{Cv} + I_p\omega^2{}_c\theta_{Ah} + I_d\omega_s^2\theta_{Av} \\
{}_sM_{Ah} &= {}_sM_{Ch} - I_p\omega^2{}_c\theta_{Av} + I_d\omega_s^2\theta_{Ah}
\end{aligned} \qquad (4.63)$$

Substitution of Equations (4.63) into Equations (4.60) then gives the following matrix equation, see page 125.

Equation (4.64) may be written more simply for the shaft element between nodes 0 and 1, where location C on the element is at node 0 and location B at node 1, as

$$\begin{Bmatrix} \{\bar{F}_0\} \\ \{F_1\} \end{Bmatrix} = \begin{bmatrix} [m_{00}] & [m_{01}] \\ [m_{10}] & [m_{11}] \end{bmatrix} \begin{Bmatrix} \{d_0\} \\ \{d_1\} \end{Bmatrix} = \begin{bmatrix} M_0 \end{bmatrix} \begin{Bmatrix} \{d_0\} \\ \{d_1\} \end{Bmatrix} \qquad (4.65)$$

where the subscripts refer to node numbers and the matrix $[M_0]$ is the dynamic stiffness matrix for element 0. A similar expression may be

$$
\begin{Bmatrix} {}_cM_{Cv} \\ {}_cQ_{Cv} - {}_cU_v + {}_cB_v \\ {}_sQ_{Cv} - {}_sU_v + {}_sB_v \\ \hline {}_cM_{Ch} \\ {}_cQ_{Ch} - {}_cU_h + {}_cB_h \\ {}_sQ_{Ch} - {}_sU_h + {}_sB_h \\ \hline \{F_{Bv}\} \\ \{F_{Bh}\} \end{Bmatrix}
=
\left[ \begin{array}{ccc|ccc|c|c}
k_{11}+m\omega^2 & k_{12}-I_d\omega^2 & 0 & 0 & 0 & 0 & & \\
k_{21} & k_{22} & 0 & 0 & 0 & 0 & & \\
0 & 0 & k_{11}+m\omega^2 & 0 & 0 & 0 & [u_{1,2}] & 0 \\
\hline
0 & 0 & 0 & I_p\omega^2 & 0 & 0 & & \\
0 & I & 0 & 0 & 0 & 0 & & \\
0 & 0 & 0 & 0 & k_{11} & k_{12}-I_d\omega^2 & 0 & [u_{1,2}] \\
& & & & k_{21}+m\omega^2 & k_{22} & & \\
\hline
 & u & 0 & -I_p\omega^2 & 0 & 0 & [u_{22}] & 0 \\
\hline
 & 0 & & u & 0 & 0 & 0 & [u_{22}]
\end{array} \right]
\begin{Bmatrix} d_{Cv} \\ \\ \\ d_{Ch} \\ \\ \\ \{d_{Bv}\} \\ \{d_{Bh}\} \end{Bmatrix}
$$

(4.64)

developed for element 1 which will be of the form

$$\begin{Bmatrix} \{\bar{F}_1\} \\ \{F_2\} \end{Bmatrix} = \begin{bmatrix} M_1 \end{bmatrix} \begin{Bmatrix} \{d_1\} \\ \{d_2\} \end{Bmatrix} \tag{4.66}$$

where $\{\bar{F}_1\}$ is similar to $\{F_1\}$ but also contains imbalance forcing terms and bearing force terms (as does $\{\bar{F}_0\}$). Equations (4.65) and (4.66) may then be combined to eliminate the internal force and moment terms of the matrix $\{\bar{F}_1\}$ and to give an overall equation, for elements 0 and 1 together, of the form

$$\begin{Bmatrix} \{\bar{F}_0\} \\ \{F_1\}^* \\ \{F_2\} \end{Bmatrix} = \begin{bmatrix} \begin{bmatrix} M_0 \end{bmatrix} & \\ & \begin{bmatrix} M_1 \end{bmatrix} \end{bmatrix} \begin{Bmatrix} \{d_0\} \\ \{d_1\} \\ \{d_2\} \end{Bmatrix} \tag{4.67}$$

where $\{F_1\}^*$ now contains only the imbalance and bearing force terms of matrix $\{\bar{F}_1\}$. Equation (4.67) may be extended for any number of system elements to give an overall system matrix equation of the form

$$\begin{Bmatrix} \{\bar{F}_0\} \\ \{F_1\}^* \\ \{F_2\}^* \\ \ldots \\ \{F_{n-1}\}^* \\ \{F_n\} \end{Bmatrix} = \begin{bmatrix} [\quad] & & & 0 \\ & [\quad] & & \\ & & \ddots & \\ 0 & & & [\quad] \end{bmatrix} \begin{Bmatrix} \{d_0\} \\ \{d_1\} \\ \{d_2\} \\ \ldots \\ \{d_{n-1}\} \\ \{d_n\} \end{Bmatrix} \tag{4.68}$$

which may be written more simply as

$$\{P\} = [Z]\{S\} \tag{4.69}$$

The left-hand side of Equation (4.68) now contains only known imbalance forcing terms, known applied forces and moments at the shaft ends (usually zero), and unknown bearing reaction forces. The bearing reaction forces can be found by multiplying both sides of Equation (4.68) by the inverse of the system dynamic stiffness matrix to give

$$\{S\} = [Z]^{-1}\{P\} \tag{4.70}$$

from which can be obtained an expression for displacement at each bearing location. In the case of bearings which behave as pinned supports these expressions can be equated to zero and solved simultaneously to give values for the bearing reaction forces. Back-substitution into Equations (4.69) then enables shaft displacements at all other locations to be evaluated.

It is noteworthy that the system dynamic stiffness matrix $[Z]$ is banded about the leading diagonal; this can be made use of when storing the matrix in the computer, since only the non-zero values need to be saved.

## 4.6 Effects of flexible bearings and pedestals

Machines are frequently designed to run in bearings which do not behave as simple supports, as has been assumed in the previous sections, but instead are themselves flexible. In addition the pedestal in which the bearing itself is mounted may also be flexible. In order to allow for these features the analytical methods described in the preceding section have to be modified accordingly.

### 4.6.1 Effects on the influence coefficients

In the case of the influence coefficient method, the influence coefficients as introduced in Equation (4.5) are evaluated from consideration of only the shaft deformation using elementary beam theory. If the bearings are not rigid, however, the total displacement of a point on the shaft is made up of that due to shaft bending, as already described, plus that which is attributable to movement of the shaft in the bearings. The latter may be calculated by first establishing the component of the applied load which is carried by each bearing, and representing the subsequent displacements at the bearings using the relationship

$$\begin{Bmatrix} \begin{pmatrix} M_1 \\ M_2 \\ N_1 \\ N_2 \end{pmatrix}_A \\ \begin{pmatrix} M_1 \\ M_2 \\ N_1 \\ N_2 \end{pmatrix}_B \end{Bmatrix} = \begin{bmatrix} \begin{bmatrix} K_{xx} & -\omega C_{xx} & -\omega C_{xy} & K_{xy} \\ \omega C_{xx} & K_{xx} & K_{xy} & \omega C_{xy} \\ \omega C_{yx} & K_{yx} & K_{yy} & \omega C_{yy} \\ K_{yx} & -\omega C_{yx} & -\omega C_{yy} & K_{yy} \end{bmatrix}_A^{-1} & 0 \\ 0 & \begin{bmatrix} K_{xx} & -\omega C_{xx} & -\omega C_{xy} & K_{xy} \\ \omega C_{xx} & K_{xx} & K_{xy} & \omega C_{xy} \\ \omega C_{yx} & K_{yx} & K_{yy} & \omega C_{yy} \\ K_{yx} & -\omega C_{yx} & -\omega C_{yy} & K_{yy} \end{bmatrix}_B \end{bmatrix}$$

$$\times \begin{Bmatrix} \begin{pmatrix} F_{bx1} \\ F_{bx2} \\ F_{by1} \\ F_{by2} \end{pmatrix}_A \\ \begin{pmatrix} F_{bx1} \\ F_{bx2} \\ F_{by1} \\ F_{by2} \end{pmatrix}_B \end{Bmatrix} \quad (4.71)$$

## MDOF SYSTEMS

which is obtained by rearranging Equation (3.37). This may be written more simply as

$$\{v\} = [I]\{F_b\} \tag{4.72}$$

Having thus obtained the displacement at the bearing location, it is then a relatively simple matter to determine the corresponding net shaft displacement at any particular location. For example, in the case of the shaft supported on two bearings a distance $l$ apart, if the location of interest $p$ is a distance $d$ from the bearing under consideration then the net displacement of $p$ is that due to shaft deformation plus a linear displacement

$$\{v'\} = \frac{(l-d)}{l}\{v\}$$

and an angular displacement

$$\{\phi'\} = \frac{\{v\}}{l}$$

This additional displacement of the point under consideration must be evaluated for displacement of the shaft at each bearing, and added to that due to shaft bending to determine the net displacement of the point. Thus the influence coefficients evaluated as described in Section 4.2 for the shaft on rigid bearings become modified to allow for bearing flexibility.

### 4.6.2 *Flexible support effects on transfer matrix types of analyses*

In the case of the analysis which makes use of the transfer matrix method the theory is modified to allow for flexible bearings by replacing the point

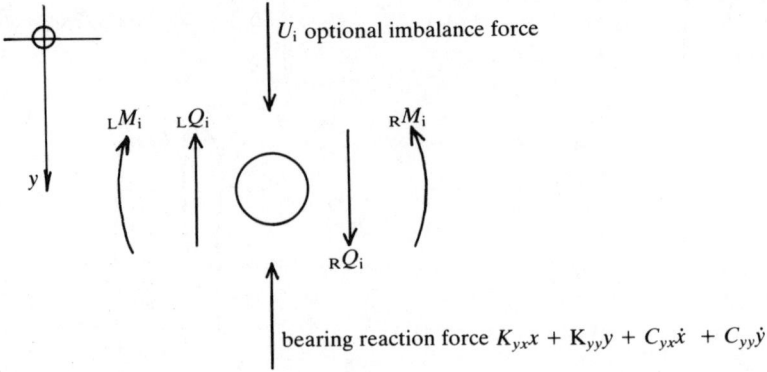

**Figure 4.12** A concentrated shaft mass with reaction forces and bearing forces acting on it. See text for details.

matrix $[P']$, representing the location on the shaft coincident with a bearing, by a bearing transfer matrix $[B]$. The bearing transfer matrix can be developed by considering the forces applied to the shaft lumped mass located at the bearing position (Kramer 1977, Rao 1983), as shown in Figure 4.12. For the mass under consideration the equations of motion in the horizontal and vertical directions are

$$_R Q_{hi} - {_L Q_{hi}} - K_{xx}x_i - K_{xy}y_i - C_{xx}\dot{x}_i - C_{xy}\dot{y}_i = m\ddot{x}_i \quad (4.74)$$
$$_R Q_{vi} - {_L Q_{vi}} - K_{yx}x_i - K_{yy}y_i - C_{yx}\dot{x}_i - C_{yy}\dot{y}_i = m\ddot{y}_i$$

respectively. If required the imbalance force can also be included in the equations as shown earlier. Also, since the values of displacement, slope and bending moment on each side of the mass are equal (assuming for the moment that the effects of shaft polar and transverse moment of inertia are negligible) we may write

$$\begin{aligned} _R x_i &= {_L x_i} \\ _R y_i &= {_L y_i} \\ _R \theta_{hi} &= {_L \theta_{hi}} \\ _R \theta_{vi} &= {_L \theta_{vi}} \\ _R M_{hi} &= {_L M_{hi}} \\ _R M_{vi} &= {_L M_{vi}} \end{aligned} \quad (4.75)$$

Remembering that the displacements are given, in general, by Equations (3.18), which may be differentiated to give expressions for velocity and acceleration (as in Equations 3.19 and 3.20), and that the slopes, bending moments and shear forces also have in-phase and quadrature terms generally and so take the form

$$\begin{aligned} \theta_h &= \theta_{h1} \sin \omega t + \theta_{h2} \cos \omega t \\ \theta_v &= \theta_{v1} \sin \omega t + \theta_{v2} \cos \omega t \\ M_h &= M_{h1} \sin \omega t + M_{h2} \cos \omega t \\ M_v &= M_{v1} \sin \omega t + M_{v2} \cos \omega t \\ Q_h &= Q_{h1} \sin \omega t + Q_{h2} \cos \omega t \\ Q_v &= Q_{v1} \sin \omega t + Q_{v2} \cos \omega t \end{aligned} \quad (4.76)$$

then on substitution of Equations (4.76), (3.18), (3.19) and (3.20) into Equations (4.75) and (4.74), and comparing coefficients of $\sin \omega t$ and $\cos \omega t$, the following matrix equation is obtained:

$$_R\{S'\} = [B] {_L\{S'\}} \quad (4.77)$$

where the matrix $\{S'\}$ is defined in Equation (4.23) and $[B]$ is a $17 \times 17$

matrix whose elements are all zero except for

$$b_{11} = b_{22} = b_{33} = \cdots = b_{17,17} = 1$$
$$b_{41} = b_{85} = -K_{xx} + m\omega^2$$
$$b_{45} = -b_{81} = \omega C_{xx}$$
$$b_{49} = -b_{8,13} = \omega C_{xy}$$
$$b_{4,13} = b_{89} = -K_{xx} \qquad (4.78)$$
$$b_{12,1} = b_{16,5} = -K_{yx}$$
$$b_{12,5} = -b_{16,1} = \omega C_{yx}$$
$$b_{12,9} = -b_{16,13} = \omega C_{yy}$$
$$b_{12,13} = b_{16,9} = -K_{yy} + m\omega^2$$

The effects of imbalance at the bearing station can be included by inserting further non-zero values in column 17 of the matrix, as in Equation (4.23). The matrix $[B]$ is now used in place of the matrix $[P']$ (see Equation 4.24) at bearing locations or, alternatively, when shaft moment of inertia is to be accounted for, the non-zero elements of the $[B]$ matrix are substituted for the corresponding elements of the $[P']$ matrix; the analysis proceeds otherwise as before.

EXAMPLE 4.4

If the shaft described in Example 4.2 is mounted on oil-film journal bearings whose stiffness and damping coefficients in the vicinity of the critical speeds take the values indicated in Table 4.1, the transfer matrix method may be used to calculate the system response as indicated in Figure 4.13.

### 4.6.3 Impedance methods

Where an impedance or receptance approach is used to model the system, the shaft impedance matrix may first be established as described by Equation (4.52). The bearing force–displacement relationship may then also be written by writing Equations 8.9 (see Chapter 8) for both bearings as

$$-\begin{Bmatrix} f_{xA} \\ f_{xB} \\ f_{yA} \\ f_{yB} \end{Bmatrix} = \begin{bmatrix} z_{xxA} & 0 & z_{xyA} & 0 \\ 0 & z_{xxB} & 0 & z_{xyB} \\ z_{yxA} & 0 & z_{yyA} & 0 \\ 0 & z_{yxB} & 0 & z_{yyB} \end{bmatrix} \begin{Bmatrix} x_A \\ x_B \\ y_A \\ y_B \end{Bmatrix} \qquad (4.79)$$

(the negative sign indicating that the force applied by the lubricant film to the shaft is opposite in direction to that applied to the bearing). In the case of flexible foundations, the bearing impedance matrix is modified as described in Chapter 3. Substituting for the bearing forces from Equations

# EFFECTS OF BEARING FLEXIBILITY

**Table 4.1** Oil film bearing stiffness and damping coefficients for the system described in Example 4.4

| | Approximate values of bearing coefficients in the region of | | | | |
|---|---|---|---|---|---|
| | 1st critical 0–1260 | 2nd critical 1280–3000 | 3rd critical 3200–6000 | 4th critical 6200–10000 | rev. min$^{-1}$ |
| $K_{xx}$ | $2.45 \times 10^6$ | $2.75 \times 10^6$ | $3.1 \times 10^6$ | $3.23 \times 10^6$ | N.m |
| $K_{xy}$ | $0.46 \times 10^6$ | $5.0 \times 10^3$ | $-0.46 \times 10^6$ | $-2.3 \times 10^6$ | N.m |
| $K_{yx}$ | $7.7 \times 10^6$ | $6.93 \times 10^6$ | $6.15 \times 10^6$ | $6.15 \times 10^6$ | N.m |
| $K_{yy}$ | $10.0 \times 10^6$ | $7.15 \times 10^6$ | $4.3 \times 10^6$ | $2.75 \times 10^6$ | N.m |
| $C_{xx}$ | $12.85 \times 10^3$ | $9.9 \times 10^3$ | $6.95 \times 10^3$ | $7.2 \times 10^3$ | N.s.m |
| $C_{xy}$ | $25.75 \times 10^3$ | $16.54 \times 10^3$ | $7.3 \times 10^3$ | $3.81 \times 10^3$ | N.s.m |
| $C_{yx}$ | $27.0 \times 10^3$ | $17.15 \times 10^3$ | $7.3 \times 10^3$ | $3.81 \times 10^3$ | N.s.m |
| $C_{yy}$ | $89.5 \times 10^3$ | $51.65 \times 10^3$ | $13.8 \times 10^3$ | $11.7 \times 10^3$ | N.s.m |

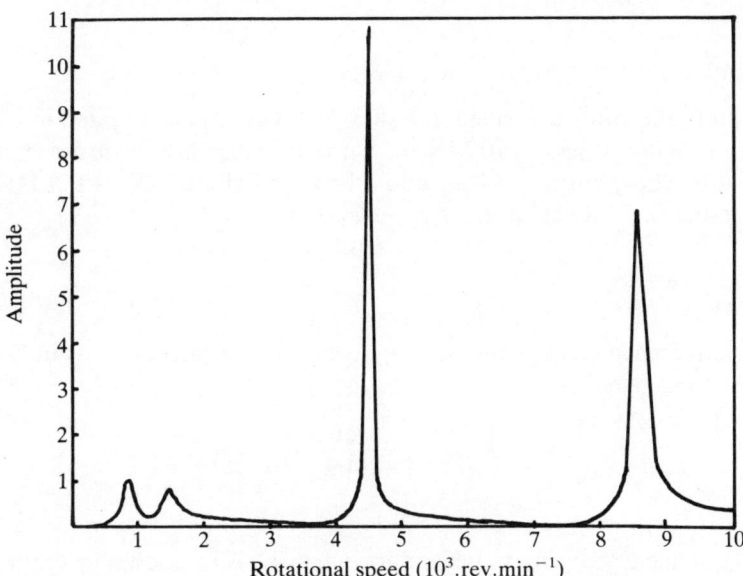

**Figure 4.13** Rotor response at rotor mass (4) when running in oil film bearings. See Example 4.4.

(4.79) into Equations (4.52) gives

$$\begin{Bmatrix} 0 \\ 0 \\ F_{xC} \\ 0 \\ 0 \\ F_{yC} \end{Bmatrix} = \begin{bmatrix} z_{AA}+z_{xxA} & z_{AB} & z_{AC} & z_{xyA} & 0 & 0 \\ z_{BA} & z_{BB}+z_{xxB} & z_{BC} & 0 & z_{xyB} & 0 \\ z_{CA} & z_{CB} & z_{CC} & 0 & 0 & 0 \\ z_{yxA} & 0 & 0 & z_{AA}+z_{yyA} & z_{AB} & z_{AC} \\ 0 & z_{yxB} & 0 & z_{BA} & z_{BB}+z_{yyB} & z_{BC} \\ 0 & 0 & 0 & z_{CA} & z_{CB} & z_{CC} \end{bmatrix} \begin{Bmatrix} x_A \\ x_B \\ x_C \\ y_A \\ y_B \\ y_C \end{Bmatrix}$$

(4.80)

MDOF SYSTEMS

It is noteworthy that the system impedance matrix in the above equation is determined simply by adding the bearing impedances to the shaft impedances, element for element. The system natural frequencies are then given by setting the determinant of the impedance matrix to zero (this can be made more simple if the data relating to forcing point C is not included; its omission will not affect the system natural frequencies). Alternatively the system forced response is obtained by pre-multiplying both sides of Equation (4.80) by the inverse of the system impedance matrix.

When establishing the effect of various support systems on critical speeds it is sometimes helpful to plot both the shaft impedance and minus the support impedance on common axes; Example 4.5 demonstrates the practice.

Further applications of impedance methods in analysis of rotor dynamics are discussed in Morton (1965, 1974).

EXAMPLE 4.5

The shaft and rotor described in Example 4.3 are to be mounted in bearings whose radial stiffness is 20 MN.m$^{-1}$; the bearings are in turn mounted in pedestals whose mass is 10 kg and whose radial stiffness is 0.9 MN.m$^{-1}$. Determine the critical speeds for the system.

SOLUTION

The shaft impedances at the bearing locations are calculated as in Example 4.3 and are given by

$$Z_{sym} = \frac{50 \times \omega^2}{1 - 14.2 \times 10^{-6}\omega^2}$$

$$Z_{asym} = 0$$

The equivalent support structure arrangement is as shown in Figure 4.14. The net support structure impedance is given by adding the impedances of the support structure components which are arranged in parallel (and adding receptances of series connected components to give a net receptance value) and is therefore given by

$$Z = \frac{1}{(1/K_B) + (1/(K_p - m_p\omega^2))}$$

$$= \frac{20 \times 10^6 \times (0.09 - \omega^2 \times 10^{-6})}{(2.09 - \omega^2 \times 10^{-6})}$$

At the bearings, the shaft and supports are effectively connected in parallel, and so the net system impedance is given by the sum of that of the shaft and

# EFFECTS OF BEARING FLEXIBILITY

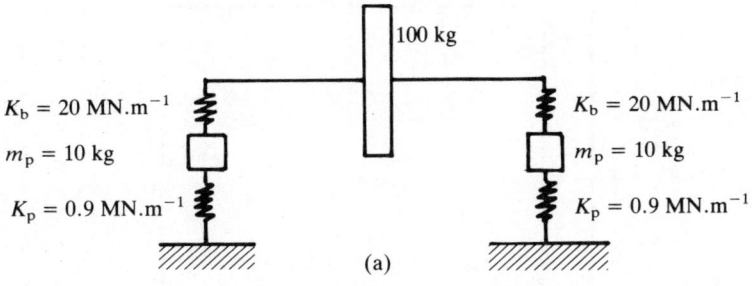

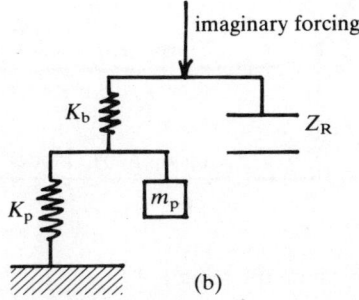

**Figure 4.14** Equivalent system for impedance calculations for the machine described in Example 4.5, (a) physical arrangement of shaft and support system; (b) equivalent system for purposes of analysis. $Z_R$, rotor impedance as measured at one bearing.

that of the supports, and is

$$Z = Z_s + Z_f$$

At resonance $Z$ is equal to zero (so long as the bearing locations are not at a node), that is $Z_s = -Z_f$. A plot of $Z_s$ and of $-Z_f$ against frequency will therefore indicate the whereabouts of the system critical speeds, as shown in Figure 4.15.

## 4.6.4 Allowance for flexible bearings in the dynamic stiffness matrix

Flexible bearings may be allowed for when using the dynamic stiffness matrix method by writing the bearing reaction forces in Equations (4.61) in terms of the bearing stiffness and damping coefficients. That is,

$$\begin{aligned} B_v &= K_{yy}y + K_{yx}x + C_{yy}\dot{y} + C_{yx}\dot{x} \\ B_h &= K_{xx}x + K_{xy}y + C_{xx}\dot{x} + C_{xy}\dot{y} \end{aligned} \quad (4.81)$$

The analysis then proceeds as described in Section 4.5, substituting from Equations (4.57) for $y$ and, on differentiation, $\dot{y}$ (and similarly for $x$ and $\dot{x}$)

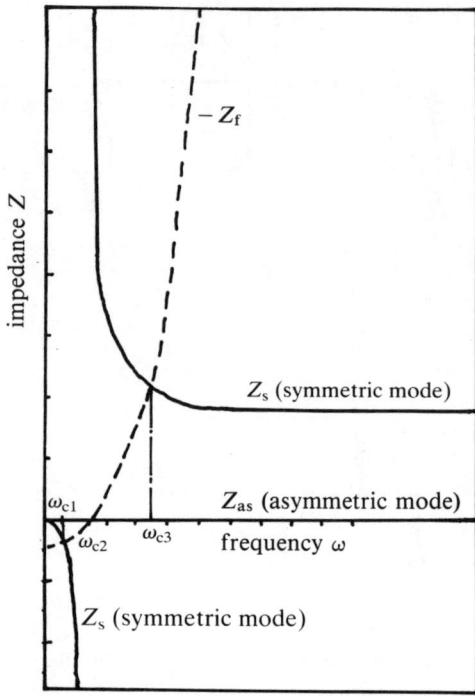

**Figure 4.15** Impedance matching to calculate system natural frequencies (Example 4.5). $\omega_{c1}$, $\omega_{c2}$, $\omega_{c3}$ are critical speeds at the intercepts between $-Z_f$ and $Z_s$ curves.

into the expressions for $B_v$ and $B_h$ used now in Equations (4.61). The subsequent equations (4.64) and (4.65) are then of a similar form to the pinned-bearing case except that $\{\bar{F}_0\}$ and $[m_{00}]$ become modified as shown on page 135.

The analysis then continues as before except that the subsequent equation (4.70) no longer contains any unknown terms (except the shaft displacements) and so the machine response can be determined directly without needing to first establish the bearing reaction forces.

## 4.7 Magnetic effects

In some types of electrical machinery, for example turbogenerators, electrical currents flowing through winding on the rotor set up magnetic fields during the normal course of operation of the machine. In turbogenerators the rotor becomes a gigantic rotating electromagnet. As a consequence of this situation any deflection of the rotor in the radial direction will result in an increased magnetic attractive force between the rotor and the surrounding metal machine stator; this force acts on the rotor

$$\{\bar{F}_0\} = \begin{Bmatrix} {}_cM_{cv} \\ {}_cQ_{cv} - {}_cU_v \\ {}_sM_{cv} \\ {}_sQ_{cv} - {}_sU_v \\ {}_cM_{ch} \\ {}_cQ_{ch} - {}_cU_h \\ {}_sM_{ch} \\ {}_sQ_{ch} - {}_sU_h \end{Bmatrix} \quad (4.82)$$

$$[m_{00}] = \begin{bmatrix} k_{11} & k_{12} - I_d\omega^2 & 0 & 0 & 0 & 0 & I_p\omega^2 & 0 \\ k_{21} - k_{yy} + m\omega^2 & k_{22} & -\omega C_{yy} & 0 & -K_{yx} & 0 & 0 & 0 \\ 0 & 0 & k_{11} & k_{12} - I_d\omega^2 & 0 & -I_p\omega^2 & 0 & 0 \\ \omega C_{yy} & 0 & k_{21} - k_{yy} + m\omega^2 & k_{22} & \omega C_{yx} & 0 & 0 & 0 \\ 0 & 0 & 0 & -I_p\omega^2 & k_{11} & k_{12} - I_d\omega^2 & 0 & -K_{yx} \\ -K_{xy} & 0 & -\omega C_{xy} & 0 & k_{21} - k_{xx} + m\omega^2 & k_{22} & -\omega C_{yx} & 0 \\ 0 & I_p\omega^2 & 0 & 0 & 0 & 0 & k_{11} & k_{12} - I_d\omega^2 \\ \omega C_{xy} & 0 & -K_{xy} & 0 & C & 0 & k_{21} - k_{xx} + m\omega^2 & k_{22} \end{bmatrix} \quad (4.83)$$

in the direction of its original deflection and is known as 'unbalanced magnetic pull'. Because unbalanced magnetic pull increases with rotor deflection, its effect is as though the shaft were acted upon by a spring of negative stiffness.

The variation of unbalanced magnetic pull with rotor deflection is in fact non-linear, but for the purposes of analysis it may be approximated to a linear relationship. The value of the effective negative spring stiffness is often expressed as force per 10% of air gap between rotor and stator, and is assumed not to vary significantly within this range of rotor displacement (see Figure 4.16). The effect of unbalanced magnetic pull on the deflection of the rotor under a static load (for example that due to the weight of the rotor) may be determined from consideration of force equilibrium. In the absence of magnetic effects the rotor deflection is

$$y = \frac{Mg}{k_s} \qquad (4.84)$$

where $M$ is the rotor mass, $k_s$ is the shaft stiffness and $g$ is gravitational acceleration. When unbalanced magnetic pull is present, a consideration of force equilibrium reveals that

$$Mg + k_u y_u = k_s y_u \qquad (4.85)$$

where $k_u$ is the unbalanced magnetic pull stiffness and $y_m$ is the rotor deflection in the presence of unbalanced magnetic pull. Rearranging

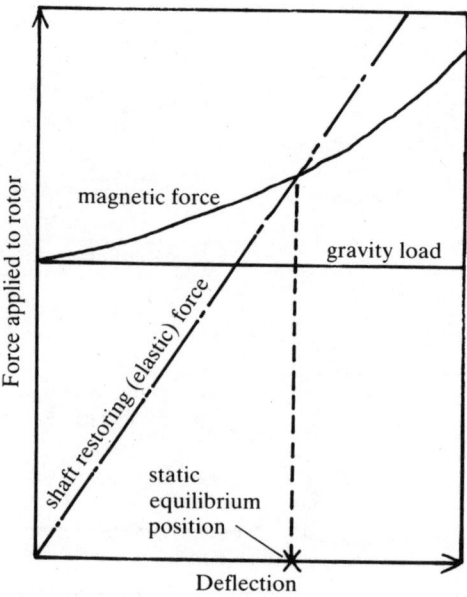

**Figure 4.16** Variation of unbalanced magnetic pull with rotor displacement.

## MAGNETIC EFFECTS

Equation (4.85) and substituting for $Mg$ from Equation (4.84) gives

$$y_u = \frac{Mg}{k_s - k_u} \qquad (4.86)$$

$$= \left(\frac{k_s}{k_s - k_u}\right) y \qquad (4.87)$$

Allowance for magnetic effects in an analysis of MDOF systems may be made by including an additional bearing station at the site of the unbalanced magnetic pull, and assigning negative stiffnesses to the bearings (together with zero damping and cross-coupling). Equations (4.86) and (4.87) indicates that if the system already possesses some initial spring stiffness property then the negative unbalanced magnetic pull stiffness is simply added to it. Alternatively, the following example shows how magnetic effects can be allowed for by modifying the system influence coefficients.

EXAMPLE 4.6.

Determine the effect of the unbalanced magnetic pull at C on the influence coefficients for the systems shown in Figures 4.17.

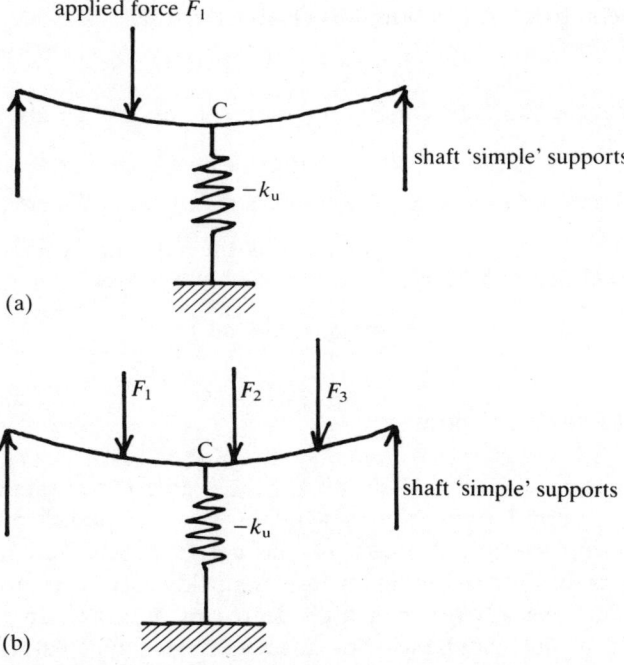

**Figure 4.17** Single mass and multi-mass systems subjected to magnetic pull at location C (Example 4.6), (a) one applied load, $F$; (b) several applied loads, $F_1, ..., F_3$.

SOLUTION

If the deflection at C, in the system shown in Figure 4.17(a), is $y_C$ then the resulting magnetic force at C is $k_u y_C$. If $a_{ij}$ is the influence coefficient relating the deflection at position $i$ due to unit force at location $j$ (see also Section 4.2), then

$$y_C = a_{C1} F_1 + \frac{k_u y_C}{k_C}$$

where $k_C$ is the shaft stiffness measured at location C. The above equation may be rearranged to give

$$y_C = \frac{a_{C1} F_1 k_C}{k_C - k_u}$$

The magnetic force acting at location C is then given by

$$U_{C1} = k_u y_C$$

$$= \frac{a_{C1} k_C k_u F_1}{k_c - k_u}$$

$$= a_{C1} K F_1 \quad \text{where } K = k_C k_u/(k_C - k_u)$$

When $n$ forces $F$ act on the shaft, at various locations (Figure 4.17b) the total magnetic force at location C is clearly given by

$$U_C = U_{C1} + U_{C2} + \cdots + U_{Cn}$$

In such cases the shaft deflection, at position 1 say, is then given by

$$y_1 = a_{11} F_1 + a_{12} F_2 + \cdots + a_{1n} F_n + a_{1C}(U_{C1} + U_{C2} + \cdots + U_{Cn})$$
$$= (a_{11} + K a_{1C}^2) F_1 + (a_{12} + K a_{1C} a_{C2}) F_2 + \cdots + (a_{1n} + K a_{1C} a_{Cn}) F_n$$

and similarly for $y_2$, $y_3$, etc. In general the influence coefficient $a_{ij}$ is therefore modified and becomes

$$_m a_{ij} = (a_{ij} + K a_{iC} a_{Cj})$$

## 4.8 Dunkerley's formula

In some situations it may be desirable for the engineer to obtain a value for the machine natural frequency in a relatively short period of time, for example during the initial stages of the design procedure. Under such circumstances the need to obtain an answer quickly may be of more concern than obtaining the answer to a high degree of accuracy. In these cases Dunkerley's formula can be used to calculate the machine natural frequencies without recourse to the numerical methods discussed in the earlier sections of this chapter; the calculation can, if necessary, be done by hand.

# DUNKERLEY'S FORMULA

Dunkerley's formula is discussed in most general vibration texts (see for example Thomson 1978); it was developed empirically, but its mathematical foundation can be explained in the context of Equation (4.13). For example, in the case of a system with three degrees of freedom, the first two terms of Equation (4.13) become

$$-\left(\frac{1}{\omega^2}\right)^3 + (a_{11}m_1 + a_{22}m_2 + a_{33}m_3)\left(\frac{1}{\omega^2}\right)^2 + \cdots \qquad (4.88)$$

But for a polynomial whose first coefficient is unity, the second coefficient is equal to minus the sum of the roots of the equation. Therefore we may write, for the three-degree-of-freedom system referred to,

$$\frac{1}{\omega_1^2} + \frac{1}{\omega_2^2} + \frac{1}{\omega_3^2} = a_{11}m_1 + a_{22}m_2 + a_{33}m_3$$

$$= \frac{m_1}{k_1} + \frac{m_2}{k_2} + \frac{m_3}{k_3}$$

$$= \frac{1}{\omega_{11}^2} + \frac{1}{\omega_{22}^2} + \frac{1}{\omega_{33}^2} \qquad (4.89)$$

where $\omega_{11}$, $\omega_{22}$ and $\omega_{33}$ are the natural frequencies of the system with only mass $m_1$, $m_2$ or $m_3$ present, respectively. In most instances the fundamental frequency $\omega_1$ will be much lower than the other natural frequencies, so that Equation (4.89) may be approximated, in general, to

$$\frac{1}{\omega_1^2} = \frac{1}{\omega_{11}^2} + \frac{1}{\omega_{22}^2} + \frac{1}{\omega_{33}^2} + \cdots \qquad (4.90)$$

which was first suggested by Dunkerley (1895). It is noteworthy that the formula always gives a value for fundamental frequency which is slightly lower than the true value, by virtue of the approximation involved.


EXAMPLE 4.7

Use Dunkerley's formula to calculate the fundamental critical speed of the system described in Example 4.1 (see also Figure 4.18).


SOLUTION

If $\omega_{11}$ is the system critical speed when only mass $A$ is mounted on the shaft, and $\omega_{22}$ is that when only mass $B$ is mounted on the shaft, and $a_{ij}$ is an influence coefficient relating displacement at $i$ due to unit force at $j$, then Dunkerley's formula may be developed to give the system fundamental critical

# MDOF SYSTEMS

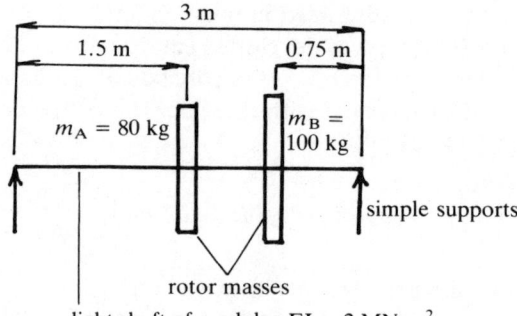

**Figure 4.18** A two-rotor system with a light flexible shaft of modulus $EI = 2$ MN.m$^2$ (Example 4.7).

speed $\omega_1$ as

$$\frac{1}{\omega_1^2} = \frac{1}{\omega_{11}^2} + \frac{1}{\omega_{22}^2}$$

$$= m_A a_{AA} + m_B a_{BB}$$

where $a_{AA}$ and $a_{BB}$ are the influence coefficients whose values are determined in Example 4.1. Then

$$\frac{1}{\omega_1^2} = 80 \times 0.281 \times 10^{-6} + 100 \times 0.158 \times 10^{-6}$$

from which

$$\omega_1 = 161.6 \text{ rad.s}^{-1}$$

## 4.9 Rayleigh's method

This is an alternative simple method of predicting system natural frequencies. The calculations involved are usually simple enough to be carried out by hand, whilst also allowing solutions to be obtained relatively quickly. The method is particularly suited to continuous systems, for example in the case of long flexible shafts which do not carry concentrated rotor masses at any point.

Rayeigh's method (1894) is founded upon equating the system maximum kinetic energy to the system maximum potential energy. That is, when a vibrating shaft reaches the limit of its displacement in one direction it has a certain amount of energy stored as strain (or potential) energy within; all of this energy is converted to kinetic energy at the instant of zero displacement from the equilibrium position, as the shaft travels back in the opposite direction. These particular moments in the vibration cycle are shown in Figure 4.19.

# RAYLEIGH'S METHOD

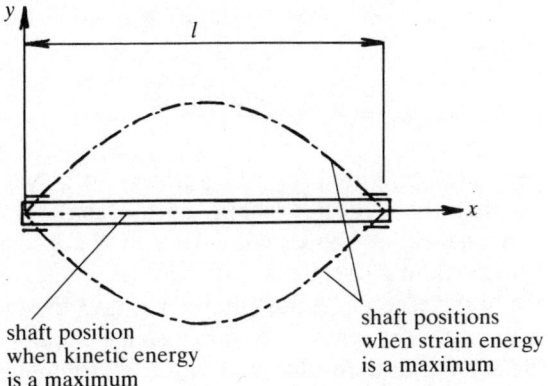

**Figure 4.19** Shaft positions when strain and kinetic energies take maximum values.

The method is most easily demonstrated for the simply supported shaft shown, where the maximum strain energy stored is given by basic mechanics theory as

$$E = \frac{1}{2} \int \frac{M^2}{EI} \, dx \tag{4.91}$$

But for a beam in bending,

$$M = \frac{EI}{R} \tag{4.92}$$

and

$$\frac{1}{R} = \frac{d\theta}{dx} = \frac{d^2 y}{dx^2} \tag{4.93}$$

Substituting Equations (4.92) and (4.93) into (4.91) gives

$$E = \frac{1}{2} \int EI \left(\frac{d^2 y}{dx^2}\right)^2 dx \tag{4.94}$$

A second expression may be written, for the kinetic energy of the system associated with vibration. For a small element of mass $dm$ at some location on the shaft, its kinetic energy is

$$E = \tfrac{1}{2} \, dm (\dot{y})^2 \tag{4.95}$$

But for sinusoidal vibrations $|\dot{y}| = \omega y$ and so Equation (4.95) leads to the following expression for the maximum kinetic energy of the whole system:

$$E = \tfrac{1}{2} \omega^2 \int y^2 m' \, dx \tag{4.96}$$

141

where $m'$ is the mass per unit length of shaft. Equating Equations (4.94) and (4.96) then gives

$$\omega^2 = \frac{\int EI(\mathrm{d}^2 y/\mathrm{d}x^2)^2 \, \mathrm{d}x}{\int y^2 m' \, \mathrm{d}x} \tag{4.97}$$

which may be evaluated provided that the variation of $y$ with $x$ is known, that is provided that the deflected shape of the shaft is known. Most reasonable assumed deflection shapes will give a good approximation to the true value of system natural frequency; usually the shaft static deflection shape is used. Only the true deflection shape will give the precise value of the true system natural frequency, any other shape implies an additional constraint to the system and results in a value of frequency which is in excess of the actual value.

EXAMPLE 4.8

Use Rayleigh's method to calculate the fundamental natural frequency of vibration for the system described in Examples 4.1 and 4.7.

SOLUTION

In the case of lumped masses an alternative expression for the system strain energy, formed by equating it to the work done by the masses as the shaft deforms under static loading conditions, is sometimes preferred. In the system under consideration work done is therefore

$$E = \sum \tfrac{1}{2} mgy$$

where $y$ is the deflection of the mass $m$, and $g$ is gravitational acceleration. Equating this energy with the maximum kinetic energy then gives a frequency equation of the form

$$\omega_1^2 = \frac{g \sum my}{\sum my^2}$$

Under static loading conditions the deflections of the two masses are

$$\begin{aligned} y_A &= a_{AA} m_A g + a_{AB} m_B g \\ &= 0.281 \times 10^{-6} \times 80 \times 9.81 + 0.193 \times 10^{-6} \times 100 \times 9.81 \\ &= 0.41 \times 10^{-3} \text{ m} \end{aligned}$$

and

$$\begin{aligned} y_B &= a_{BA} m_A g + a_{BB} m_B g \\ &= 0.193 \times 10^{-6} \times 80 \times 9.81 + 0.158 \times 10^{-6} \times 100 \times 9.81 \\ &= 0.306 \times 10^{-3} \text{ m} \end{aligned}$$

Substitution into the frequency equation then gives a fundamental critical speed of

$$\omega_1^2 = \frac{9.81(80 \times 0.41 \times 10^{-3} + 100 \times 0.306 \times 10^{-3})}{(80 \times 0.41^2 \times 10^{-6} + 100 \times 0.306^2 \times 10^{-6})}$$

that is

$$\omega_1 = 165.1 \text{ rad.s}^{-1}$$

## References

Bishop, R. E. D. & D. C. Johnson 1960. *The mechanics of vibration*. Cambridge: Cambridge University Press.

Dunkerley, S. 1895. On the whirling and vibration of shafts. *Phil. Trans. Roy. Soc. London A* **185**, 269–360.

Gladwell, G. M. L. & R. E. D. Bishop 1959. The receptances of uniform and non-uniform rotating shafts. *J. Mech. Eng. Sci.* **1**, 78–91.

Gunter, E. J., Choy, K. C. & P. E. Allaire 1978. Modal analysis of turborotors using planar modes theory. *J. Franklin Inst.* **305**, 221f.

Kikuchi, K. 1970. Analysis of unbalance vibration of rotating shaft systems with many bearings and disks. *Bull. JSME* **13**, 61, 864–872.

Kramer, E. 1977. Computation of unbalance vibrations of turbo rotors. *Trans. ASME.* Design Engineering Transactions, 13, Paper 77-DET-13.

Lund, J. W. 1974a. Modal response of a flexible rotor in fluid film bearings. *Trans. ASME., J. Eng. Ind.* **96**, May, 525–533.

Lund, J. W. 1974b. Stability and damped critical speeds of a flexible rotor in fluid film bearings. *Trans. ASME., J. Eng. Ind.* May, 509–517.

Lund, J. W. & F. K. Orcutt 1967. Calculations and experiments on the unbalance response of a flexible rotor. *Trans. ASME., J. Eng. Ind.* November, 785–796.

Morton, P. G. 1965. On the dynamics of large turbo-generator rotors. *Proc. IMechE.* **180**, 295–329.

Morton, P. G. 1974. The derivation of bearing characteristics by means of transient excitation applied directly to a rotating shaft. *Proc. IUTAM Symp. on Dynamics of Rotors.* Copenhagen, October, 350–379.

Morton, P. G. 1975. Dynamic characteristics of bearings: measurement under operating conditions. *GEC J. Sci. Techn.* **42**, 1, 37–47.

Myklestad, N. O. 1944. A new method of calculating natural modes of uncoupled bending vibrations of aeroplane wings and other types of beams. *J. Aero. Sci.* **11**, 2, 153–162.

Nelson, H. D. & J. M. McVaugh 1976. The dynamics of rotor bearing systems using finite elements. *Trans. ASME., J. Eng. Ind.* **98**, 593ff.

Nicholas, J. C. 1986. Improving critical speed calculations using flexible support modal analysis compliance data. *Proc. 15th Turbomach. Symp., Texas A&M Univ.*, November.

Pestel, E. C. & F. A. Leckie 1963. *Matrix methods in elastomechanics*. New York: McGraw-Hill.

Prohl, M. A. 1945. A general method for calculating critical speeds of flexible rotors. *Trans. ASME., J. Appl. Mech.*, September, 48.

Rao, J. S. 1983. *Rotor dynamics* New Delhi: Wiley Eastern.

Rayleigh, J. W. S. 1894. *The theory of sound*. New York: Dover (Reprinted) 1945.

Rieger, N. F., Thomas, C. B. & W. W. Walter 1976. Dynamic stiffness matrix approach for rotor-bearing system analysis. *Proc. IMechE. Conf. Vibrations in Rotating Machinery.* Cambridge. Paper C187/76.

Rouch, K. E. & J. S. Kao 1980. Dynamic reduction rotor dynamics by the finite element method. *Trans. ASME., J. Mach. Des.* **102**, 360ff.

Thomson, W. T. 1978. *Vibration theory and applications.* London: George Allen & Unwin.

Vance, J. M. 1988. *Rotordynamics of turbomachinery.* New York: Wiley.

# 5 Torsional vibrations

## 5.1 Introduction

Most rotating machinery is susceptible to some source of torsional vibration forcing. Examples include the inertia forces of reciprocating mechanisms (such as the pistons in internal combustion engines), impulsive loads occurring during a normal machine cycle (for example during the operation of punch presses), shock loads applied to electrical machinery (such as a generator line fault followed by fault removal and automatic reclosure), and torques related to gear mesh frequencies.

Machines in which such sources constitute a potential design problem area are those with massive rotors and flexible shafts, where the system natural frequencies of torsional vibration may lie close to, or within, the source frequency range during normal operation of the machine. In these instances it is essential that the designer should accurately predict the value of the machine natural frequencies, and should ensure that they do not coincide with the frequency of any torsional load fluctuations which might be present.

This chapter introduces the reader to the concept of torsional vibration and shows how the torsional natural frequencies of simple systems can be calculated by hand. The next section describes how more complicated structures, with many rotor masses, can be dealt with using numerical methods. Later, methods of dealing with geared and branched systems are discussed, while the final section describes the effects of torsional dampers on machine behaviour, and how this may be allowed for in design calculations.

## 5.2 Simple systems with one or two rotor masses

The simplest system in which to consider torsional oscillations is that with only one rotor mass attached to the end of a flexible shaft, the other end of the shaft being rigidly supported. Such a system is shown in Figure 5.1. The rotor polar moment of inertia is $I$, and the shaft torsional stiffness is $k_t$ which is defined as

$$k_t = \frac{T}{\theta} = \frac{GJ}{l} \qquad (5.1)$$

where $G$ is the shaft material modulus of rigidity, $J$ is the shaft polar second moment of area, $l$ is the shaft length and $\theta$ is the angular displacement of

## TORSIONAL VIBRATIONS

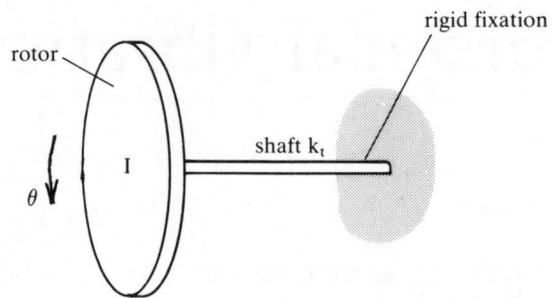

**Figure 5.1** A single-rotor torsional system. See text for details.

the rotor. With no externally applied loads the equation of motion of the rotor is

$$-k_t\theta = I\ddot{\theta} \qquad (5.2)$$

where the rotor angular displacement $\theta$ takes the form

$$\theta = \hat{\theta}\sin\omega t \qquad (5.3)$$

Differentiating Equation (5.3) with respect to time to obtain an expression for $\ddot{\theta}$, and substituting into Equation (5.2) gives, on rearranging,

$$\omega_n^2 = k_t/I \qquad (5.4)$$

where $\omega_n$ is the natural frequency of free torsional vibrations of the system.

A system with two rotor masses is shown in Figure 5.2. In this case the whole of the system is free to rotate, the shaft being mounted in frictionless bearings. For free vibrations the equations of motion for the two masses may be written as

$$k_t(\theta_2 - \theta_1) = I_1\ddot{\theta}_1, \qquad k_t(\theta_1 - \theta_2) = I_2\ddot{\theta}_2 \qquad (5.5)$$

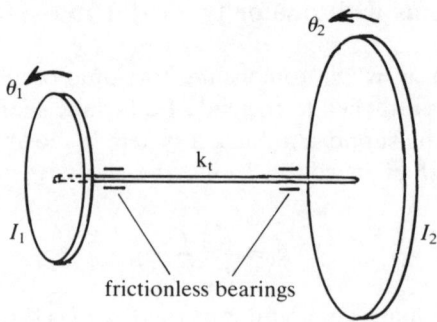

**Figure 5.2** A two-rotor torsional system.

## SIMPLE SYSTEMS

The solutions to Equations (5.5) will take the form

$$\theta_1 = \hat{\theta}_1 \sin \omega t, \qquad \theta_2 = \hat{\theta}_2 \sin \omega t \qquad (5.6)$$

which may be differentiated to obtain expressions for $\ddot{\theta}_1$ and $\ddot{\theta}_2$ as before. Back-substitution into Equations (5.5) then gives, on rearranging,

$$\begin{bmatrix} k_t - I_1\omega_n^2 & -k_t \\ -k_t & k_t - I_2\omega_n^2 \end{bmatrix} \begin{Bmatrix} \hat{\theta}_1 \\ \hat{\theta}_2 \end{Bmatrix} = \begin{Bmatrix} 0 \\ 0 \end{Bmatrix} \qquad (5.7)$$

which may be written more simply as

$$[K]\{\theta\} = \{0\} \qquad (5.8)$$

The non-trivial solution to Equation (5.8) is

$$|K| = 0 \qquad (5.9)$$

that is

$$I_1 I_2 \omega_n^4 - (I_1 + I_2) k_t \omega_n^2 = 0 \qquad (5.10)$$

for which the roots are

$$\omega_{n1} = 0, \qquad \omega_{n2} = \left[\frac{(I_1 + I_2)k_t}{I_1 I_2}\right]^{1/2} \qquad (5.11)$$

This approach to solving the torsional vibration problem can also be extended to deal with forced vibration, and with systems incorporating torsional damping elements (Vance 1988). It can also be extended for use with MDOF systems; in these cases a methodical approach will be required to solve Equation (5.9) to find the system natural frequencies (Moore 1967, Carnahan et al. 1969), and to find the forced response by solving Equation (5.8), where the right-hand side is no longer zero (Collar and Simpson 1987; Pipes 1963).

The first of the roots in Equation (5.11) represents the case when both rotor inertias simply roll, together, in phase with one another. It is of little practical significance. The second root represents the case when both masses vibrate in anti-phase with one another, as indicated by Figure 5.3. Since both masses are always vibrating in opposite directions, it follows that there must be some location on the shaft where no torsional oscillation is evident – a torsional node. The location of the node may be established by treating each end of the real system as a separate single-rotor system of the type discussed above, the node being treated as the point where the shaft is rigidly fixed. Since the value of the natural frequency is known we may write

$$\omega_n^2 = \frac{k_{t1}}{I_1} = \frac{k_{t2}}{I_2} \qquad (5.12)$$

whereupon $k_{t1}$ and $k_{t2}$, and hence $l_1$ and $l_2$, may be found. The relative

TORSIONAL VIBRATIONS

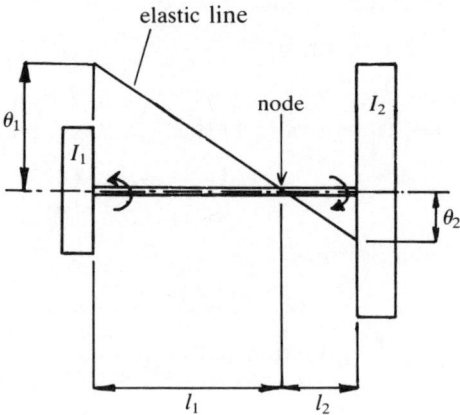

**Figure 5.3** Vibration mode of a two-rotor system, showing rotor inertias vibrating in opposite directions.

amplitudes of vibration of the two rotor masses, $\hat{\theta}_1/\hat{\theta}_2$, may be found using either of Equations (5.7) once the value of $\omega_n$ is known.

In some systems the shaft may be stepped such that its diameter is different at different axial locations, as shown in Figure 5.4. In these cases the actual shaft should be replaced by an unstepped equivalent shaft for the purposes of analysis; frequently the equivalent shaft diameter will be the same as the smallest diameter of the real shaft. The equivalent shaft must have the same torsional stiffness as the real shaft, so that an equivalent shaft length is given by multiplying the real shaft length by the ratio $J_e/J$, where $J$ is the polar second moment of area of the real shaft and $J_e$ that of the equivalent shaft. For the system shown in Figure 5.4 the equivalent shaft length is given by

$$l_e = l_{e1} + l_{e2} + l_{e3}$$
$$= l_1 J_e/J_1 + l_2 J_e/J_2 + l_3 J_e/J_3 \qquad (5.13)$$

The remainder of the analysis now proceeds as before, except that the final location of the node in the real system is given by proportion along the length of shaft in which the node occurs. For example, in the system shown in Figure 5.4 the node is located such that $a'/b' = a/b$ where $a$ and $b$ are known once the node has been located in the equivalent system.

EXAMPLE 5.1

A ship's propeller has a mass of 1000 kg and a radius of gyration of 0.6 m. It is connected by a stepped solid shaft to an engine whose rotating

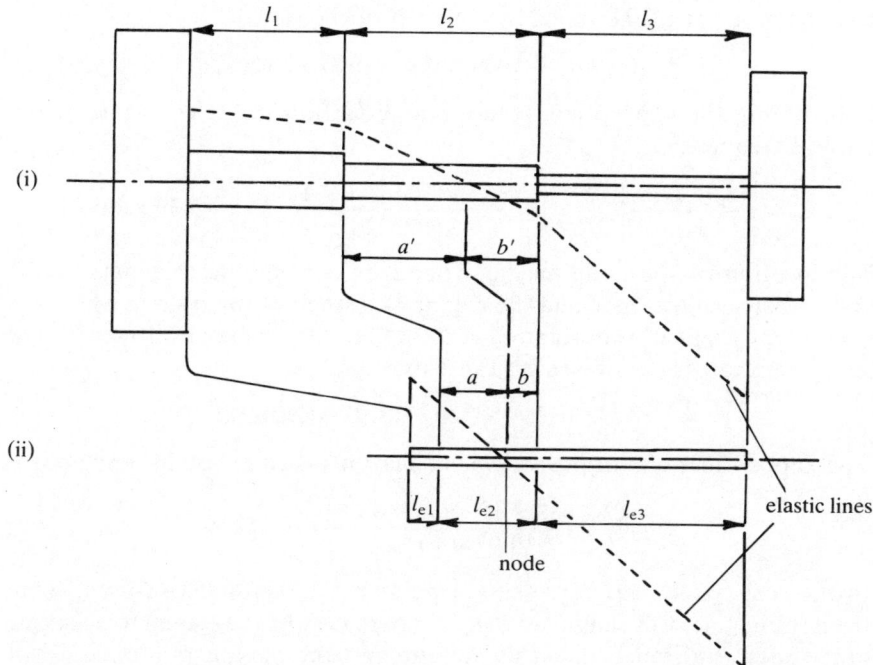

**Figure 5.4** Reducing a stepped shaft to an equivalent single diameter shaft, (i) real system (stepped) and (ii) equivalent system (unstepped).

components have an effective moment of inertia of 720 kg.m.$^2$ The drive shaft is 11 m long and its diameter is 300 mm between the engine and a point 8 m from the engine, and 250 mm for the remainder of its length to the propeller. The shaft material modulus of rigidity is $G = 83$ GN.m$^{-2}$. Calculate the natural frequency of torsional oscillations of the system, the location of the node, and the ratio of amplitudes of the two rotors.

SOLUTION

For the purposes of analysis the stepped shaft may be replaced by an equivalent unstepped shaft whose diameter is 250 mm, and whose torsional stiffness is the same as that of the real shaft. The length of the equivalent shaft is given by Equation (5.13) as

$$l_e = \frac{8\pi \times 0.25^4/32}{\pi \times 0.3^4/32} + 3 = 6.858 \text{ m}$$

The torsional stiffness of the shaft is then given by

$$k_t = \frac{T}{\theta} = \frac{GJ}{l} = \frac{83 \times 10^9 \times \pi \times 0.25^4}{6.858 \quad 32} = 4.641 \text{ MN.m.rad.}^{-1}$$

and the moment of inertia of the ship's propeller is

$$I = mr_g^2 = 1000 \times 0.6^2 = 360 \text{ kg.m}^2$$

Substituting the appropriate values into Equation (5.11) gives the system natural frequency as

$$\omega_n = \left[\frac{(I_p + I_e)k_t}{I_p I_e}\right]^{1/2} = \left[\frac{(360 + 720)4.641 \times 10^6}{360 \cdot 720}\right]^{1/2} = 139 \text{ rad.s}^{-1}$$

The location of the node may be found by considering that part of the system between the node and the engine as a single-rotor system, where the shaft is constrained from rotating at the node. The torsional stiffness of that length of shaft is then given by Equation (5.12) as

$$k_e = \omega_n^2 I_e = 139^2 \times 720 = 13.92 \text{ MN.m.rad}^{-1}$$

whereupon the length of the equivalent shaft between the engine and node is

$$l_e = \frac{GJ}{k} = \frac{83 \times 10^9 \times \pi \times 0.25^4}{13.92 \times 10^6 \times 32} = 2.287 \text{ m}$$

In the real system this represents a point $8 \times 2.287/(6.858 - 3) = 4.742$ m from the engine. Continuing to consider each part of the system to each side of the node as an independent single rotor system, the ratio of amplitudes of the two rotors is

$$\frac{\hat{\theta}_e}{\hat{\theta}_p} = \frac{GJ/l_p}{GJ/l_e} = \frac{l_e}{l_p} = \frac{2.287}{(6.858 - 2.287)} = 0.5$$

This answer may also be obtained using either of Equations (5.7).

## 5.3 MDOF systems – transfer matrix methods

In many machines there are more than one or two rotor masses mounted on the shaft, or the mass of the shaft itself may be significant. In these situations the preceding theoretical analyses are inadequate for the purposes of modelling the system; although they could be extended to allow for more rotor masses the resulting mathematics becomes somewhat cumbersome and an alternative approach is favoured. Two of the most commonly used techniques are the transfer matrix method (Pestel and Leckie 1963, Rao 1983), and the method of mechanical impedances. The former of these is described below; the latter is discussed in section 5.6.

Consider the three-rotor system shown in Figure 5.5(a). The instantaneous torques applied to rotor inertia 2 and shaft section 2 are as shown in Figure 5.5(b). The equation of motion for the rotor is then given by

$$_R T_2 - {}_L T_2 = I_2 \ddot{\theta}_2 \qquad (5.14)$$

# MDOF SYSTEMS

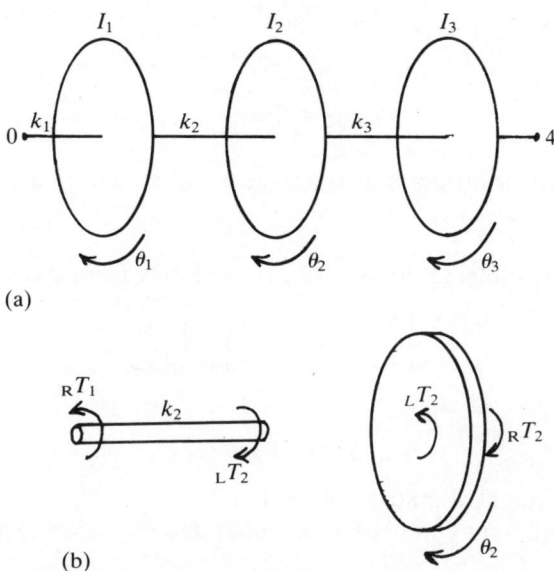

(a)

(b)

**Figure 5.5** Elements in a three-rotor system, (a) overall arrangement; (b) torques and displacements relating to rotor inertia 2 and shaft section 2.

The angular oscillations of the rotor will take the form

$$\theta_2 = \hat{\theta}_2 \sin \omega_2 t \tag{5.15}$$

which may be differentiated with respect to time to obtain an expression for $\ddot{\theta}_2 = -\omega^2 \theta_2$. On substituting back into Equation (5.14) this gives

$$_R T_2 - {}_L T_2 = -\omega_n^2 I_2 \theta_2 \tag{5.16}$$

Also it is clear that the angular displacement of the rotor is the same when measured on either side, that is

$$_R \theta_2 = {}_L \theta_2 \tag{5.17}$$

Equations (5.16) and (5.17) may now be combined and written in matrix form as

$$_R \begin{Bmatrix} \theta \\ T \end{Bmatrix}_2 = \begin{bmatrix} 1 & 0 \\ -\omega^2 I & 1 \end{bmatrix}_2 {}_L \begin{Bmatrix} \theta \\ T \end{Bmatrix}_2 \tag{5.18}$$

which may be written more simply as

$$_R \{S\}_2 = [P]_2 {}_L \{S\}_2 \tag{5.19}$$

where $\{S\}_2$ is the state vector at station 2, and $[P]_2$ is the point matrix for station 2.

If we now consider the shaft section 2, we may note that its angle of twist is related to its torsional stiffness, and to the torque which is transmitted through it, by

$$\theta_2 - \theta_1 = \frac{T}{k_2} \tag{5.20}$$

whilst the torque transmitted is the same at either end of the shaft, so that

$$_L T_2 = {}_R T_1 \tag{5.21}$$

Combining Equations (5.20) and (5.21) in matrix form gives

$$\left\{ \begin{matrix} \theta \\ T \end{matrix} \right\}_{L\,2} = \begin{bmatrix} 1 & 1/k \\ 0 & 1 \end{bmatrix}_2 {}_R \left\{ \begin{matrix} \theta \\ T \end{matrix} \right\}_1 \tag{5.22}$$

which may be written as

$$_L \{S\}_2 = [F]_2 \,_R \{S\}_1 \tag{5.23}$$

where $[F]_2$ is the field matrix for station 2.

Equation (5.23) may now be substituted into Equation (5.19) to give

$$_R \{S\}_2 = [P]_2 \,_L \{S\}_2$$
$$= [P]_2 [F]_2 \,_R \{S\}_1$$
$$= [U]_2 \,_R \{S\}_1 \tag{5.24}$$

where $[U]_2$ is the transfer matrix relating to station 2. (An alternative form of the $[U]$ matrix is sometimes used, when the rotor mass is uniformly distributed along the shaft length, which enables a reduction in the number of nodes used in the analysis, Rao *et al.* 1980.) Similarly the transfer matrices for all other stations in the system may be established. They are all related to one another by the expressions

$$\begin{aligned}
{}_R \{S\}_1 &= [U]_1 \,_R \{S\}_0 \\
{}_R \{S\}_2 &= [U]_2 \,_R \{S\}_1 = [U]_2 [U]_1 \,_R \{S\}_0 \\
{}_R \{S\}_3 &= [U]_3 \,_R \{S\}_2 = [U]_3 [U]_2 [U]_1 \,_R \{S\}_0 \\
{}_R \{S\}_4 &= [U]_4 \,_R \{S\}_3 = [U]_4 [U]_3 [U]_2 [U]_1 \,_R \{S\}_0 \\
&= [T] \,_R \{S\}_0
\end{aligned} \tag{5.25}$$

where $[T]$ is the overall system transfer matrix.

Equation (5.25) may now be written more explictly as

$$\left\{ \begin{matrix} \theta \\ T \end{matrix} \right\}_{R\,4} = \begin{bmatrix} t_{11} & t_{12} \\ t_{21} & t_{22} \end{bmatrix} {}_R \left\{ \begin{matrix} \theta \\ T \end{matrix} \right\}_0 \tag{5.26}$$

where, for free torsional vibrations, conditions which must be satisfied are that at each end of the machine the torque transmitted through the shaft is zero, that is

$$_R T_4 = {}_R T_0 = 0 \tag{5.27}$$

## MDOF SYSTEMS

which when substituted in Equation (5.26) gives

$$t_{21} = 0 \tag{5.28}$$

which is satisfied for any value of $\omega$ which is a system natural frequency. These roots of $\omega_n$ may be found by any root-searching technique, for example by using trial substitutions of $\omega$ in the equation and monitoring the change in the value of $t_{21}$.

For each value of natural frequency thus obtained, the system mode shape may be determined by substituting $T = 0$ and $\theta = 1$ into Equations (5.25), and so obtaining the state vector $\{S\}$ at all other stations in the system.

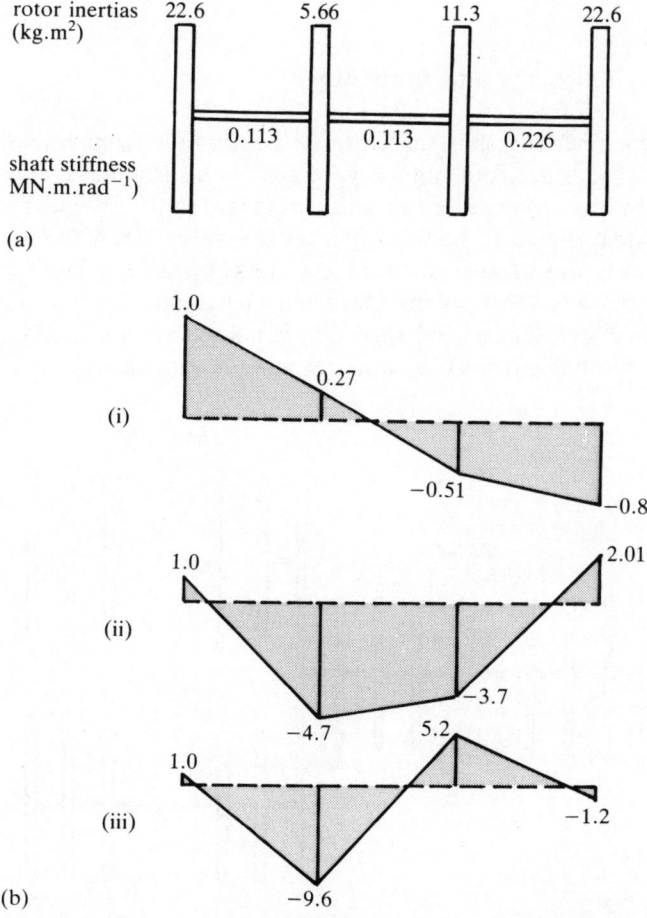

**Figure 5.6** Natural frequencies and mode shapes for a four-rotor system (Example 5.2), (a) system layout; (b) natural frequencies and mode shapes. The figures indicate relative amplitudes. (i) 10 Hz, (ii) 27 Hz and (iii) 37 Hz.

# TORSIONAL VIBRATIONS

EXAMPLE 5.2

Determine the natural frequencies and mode shapes of the system shown in Figure 5.6(a).

SOLUTION

Following the procedure described in Section 5.3, using a computer program to carry out the calculations, the system natural frequencies are found to be at 10 Hz, 27 Hz and 37 Hz. For each of these frequencies the system torsional mode shape may be computed and found to be as shown in Figures 5.6(b), where the amplitudes are normalized to give a value of unity at one end of the system.

## 5.4 Geared systems and branching

In some machines the shaft may not be continuous from one end of the machine to the other, but may have a gear box installed at one or more locations. In cases where there are more than two shafts attached to the gear box the system is said to be branched, as shown in Figure 5.7.

For the purposes of analysis, geared systems must be reduced to systems with a continuous shaft so that they may be treated as described in the preceding section. Figures 5.8 show an actual system where the gear box, shaft of torsional stiffness $k_2$ and rotor with moment of inertia $I_2$ are

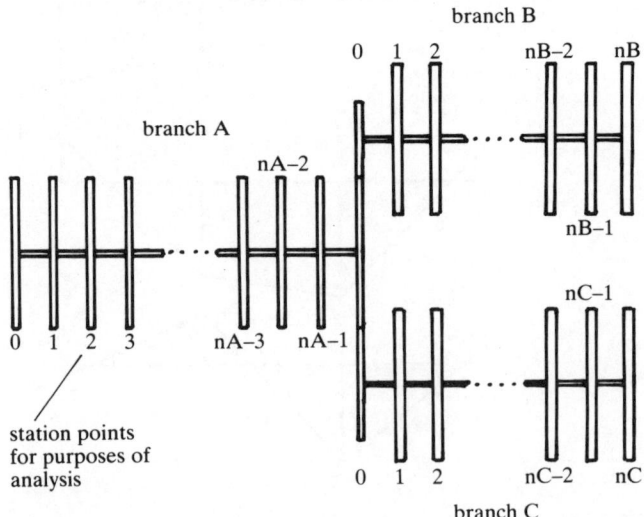

**Figure 5.7** A branched system with many rotor masses. 0, 1, 2, 3 etc. are station points for the purpose of analysis.

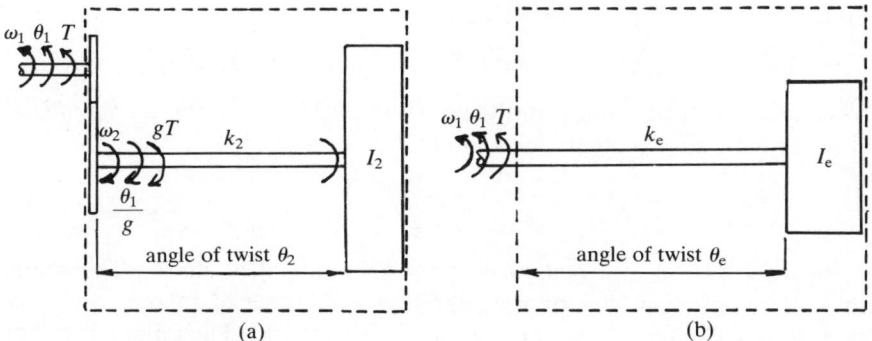

**Figure 5.8** A gearbox, shaft, and rotor, (a) actual system; (b) equivalent system without a gearbox. See text for details.

replaced in the equivalent system with a shaft of torsional stiffness $k_e$ and rotor of inertia $I_e$. When vibrations of the system take place kinetic energy of the rotor is transformed to strain energy in the shaft. These energy values must be the same in both the real and equivalent systems for the theoretical model to be valid.

Considering strain energy first, by imagining the rotor $I_2$ to be held rigidly whilst shaft 1 is rotated through some angle $\theta_1$ at the gear box, it may be noted that shaft 2 is rotated through an angle $\theta_1/g$ at the gear box, where $g$ is the gear ratio. The strain energy stored in shaft 2 must then be given by

$$E = \tfrac{1}{2} k_2 (\theta_1/g)^2 \tag{5.29}$$

whilst applying the same input to the left-hand side of the equivalent system results in the strain energy

$$E = \tfrac{1}{2} k_e \theta_1^2 \tag{5.30}$$

being stored in the equivalent shaft. By equating these two values of strain energy we may note that

$$k_e = \frac{k_2}{g^2} \tag{5.31}$$

If consideration is now given to the kinetic energies of both the real and equivalent systems, which must also be equal, we may write

$$\tfrac{1}{2} I_2 \omega^2 = \tfrac{1}{2} I_e \omega_e^2 \tag{5.32}$$

where $\omega$ and $\omega_e$ are the angular frequencies of inertias $I_2$ and $I_e$ respectively. Alternatively this may be written as

$$\tfrac{1}{2} I_2 (\omega_2 + d\theta_2/dt)^2 = \tfrac{1}{2} I_e (\omega_1 + d\theta_e/dt)^2 \tag{5.33}$$

where $\omega_1$ and $\omega_2$ are the angular frequencies of the high-speed and low-speed gear box shafts respectively, and $\theta_2$ and $\theta_e$ are the angles of twist of the actual shaft 2 and the shaft in the equivalent system respectively.

## TORSIONAL VIBRATIONS

Equation (5.33) may also be written as

$$\tfrac{1}{2}I_2(\omega_1/g + d(gT/k_2)/dt)^2 = \tfrac{1}{2}I_e(\omega_1 + d(T/k_e)/dt)^2 \qquad (5.34)$$

where $T$ is the torque input to the gear box pinion. Substituting $k_e = k_2/g^2$ from Equation (5.31) then gives, on rearranging,

$$I_e = \frac{I_2}{g^2} \qquad (5.35)$$

In the above, $k_e$ and $I_e$ are said to be the equivalent shaft stiffness and rotor moment of inertia of the geared system referred to the 'reference shaft' (the left-hand side of the gear box in Figure 5.8). The general rule, for forming the equivalent system for the purposes of analysis, is to divide all shaft stiffnesses and rotor inertias of the geared system by $g^2$, where $g$ is the speed ratio of the reference shaft to the geared shaft. When the analysis has been completed, it will be remembered that the elastic line of the real system is modified as shown in Figure 5.9, as compared with that of the equivalent system.

For some types of machinery, such as marine vessel power transmission shafts or machine-tool drives, there may be many rotor inertias in the system and the gear box may be a branch point where more than two shafts are linked together, as shown in Figure 5.7 for example. In these cases a slightly different approach is required in order to determine the system natural frequencies. First of all the transfer matrix equations are written, as described earlier, for each branch separately; for the system shown in Figure 5.7 these are

$$\{S\}_{nA} = [A]\{S\}_{0A} \qquad (5.36)$$

$$\{S\}_{nB} = [B]\{S\}_{0B} \qquad (5.37)$$

$$\{S\}_{nC} = [C]\{S\}_{0C} \qquad (5.38)$$

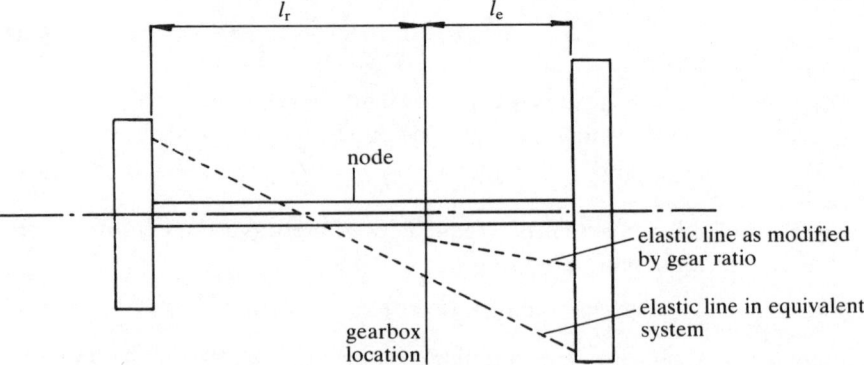

**Figure 5.9** Modification of the elastic line in the equivalent system to allow for the gearbox. $l_r$, shaft length on reference side of gearbox; $l_e$, shaft length whose stiffness is $K_e$.

Recognizing that the boundary condition at the free end of branch A is $T_{0A} = 0$, and imposing a unit angular displacement $\theta_{0A} = 1$ there, gives, on expanding equation (5.36),

$$\begin{Bmatrix} \theta \\ T \end{Bmatrix} = \begin{bmatrix} a_{11} & a_{12} \\ a_{21} & a_{22} \end{bmatrix} \begin{Bmatrix} 1 \\ 0 \end{Bmatrix} \qquad (5.39)$$

where $a_{11}$, $a_{12}$ etc. are elements of the matrix $[A]$, and which leads to $T_{nA} = a_{21}$ and $\theta_{nA} = a_{11}$. Noting that at the branch point $\theta_{0B} = \theta_{nA}/g_{AB}$, where $g_{AB}$ is the gear ratio between shafts A and B, and that $T_{nB} = 0$ at the free end of branch B is another boundary condition, Equation (5.37) may now be expanded to give

$$\begin{Bmatrix} \theta \\ 0 \end{Bmatrix} = \begin{bmatrix} b_{11} & b_{12} \\ b_{21} & b_{22} \end{bmatrix} \begin{Bmatrix} a_{11}/g_{AB} \\ T \end{Bmatrix} \qquad (5.40)$$

where $b_{11}$, $b_{12}$, etc. are elements of the matrix $[B]$, whereupon $T_{0B} = -b_{21}a_{11}/b_{22}g_{AB}$. Two further boundary conditions to be satisfied are that at the branch point $T_{nA} = T_{DB}/g_{AB} + T_{DC}/g_{AC}$ and $\theta_{DC} = \theta_{NA}/g_{AC}$, where $g_{AC}$ is the gear ratio between shafts A and C. Thus both $T_{0C}$ and $\theta_{0C}$ may now be evaluated and substituted into the expanded form of Equation (5.38) to give

$$\begin{Bmatrix} \theta \\ 0 \end{Bmatrix} = \begin{bmatrix} c_{11} & c_{12} \\ c_{21} & c_{22} \end{bmatrix} \begin{Bmatrix} a_{11}/g_{AC} \\ g_{AC}a_{21} + b_{21}a_{11}g_{AC}/(b_{22}g_{AB}^2) \end{Bmatrix} \qquad (5.41)$$

where $T_{nC} = 0$ is the boundary condition describing the free end of branch C. The second of Equations (5.41) is

$$0 = \frac{c_{21}a_{11}}{g_{AC}} + g_{AC}c_{22}a_{21} + \frac{c_{22}b_{21}a_{11}g_{AC}}{b_{22}g_{AB}^2} \qquad (5.42)$$

the roots of which are the system natural frequencies.

As before, these frequencies may then be substituted back into the transfer matrices for each station considered, whereupon the state vector at each station may be evaluated. A plot of angular displacement against shaft position then indicates the system mode shapes. Using this method, there is of course no need to modify the elastic line of the system to allow for gear box ratios since these have now already been allowed for in the analysis.

Torsional oscillations in branched systems are also discussed in Rao (1983), Vance (1988), and Murphy (1984).

EXAMPLE 5.3

A turbine whose moment of inertia is 0.5 kg.m$^2$ is coupled to a 3:1 speed reduction gear box via a 50 mm diameter solid steel shaft of length 0.8 m. The output shaft of the gear box is 0.6 m long and 75 mm in diameter, and drives a generator whose moment of inertia is 2.7 kg.m$^2$. Determine the torsional natural frequency of the system, and the amplitude of oscillations at the turbine when those measured at the generator are 0.1 10$^{-3}$ rad. Ignore gear inertia. The shaft material modulus of rigidity is 80 GN.m$^{-2}$.

SOLUTION

The turbine shaft torsional stiffness is

$$k_t = \frac{T}{\theta} = \frac{GJ}{1} = \frac{80 \times 10^9 \times \pi \times 0.05^4}{0.8 \times 32} = 61.4 \text{ kN.m.rad}^{-1}$$

That of the generator shaft is

$$k_g = \frac{GJ}{1} = \frac{80 \times 10^9 \times \pi \times 0.075^4}{0.6 \times 32} = 414 \text{ kN.m.rad}^{-1}$$

which when referred to the tubine shaft is

$$k_g' = 414/3^2 = 46 \text{ kN.m.rad}^{-1}$$

(using Equation 5.31). Also, the generator moment of inertia referred to the turbine shaft is, from Equation (5.35),

$$I_g' = 2.7/3^2 = 0.3 \text{ kg.m}^2$$

The turbine shaft, and the generator shaft referred to the turbine shaft axis, are connected in series, so that the overall shaft torsional stiffness between turbine and generator is

$$k_e = \left[\frac{1}{k_t} + \frac{1}{k_g'}\right]^{-1} = 26.3 \text{ kN.m.rad}^{-1}$$

whereupon the system torsional natural frequency may be evaluated using Equation (5.11) as

$$\omega_n = \left[\frac{(I_t + I_g')k_e}{I_t I_g'}\right]^{1/2} = \left[\frac{(0.5 + 0.3) \times 26.3 \times 10^3}{0.5 \times 0.3}\right]^{1/2} = 375 \text{ rad.s}^{-1}$$

The ratio of amplitudes of the two rotors may be determined using, say, the first of Equations (5.7) applied to the equivalent system

$$\frac{\theta_t}{\theta_g} = \frac{k_e}{(k_e - I_t \omega_n^2)} = \frac{26.3 \times 10^3}{(26.3 \times 10^3 - 0.5 \times 375^2)} = -0.6$$

But $\theta_g$, referred to the turbine shaft, is $0.1 \times 10^{-3} \times 3 = 0.3 \times 10^{-3}$ rad. Then $\theta_t$ is given by

$$\theta_t = -0.6 \times 0.3 \times 10^{-3} \qquad = -0.18 \times 10^{-3} \text{ rad}$$

(the negative sign indicating that the motion of the turbine is in anti-phase with that of the generator).

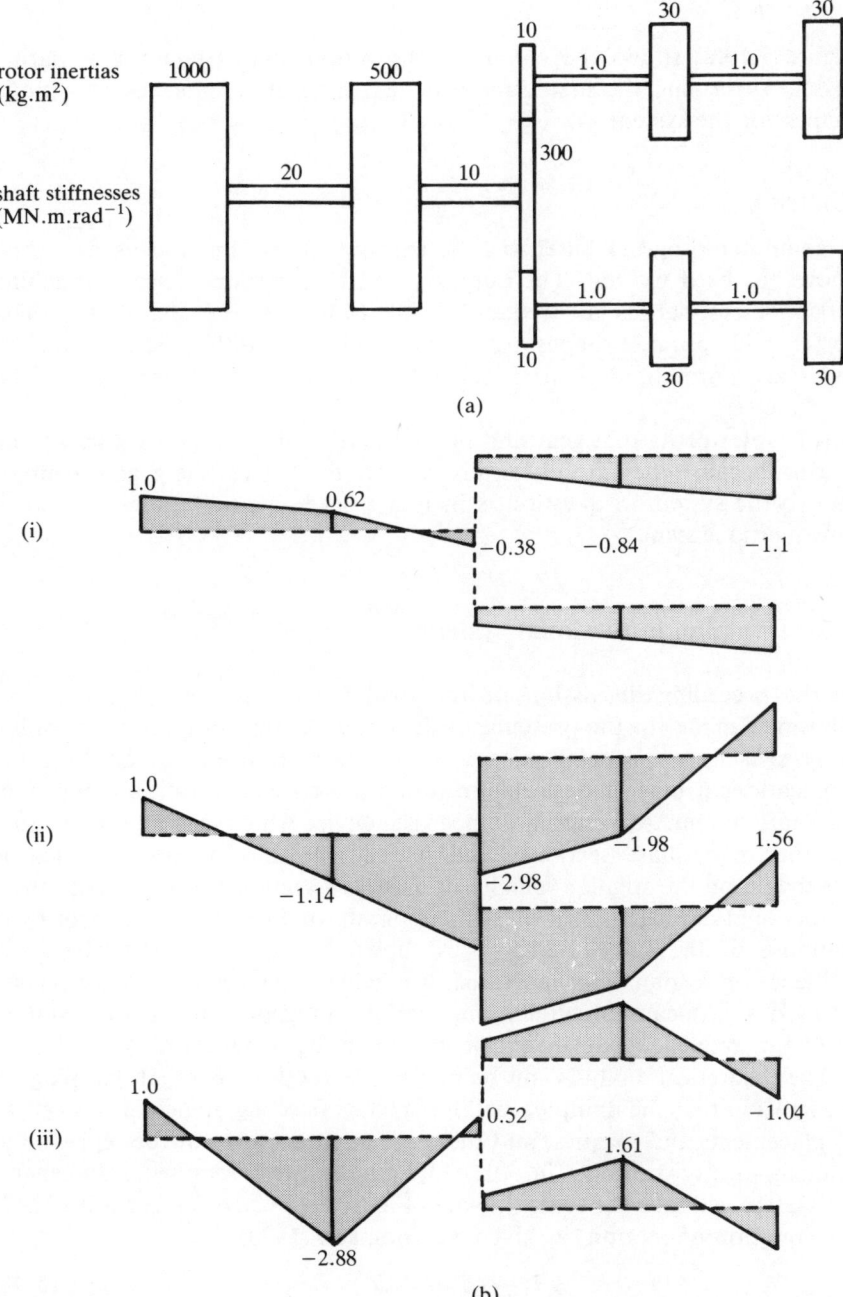

**Figure 5.10** A branched system with several rotor masses (Example 5.4), (a) system layout; (b) natural frequencies and mode shapes. The figures indicate relative amplitudes (i) 14 Hz, (ii) 33 Hz and (iii) 44 Hz.

EXAMPLE 5.4

Figure 5.10(a) shows the layout of the propulsion system for a marine vessel. Determine the first three torsional natural frequencies and mode shapes for the system.

SOLUTION

A computer program based on the transfer matrix method as described above has been written. The output from the program indicates machine torsional resonances at frequencies of 14 Hz, 33 Hz and 44 Hz. The corresponding mode shapes are shown in Figure 5.10(b) where the amplitudes are normalized so as to give a value of unity at one end of the machine.

It is noteworthy that gear and pinion inertias always act together as one inertia (because they are physically rigidly coupled via the gear meshing), and so the system in question is in fact a seven-inertia system and not a nine-inertia system.

## 5.5 Damping in torsional systems

In the preceding discussion of torsional vibrations there has been no allowance made for the presence of damping. A small amount of damping is present as a property of the shaft material itself, whilst in some applications it may be desirable to install a torsional vibration damper on the shaft system. A torsional damper is a device which may be used to join together two shaft sections, and which transmits a torque which is dependent on the angular velocity of one shaft section relative to the other. A torsional damper is shown in the diagram of Figure 5.11(a), whilst the response of the rotor inertia $I_1$ is shown in Figure 5.11(b) where the influence of the damper is indicated. It is evident that torsional dampers can be used as a means of attenuating system vibrations and to tune system resonant frequencies to suit particular operating conditions.

The theoretical analysis can be made to allow for torsional damping by recognizing that the damping will introduce phase lag angles to the system displacements and torques, and these parameters must now be represented mathematically with both in-phase and quadrature components. In general the system will take the form shown in Figure 5.12 where for rotor inertia $I_n$ the equation of motion, as given by Thomson (1978), is

$$_R T_n - {_L T_n} - C_n \dot{\theta}_n = I_n \ddot{\theta}_n \tag{5.43}$$

and where

$$_L \theta_n = {_R \theta_n} \tag{5.44}$$

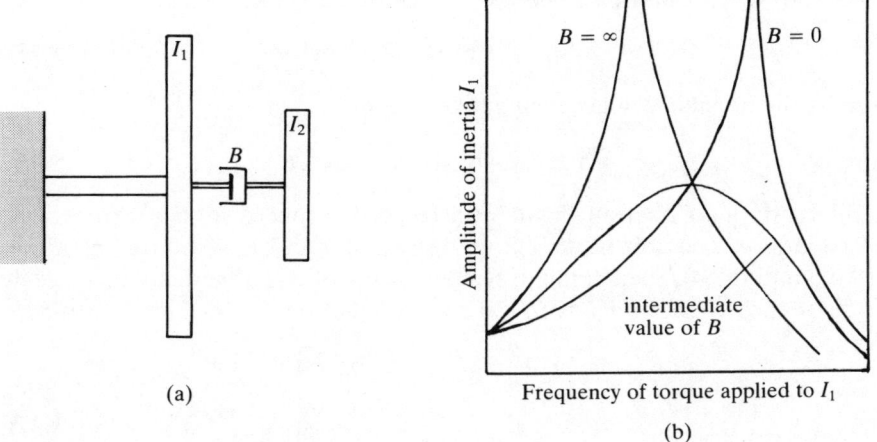

(a)

(b)

**Figure 5.11** A torsional damper in a simple system, and its effect on vibration, (a) system arrangement; (b) system frequency response.

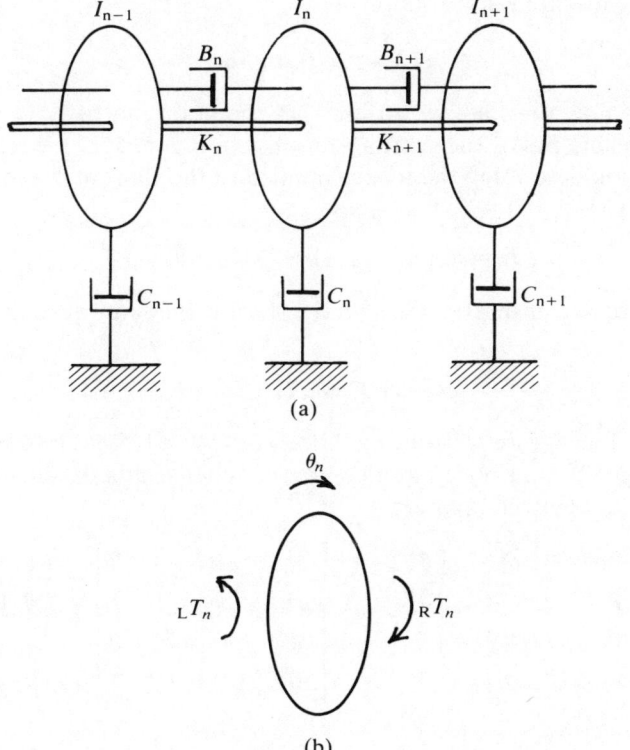

(a)

(b)

**Figure 5.12** General layout of a multi-degree-of-freedom system with damping, (a) general arrangement; (b) torques applied to left-hand and right-hand sides of inertia via shaft sections on each side.

The torques $_L T_n$ and $_R T_n$ may be written in the form

$$T_n = T_{n1} \sin \omega t + T_{n2} \cos \omega t \tag{5.45}$$

whilst the angular displacements take the form

$$\theta_n = \theta_{n1} \sin \omega t + \theta_{n2} \cos \omega t \tag{5.46}$$

Differentiating Equation (5.46) with respect to time to obtain expressions for $\dot{\theta}$ and $\ddot{\theta}$, and substituting Equations (5.45) and (5.46) into Equations (5.43) and (5.44), then comparing coefficients of $\sin \omega t$ and $\cos \omega t$ on each side of each equation leads to the matrix equation

$$_R\begin{Bmatrix} \theta_1 \\ \theta_2 \\ T_1 \\ T_2 \end{Bmatrix}_n = \begin{bmatrix} 1 & 0 & 0 & 0 \\ 0 & 1 & 0 & 0 \\ -\omega^2 I & \omega C & 1 & 0 \\ \omega C & -\omega^2 I & 0 & 1 \end{bmatrix}_n {}_L\begin{Bmatrix} \theta_1 \\ \theta_2 \\ T_1 \\ T_2 \end{Bmatrix}_n \tag{5.47}$$

which may be written more simply as

$$_R\{S\}_n = [P]_n {}_L\{S\}_n \tag{5.48}$$

where $\{S\}$ is now the state vector and $[P]$ the point matrix.

The characteristics of the shaft at station $n$ in Figure 5.12 are represented in the equation describing the torque applied to the shaft at the location of rotor $n$, that is

$$_L T_n = k_n(\theta_n - \theta_{n-1}) + B_n(\dot{\theta}_n - \dot{\theta}_{n-1}) \tag{5.49}$$

while the torque transmitted through the shaft is the same at each end, and so

$$_L T_n = {}_R T_n \tag{5.50}$$

Substituting for $T$, $\theta$ and $\dot{\theta}$ from Equations (5.45) and (5.46) into Equations (5.49) and (5.50), and once more comparing coefficients of $\sin \omega t$ and $\cos \omega t$, leads to the matrix equation

$$\begin{bmatrix} k & -\omega B & 0 & 0 \\ \omega B & k & 0 & 0 \\ 0 & 0 & 1 & 0 \\ 0 & 0 & 0 & 1 \end{bmatrix}_n {}_L\begin{Bmatrix} \theta_1 \\ \theta_2 \\ T_1 \\ T_2 \end{Bmatrix}_n = \begin{bmatrix} k & -\omega B & 1 & 0 \\ \omega B & k & 0 & 1 \\ 0 & 0 & 1 & 0 \\ 0 & 0 & 0 & 1 \end{bmatrix}_n {}_R\begin{Bmatrix} \theta_1 \\ \theta_2 \\ T_1 \\ T_2 \end{Bmatrix}_{n-1} \tag{5.51}$$

which may be written more simply as

$$[L]_n {}_L\{S\}_n = [M]_n {}_R\{S\}_{n-1} \tag{5.52}$$

## TORSIONAL IMPEDANCES

Pre-multiplying both sides of Equation (5.52) by the inverse of $[L]$ gives

$$_L\{S\}_n = [L]_n^{-1}[M]_n {}_R\{S\}_{n-1} \qquad (5.53)$$
$$= [F]_n {}_R\{S\}_{n-1}$$

where $[F]_n$ is now the field matrix at station $n$.

Substituting Equation (5.53) into Equation (5.48) then gives

$$_R\{S\}_n = [P]_n[F]_n {}_R\{S\}_{n-1} \qquad (5.54)$$
$$= [U]_n {}_R\{S\}_{n-1}$$

where $[U]_n$ is now the transfer matrix for station $n$. The transfer matrix may be so evaluated for each station and combined as before, in Equations (5.25), to give the overall system transfer matrix. If the system forcing is known, then the response at each point may be calculated as discussed previously in Section 5.3.

Further detailed studies of the effects of damping on torsional vibrations are discussed by Wilson (1967).

## 5.6 Calculation of torsional natural frequencies and forced response using the method of mechanical impedances

For MDOF systems the method of mechanical impedances may be used as an alternative to the transfer matrix method for calculating system torsional natural frequencies and forced response. The method is probably more adaptable for use in hand calculations than is the transfer matrix method, provided that the number of rotor inertias is not excessive. Application of the method is based upon the principles discussed in section 4.4 for lateral vibrations; these principles are equally applicable to torsional systems. In summary, the rules which must be applied are that when the forcing point has been decided upon, the equivalent system must be considered and impedances evaluated for each element in the equivalent system. To determine the impedance at the forcing point (the 'driving point' impedance), the impedances of elements connected in parallel to this point in the equivalent system must be added; if elements are connected in series then their receptances are added. The method is demonstrated in the following example.

### EXAMPLE 5.5

For the system shown in Figure 5.13(a) plot curves showing the variation of driving point impedance $Z_{22}$ with forcing frequency. Also, determine the system natural frequencies, the frequencies at which $I_2$ is at a node, and the amplitude of $I_1$ when $\omega = 400$ rad.s$^{-1}$.

TORSIONAL VIBRATIONS

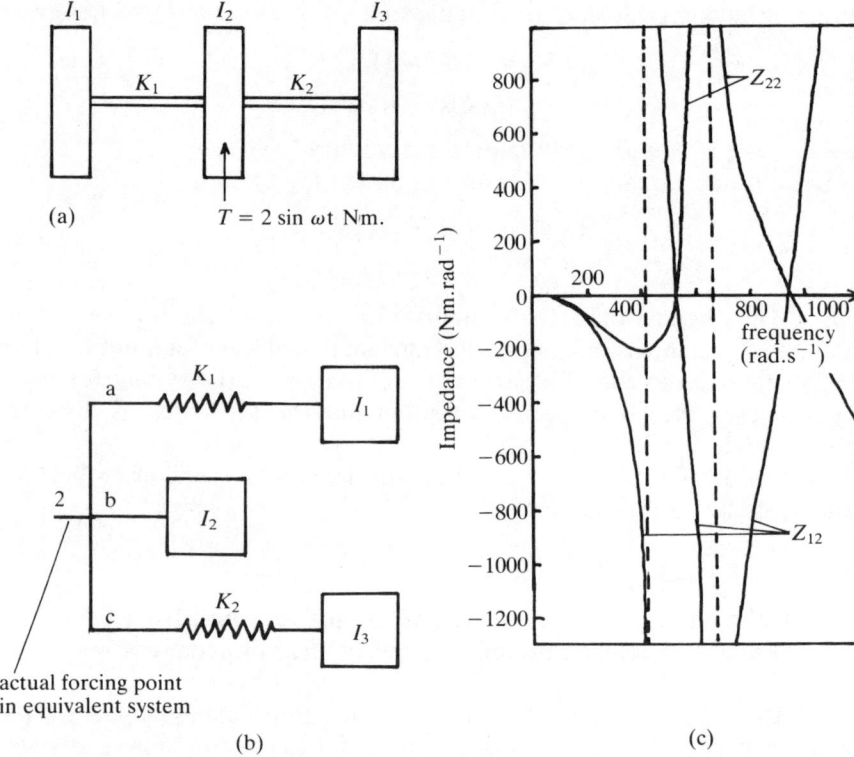

(b)

actual forcing point
in equivalent system

(c)

**Figure 5.13** Equivalent system and impedance characteristics for a three-rotor torsional system (Example 5.5), (a) actual system, $I_1 = 10^{-3}$ kg.m$^2$, $I_2 = 0.7 \times 10^{-3}$ kg.m$^2$, $I_3 = 0.5 \times 10^{-3}$ kg.m$^2$, $K_1 = 200$ Nm.rad$^{-1}$, $K_2 = 200$ Nm.rad$^{-1}$. The torque $T$ is applied to $I_2$; (b) equivalent system for purpose of analysis; $a$, $b$, $c$ is the notation for evaluating individual branch impedances; (c) variation of system impedances with forcing frequency.

SOLUTION

Since the forcing is at $I_2$, this rotor inertia is in parallel with the torsional springs and rotor inertias on each side of $I_2$ (the torsional forcing is applied directly to the point of junction of these system elements), and so the equivalent system is as shown in Figure 5.13(b). Applying the rules described above enables calculation of $Z_{22}$ as

$$Z_{22} = Z_{aa} + Z_{bb} + Z_{cc}$$

$$= \frac{1}{(R_{K_1} + R_{I_1})} + Z_{I_2} + \frac{1}{(R_{K_2} + R_{I_3})} \quad \text{(where } R \text{ is a receptance)}$$

$$= \frac{1}{(1/200 - 1/10^{-3}\omega^2)} - 0.7 \times 10^{-3}\omega^2 + \frac{1}{(1/200 - 1/0.5 \times 10^{-3}\omega^2)}$$

164

## TORSIONAL IMPEDANCES

To evaluate $Z_{12}$ consider the following:

$$\theta_1 = \frac{T_1}{Z_{I_1}}$$

$$= \frac{TZ_{aa}}{Z_{I_1}Z_{22}}$$

and so therefore

$$Z_{12} = \frac{T}{\theta_1} = \frac{Z_{I_1}Z_{22}}{Z_{aa}}$$

$Z_{22}$ and $Z_{12}$ may be plotted against $\omega$ as shown in Figure 5.13(c). The system natural frequencies occur when $Z_{22} = Z_{12} = 0$ (as discussed in Section 4.4), that is at $\omega_{n1} = 530$ rad.s$^{-1}$ and at $\omega_{n2} = 940$ rad.s$^{-1}$. Inertia $I_2$ is at a node when $Z_{22} = \infty$, that is at 420 rad.s$^{-1}$ and at 650 rad.s$^{-1}$. The amplitude of $I_1$ at 400 rad.s$^{-1}$ may be evaluated from

$$\theta_1 = \frac{T}{Z_{12}} = \frac{2}{-209} = -9.6 \times 10^{-3} \text{ rad.}$$

where the negative sign indicates that its angular displacement is in the opposite direction to the forcing at $I_2$.

It is noteworthy that the method can also be readily applied to systems with gearing and branching, recognizing that different branches are connected in parallel in the equivalent system. The geared branch must be referred to the reference shaft for the purposes of analysis by dividing all rotor inertias and shaft torsional impedances by $g^2$, where $g$ is the speed ratio of the reference shaft to the geared shaft (see also Section 5.4).

Allowance may also be made for damping in torsional systems by writing the system impedances as complex numbers. In the case of a single torsional damper the element impedance may be evaluated from

$$T = B\dot{\theta}$$
$$= j\omega B\theta \qquad (5.55)$$

(where $B$ is the torsional damping coefficient) then the element impedance is

$$Z = \frac{T}{\theta} = j\omega B \qquad (5.56)$$

Similarly, consideration of a damper in parallel with a torsional spring of stiffness $k$ enables the corresponding impedance to be evaluated as

$$Z = (k + j\omega B) \qquad (5.57)$$

# References

Carnahan, B., Luther, H. A. & J. O. Wilkes 1969. *Applied numerical methods*. New York: Wiley.
Collar, A. R. & A. Simpson 1987. *Matrices and engineering dynamics*. Chichester: Ellis Horwood.
Moore, J. B. 1967. A convergent algorithm for solving polynomial equations. *J. Assoc. Comput. Mach.* **14**, 2, 311–315.
Murphy, B. T. 1984. Eigenvalues of rotating machinery. Ph.D. Thesis, Texas A&M University.
Pestel, E. C. & F. A. Leckie 1963. *Matrix methods in elastomechanics*. New York: McGraw-Hill.
Pipes, L. A. 1963. *Matrix methods for engineering*. Englewood Cliffs, NJ: Prentice-Hall.
Rao, J. S. 1983. *Rotor dynamics*. New Delhi: Wiley Eastern.
Rao, J. S., Rao, D. K. & K. V. Bhaskara Sarma 1980. The transient response of turbo-alternator rotor systems under short circuit conditions. *Proc. IMech.E Conf. Vibrations in Rotating Machinery*, Cambridge, 271ff.
Thomson, W. T. 1978. *Vibration theory and applications*. London: George Allen and Unwin.
Vance, J. M. 1988. *Rotordynamics of turbomachinery*. New York: Wiley.
Wilson, W. J. K. 1967. *A handbook of torsional vibration analysis*. New York: Wiley.

# 6 Instability in rotating machines

## 6.1 Introduction

In certain circumstances, depending on the design, some machines may be prone to instability. This means that machine vibrations set in, even in the absence of imbalance effects, resulting in high levels of noise and component stress and a corresponding reduced fatigue life. In linear systems the magnitude of these vibrations tends towards infinity, although in practice shaft vibration amplitudes are often limited by system non-linearities. Such machine instabilities may originate from a number of sources including oil-film bearings, shaft stiffness asymmetry, internal friction between mating components, and aerodynamic forces. The designer's problem is to investigate the possibility of machine instability, and to change the appropriate machine design parameters to ensure that any potential unstable modes of operation lie outside the normal operating regime of the machine. This chapter discusses several of the more common sources of machine instability and shows how the designer might investigate them.

## 6.2 Oil whirl

Oil-film bearings are possibly the most common source of machine instability. Oil-whirl instability is also known as 'half-speed whirl' because the frequency of the vibrations which are set up is often just below half the shaft rotational frequency (typically 0.46–0.48 times the shaft rotational frequency). Oil whirl was first investigated by Newkirk and Taylor (1925); it tends to occur only in lightly loaded oil-film bearings operating at a very small eccentricity ratio. It is an extremely dangerous condition because the bearing load-carrying capacity is in effect reduced to zero as vibrations of increasing amplitude set in, and consequent destruction of the bearings is a possibility.

An indication of why the machine vibrations occur at about half the shaft rotational frequency may be understood by considering Figure 6.1, where the journal in an oil-film bearing rotates at frequency $\omega$ and whirls around the bearing clearance at frequency $n\omega$. Because the bearing is lightly loaded (and so operates at only a small eccentricity $e$ in the first instance), the

# INSTABILITY

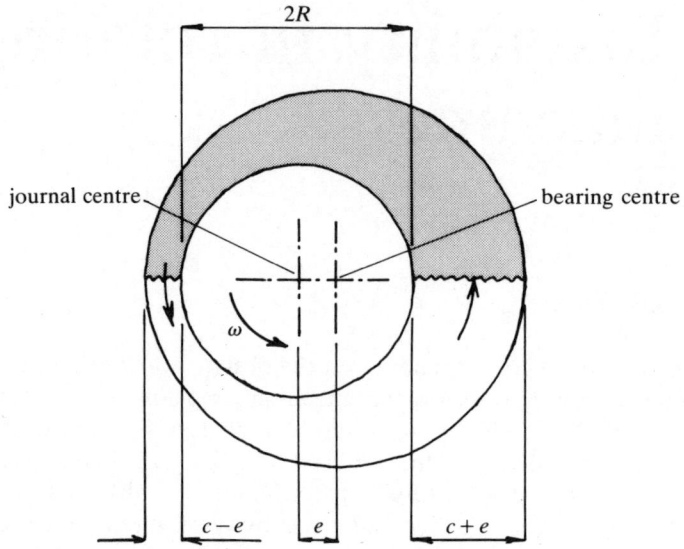

**Figure 6.1** General arrangement of journal and lubricant in a bearing during oil whirl.

variation in oil pressure around the bearing circumference may be considered to be negligible so that the only oil flow around the bearing is that which is induced by the rotation of the journal. The lubricant flow rate into, and out of, the shaded wedge-shaped area shown in Figure 6.1 is then given by basic fluid mechanics theory as

$$Q_{in} = R\omega(c+e)/2, \qquad Q_{out} = R\omega(c-e)/2 \qquad (6.1)$$

respectively, per unit length of the bearing. Since the journal is whirling within the bearing clearance with some frequency $n\omega$, the tangential velocity of the journal centre at the instant in time shown is $en\omega$, and so the volume of the shaded wedge area, per unit length of the bearing, must be increasing at a rate given by

$$Q_{vol} = 2Ren\omega \qquad (6.2)$$

This volume flow rate must be provided by the net lubricant flow into the shaded area under consideration, so that we may write

$$2Ren\omega = R\omega(c+e)/2 - R\omega(c-e)/2 \qquad (6.3)$$

which may be solved to give $n = 0.5$.

## 6.2.1 Stability analysis using linearized stiffness and damping coefficients

To investigate the likelihood of oil whirl, attention needs to be given to the bearing operating characteristics themselves. For oil-film journal bearings

## OIL WHIRL

these may be expressed in terms of the eight linearized stiffness and damping coefficients introduced in Chapter 2. The relationship between the bearing forces and the journal motion is given by the equations of motion for the journal, which in the case of a symmetrical system with a rigid rotor case are

$$K_{xx}x + K_{xy}y + C_{xx}\dot{x} + C_{xy}\dot{y} = -M\ddot{x}$$
$$K_{yx}x + K_{yy}y + C_{yx}\dot{x} + C_{yy}\dot{y} = -M\ddot{y} \quad (6.4)$$

where $x$ and $y$ are the horizontal and vertical displacements of the rotor and $M$ is half the rotor mass. For the flexible shaft case the bearing stiffness and damping coefficients can be modified to allow for shaft flexibility as described in Chapter 3. If the rotor is momentarily displaced from its equilibrium position by some random input, any subsequent free vibrations of the rotor in the horizontal and vertical directions will take the form

$$x = X_0 \exp(st)$$
$$y = Y_0 \exp(st + \phi) \quad (6.5)$$

where $X_0$ and $Y_0$ are the vibration amplitudes in the horizontal and vertical directions respectively, and $\phi$ is a phase lag angle. Equations (6.5) may be differentiated to give expressions for velocity and acceleration in the horizontal and vertical directions respectively as

$$\dot{x} = X_0 s \exp(st)$$
$$\dot{y} = Y_0 s \exp(st + \phi) \quad (6.6)$$

$$\ddot{x} = X_0 s^2 \exp(st)$$
$$\ddot{y} = Y_0 s^2 \exp(st + \phi) \quad (6.7)$$

Substituting Equations (6.5) to (6.7) into Equations (6.4), and dividing throughout by $\exp(st)$, gives the following two equations:

$$X_0(-Ms^2 - K_{xx} - C_{xx}s) = Y_0 e^{\phi}(K_{xy} + C_{xy}s)$$
$$X_0(K_{yx} + C_{yx}s) = Y_0 e^{\phi}(-Ms^2 - K_{yy} - C_{yy}s) \quad (6.8)$$

Expressing each of Equations (6.8) as a ratio $X/Ye^{\phi}$ then gives

$$\frac{X_0}{Y_0 e^{\phi}} = \frac{(K_{xy} + C_{xy}s)}{(-Ms^2 - K_{xx} - C_{xx}s)} = \frac{(-Ms^2 - K_{yy} - C_{yy}s)}{(K_{yx} + C_{yx}s)} \quad (6.9)$$

which may be rearranged to give

$$(M^2)s^4 + (C_{xx}M + C_{yy}M)s^3 + (K_{xy}M + K_{yy}M + C_{xx}C_{yy} - C_{xy}C_{yx})s^2$$
$$+ (K_{yy}C_{xx} + C_{yy}K_{xx} - K_{xy}C_{yx} - C_{xy}K_{yx})s + (K_{xx}K_{yy} - K_{xy}K_{yx}) = 0 \quad (6.10)$$

Equation (6.10) has four roots of $s$. To test for system stability one must examine the motion of the journal which follows a momentary displacement from its steady running position, that is, does the journal return to a stable equilibrium position or not? If the journal were to return to a stable equilibrium position then this would be characterized by values of displacement $x$ and $y$ which decreased with time, that is by negative values of $s$. In general the roots of $s$ will have both real and imaginary parts, indicating that the transient motion of the journal will take the form of a sine wave having a decaying amplitude when the system is stable (when the real parts of all roots of $s$ are negative). The imaginary parts of the roots of $s$ indicate the frequency of the resulting vibrations.

The circumstances under which the real parts of all roots of $s$ are negative are given by the Routh–Hurwitz stability criterion (Collar and Simpson 1987) as being when the following conditions are met:

(i) all coefficients of the characteristic equation must have the same sign
(ii) each of the following determinants must be positive:

$$R_1 = |a_1|$$

$$R_2 = \begin{vmatrix} a_1 & a_0 \\ a_3 & a_2 \end{vmatrix}$$

$$R_3 = \begin{vmatrix} a_1 & a_0 & 0 \\ a_3 & a_2 & a_1 \\ a_5 & a_4 & a_3 \end{vmatrix}$$

$$R_4 = \begin{vmatrix} a_1 & a_0 & 0 & 0 \\ a_3 & a_2 & a_1 & a_0 \\ a_5 & a_4 & a_3 & a_1 \\ a_7 & a_6 & a_5 & a_4 \end{vmatrix} \tag{6.11}$$

etc.

$$R_n = \begin{vmatrix} a_1 & a_0 & 0 & 0 & \cdots \\ a_3 & a_2 & a_1 & & \\ a_5 & a_4 & & & \\ a_7 & \cdots & & \text{etc.} & \\ & \cdots & & & \\ a_{2n-1} & a_{2n-2} & & & \end{vmatrix}$$

where the system characteristic equation takes the form

$$a_{n+1}s^{n+1} + a_n s^n + \cdots + a_4 s^4 + a_3 s^3 + a_2 s^2 + a_1 s + a_0 = 0 \tag{6.12}$$

Substitution of appropriate values into Equations (6.11) then allows the designer to check the likely stability of the bearings.

Use of the Routh–Hurwitz criterion only enables the designer to determine whether a machine is likely to be stable or unstable. It does not indicate how stable (or unstable) a machine may be. This information may

## OIL WHIRL

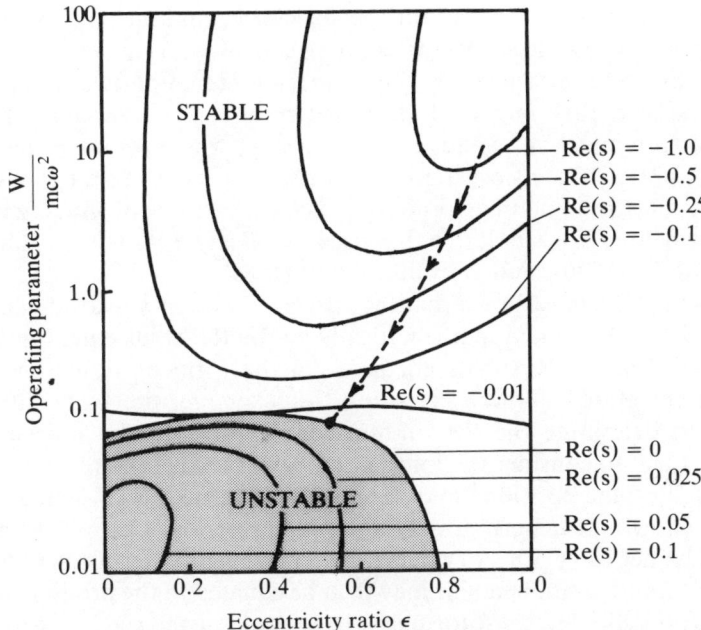

**Figure 6.2** A typical journal bearing stability map.

be vital in studies where the system is apparently stable but where the accuracy of the input data may be questionable. For this reason it may be much more useful for the designer to evaluate the magnitude of the largest real part of the roots of $s$, since this indicates the stability reserve of the system. This information is frequently recorded in the form of a system stability map (Cameron 1981, Ogrodnik et al. 1985) similar to that shown in Figure 6.2.

It will be noted from the above that system stability depends upon the stiffness and damping of the bearings. The above analysis relates to rigid-shaft machines for which a damping ratio of 1.0 maximizes stability, in the absence of cross-coupling. In the case of flexible-shaft machines, the optimum value of support damping for maximum stability depends upon the bearing and shaft stiffness (Barrett et al. 1978). It may also be possible to improve system stability by reducing the bearing stiffness (Goodwin et al. 1988).

### 6.2.2 Stability analysis allowing for oil-film non-linearity

In the previous section a method of investigating the possibility of oil whirl was presented which relied upon accurate information about the oil-film behaviour being expressed in terms of the eight linearized oil-film coefficients. Although the method is relatively easy to apply, it is sometimes criticized because it utilizes linearized oil-film coefficients which are only valid for small displacements of the journal away from its initial static equilibrium position. Since oil whirl implies large amplitude vibration, it may be argued that any analysis based upon an assumption of small vibration amplitudes is invalid. For this reason, where resources permit, a different approach to oil-whirl instability analysis, which does not assume a linear oil film, is sometimes preferred. Such a method of investigation is described below; it is similar to that used by Akers *et al.* (1971), Bannister and Makdissy (1980), and Ogrodnik *et al.* (1989).

In section 2.3 it was shown that the characteristics of a hydrodynamic oil film could be represented mathematically by the Reynolds equation. It was also noted that the Reynolds equation, in the form of Equation (2.19), could be integrated numerically using, for example, the finite difference method to determine the net oil-film force acting on the journal. It is possible, then, to consider the journal to be momentarily displaced from its static equilibrium position, and to compute the net force acting on the journal, allowing for both gravity and imbalance loads as well as oil-film forces. The net force may, of course, be represented in terms of radial and tangential force components. It may then be equated to the product of rotor acceleration and effective rotor mass, in the case of the rigid rotor system. This can be expressed mathematically as

$$-F_r + mu\omega^2 \cos \alpha - W \cos \phi = m(\ddot{e} - e\dot{\phi}^2)$$
$$-F_t - mu\omega^2 \sin \alpha + W \sin \phi = m(e\ddot{\phi} + 2\dot{e}\dot{\phi})$$
(6.13)

for journal motion in the radial and tangential directions respectively, where $F_r$ and $F_t$ are the oil-film forces, $mu$ is the rotor out-of-balance, $W$ is the gravity load acting on the bearing, and $m$ is the effective rotor mass at the bearing (equal to half the actual mass for a symmetrical rigid rotor system). Figure 6.3 shows the general arrangement of the system of forces acting on the journal.

From this point the analysis may proceed in the manner described in section 2.5.2, comparing Equations (6.13) with Equations (2.107). The only unknowns in Equations (6.13) are $e$ and $\phi$, and their derivatives. The equations may be integrated forward in time, starting with some assumed values of the unknowns compatible with some momentary disturbance of the journal away from its static equilibrium position. Two of the most commonly used methods of integrating Equations (6.13) are the Euler and Runge–Kutta methods, the former being described in section 2.5.2.

It is thus possible to determine the journal position, described by $e$ and $\phi$, at various moments in time following the initial disturbance of the system. If these positions are plotted then the path followed by the journal is effectively traced out. The resultant journal whirl orbit would normally settle down to an elliptical shape after sufficient iterations, as shown in

OIL WHIRL

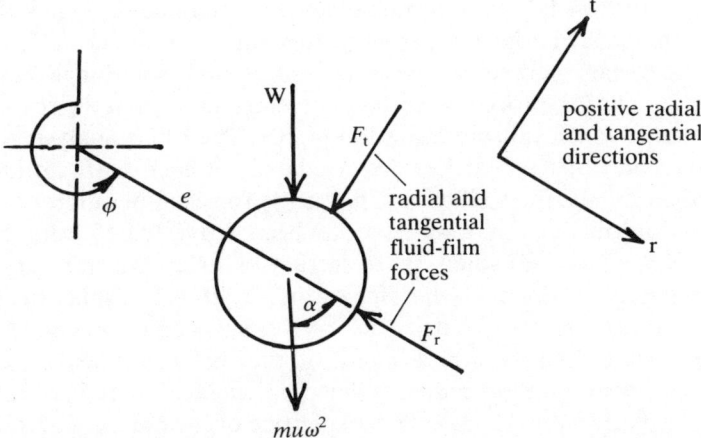

**Figure 6.3** General arrangement of forces acting on a journal. $W$, gravity load; $mu\omega^2$, imbalance force; $r$, $t$ radial and tangential directions; $F_r$, $F_t$, fluid-film forces.

Figure 6.4(a), in the case of an imbalance load being present. When no imbalance load is present the oribit collapses to a point – the static equilibrium position. If, however, the system is unstable then the orbit does not degenerate to an ellipse and may increase in amplitude with time, indicating subsequent destruction of the bearing, as shown in Figure 6.4(b).

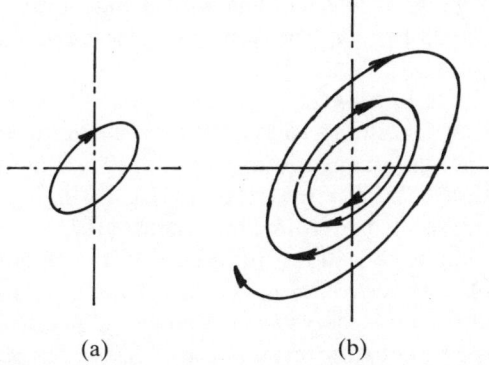

(a)　　　　　　(b)

**Figure 6.4** Some typical journal whirl orbits under stable and unstable running conditions, (a) stable system whirls orbit; (b) journal trajectory for an unstable system.

### 6.2.3 MDOF systems

A method of analysing the stability of MDOF systems has been suggested by Lund (1965, 1974). This method uses transfer matrices to represent the system, as described earlier in Chapter 4. The procedure is identical to that described in section 4.3, with the imbalance forcing omitted, except that the vibration frequency in the point and bearing matrices is defined as complex, with the imaginary part set to about half the journal rotational frequency. An iterative solution procedure is then implemented to find the real part of the eigenvalue which satisfies Equation (4.30). The Lund analysis requires that the system transfer matrices are evaluated for each shaft section, and combined together as described in Chapter 4, for each iteration.

A variation on the Lund analysis has been suggested by Murphy and Vance (1983). The individual shaft section transfer matrices are again represented as functions of the eigenvalues $s$, if the displacements are written as functions of $e$, but in this method the entire system transfer matrix is next evaluated, where many of the matrix elements are polynomials in $s$. A solution procedure is then implemented to find the values of $s$ which satisfy Equation (4.30). The magnitude of the real part of $s$ is again a measure of the stability of the system, as discussed above.

The characteristic matrix method, used in section 5.2 for calculating system torsional natural frequencies, can also be used in relation to rotor lateral vibration, and can be extended to indicate the degree of stability of MDOF systems (Goodwin *et al.* 1989). The system displacements are again represented as functions of $e$, so that the elements of the matrix $K$ in Equation (5.9) are also functions of $s$. Again, a solution procedure is required to calculate the values of $s$ which satisfy Equation (5.9); once more these will be complex quantities and the real part of $s$ will indicate the degree of stability. The disadvantage with this technique, however, is that the $K$ matrix is very large for systems with a high number of degrees of freedom, and this may present problems with storing and manipulating the data on the computer.

Another alternative approach for MDOF systems is to write down the equations of motion relating to each degree of freedom, and to solve these equations simultaneously using a 'time-marching' technique similar to that described in section 6.2.2 for the system with a single rotor mass. This method has been described by Kirk and Gunter (1974). The advantages of this method are that it can allow for system non-linearities, for example non-linear oil-film properties, and system transient displacements are computed. Its shortcoming, however, is that large quantities of computing power are required in order to calculate the system response over only a short period of real time, which may represent only a few cycles of vibration.

EXAMPLE 6.1

A symmetrical rigid rotor machine runs in plain hydrodynamic oil-film bearings which have a length/diameter ratio of unity. The bearings are

## OIL WHIRL

lubricated with oil of dynamic viscosity of 0.03 N.s.m$^{-2}$ and have a radial clearance of 120 $\mu$m. The journal diameter is 100 mm and the rotor mass is 1000 kg. Estimate whether or not the bearings are stable at 10 000 rev. min$^{-1}$.

SOLUTION

For the machine in question the bearing Sommerfeld number is given by

$$S = \frac{\mu L D}{W} \left(\frac{R}{C}\right)^2 N$$

$$= \frac{0.03 \times 0.1 \times 0.1}{1000 \times 9.81} \left(\frac{50}{0.12}\right)^2 \times 10^5 \text{ (1000kg effective mass at each bearing)}$$

$$= 0.885$$

Using the data in Figures 2.19 enables the corresponding bearing stiffness and damping coefficients to be evaluated as

$$\begin{aligned}
K_{xx} &= 155 \text{ MN.m}^{-1} & C_{xx} &= 1.01 \text{ MN.s.m}^{-1} \\
K_{xy} &= -163 \text{ MN.m}^{-1} & C_{xy} &= 0.171 \text{ MN.s.m}^{-1} \\
K_{yx} &= 326 \text{ MN.m}^{-1} & C_{yx} &= 0.171 \text{ MN.s.m}^{-1} \\
K_{yy} &= 81.6 \text{ MN.m}^{-1} & C_{yy} &= 1.01 \text{ MN.s.m}^{-1}
\end{aligned}$$

whereupon the coefficients of the characteristic equation (6.12) may be evaluated as

$$a_0 = K_{xx}K_{yy} - K_{xy}K_{yx} = 65.8 \times 10^{15} \text{ N}^2.\text{m}^{-2}$$
$$a_1 = K_{yy}C_{xx} + C_{yy}K_{xx} - K_{xy}C_{yx} - C_{xy}K_{yx} = 211 \times 10^{12} \text{ N}^2.\text{s.m}^{-2}$$
$$a_2 = K_{xx}M + K_{yy}M + C_{xx}C_{yy} - C_{xy}C_{yx} = 1.23 \times 10^{12} \text{ N}^2.\text{s}^2.\text{m}^{-2}$$
$$a_3 = C_{xx}M + C_{yy}M = 2.02 \times 10^9 \text{ N}^2.\text{s}^3.\text{m}^{-2}$$
$$a_4 = M^2 = 1.0 \times 10^6 \text{ N}^2.\text{s}^4.\text{m}^{-2}$$

Using the method described in section 6.2.1, the first condition for stability is satisfied since all coefficients are positive. The Routh determinants are:

$$R_1 = a_1 = 0.211 \times 10^{15}$$
$$R_2 = a_1 a_2 - a_3 a_0 = 126 \times 10^{21}$$
$$R_3 = a_1(a_2 a_3 - a_4 a_1) - a_0(a_3^2 - a_1 a_5) - a_3(a_0 a_3 - 0) + a_2(a_1 a_3 - 0)$$
$$\quad - a_1(a_1 a_4 - a_5 a_0) + a_5(a_0 a_1 - 0) - a_4(a_1^2 - 0) + a_3(a_1 a_2 - a_3 a_0)$$
$$= -1.29 \times 10^{37}$$

These values are not all positive, and so the bearings of the machine in question are unstable at 10 000 rev. min$^{-1}$ and oil whirl is likely to set in.

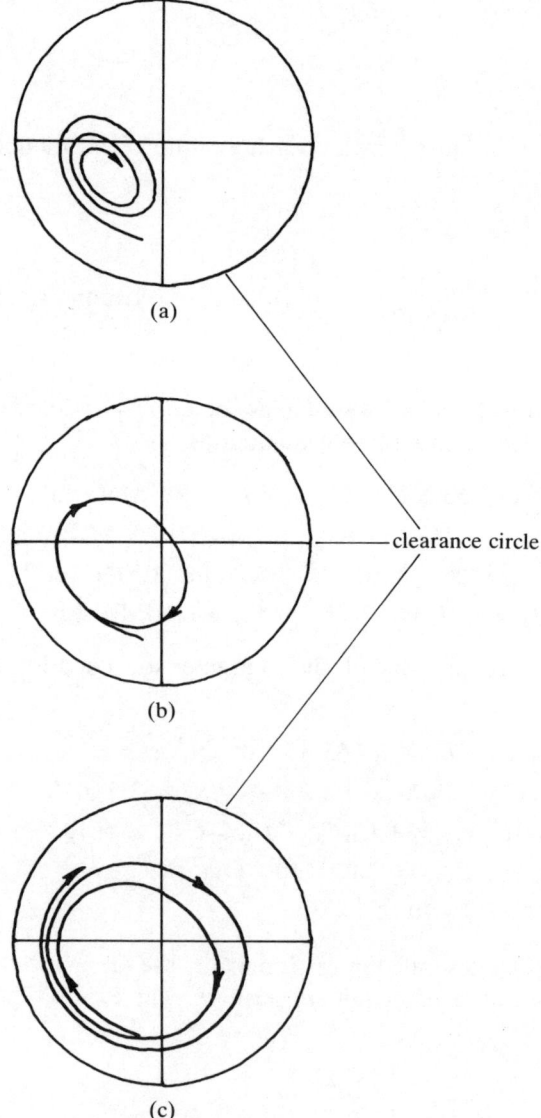

**Figure 6.5** Journal whirl orbits predicted using a time-marching technique (Example 6.2), (a) rotational speed 7000 rev.min$^{-1}$, orbit converging to the static equilibrium position (point stability); (b) rotational speed 8000 rev.min$^{-1}$, orbit maintaining a constant amplitude (orbitally stable); (c) rotational speed 9000 rev.min$^{-1}$, orbit of increasing amplitude (unstable).

EXAMPLE 6.2

A symmetrical rigid rotor machine runs in plain cylindrical hydrodynamic oil-film bearings and may be required to operate at a speed which is close to that at which oil-whirl instability is likely to set in. Because of the likelihood of large amplitude vibrations, an analysis of the situation based solely on linearized bearing coefficients, as discussed in Section 6.2.1, is deemed to be inappropriate. Instead, the limits of the stable operating regime for the bearings are to be determined by means of an analysis which allows for oil-film non-linearity effects and is based upon a time-marching technique. The machine details are as follows:

| | |
|---|---|
| rotor mass | 891 kg |
| journal diameters | 50 mm |
| bearing radial clearances | 127 $\mu$m |
| bearing lengths | 50 mm |
| lubricant dynamic viscosity | 0.069 N.s.m$^{-2}$ |

Determine the limiting speed of stable operation of the bearings.

SOLUTION

The procedure described in section 6.2.2 may be adopted to determine the journal whirl orbits at various running speeds. Typical results of such an analysis are shown in Figures 6.5; these results were obtained by modelling the bearing oil film with a finite difference mesh having nine nodes along the bearing length (including boundary nodes) and eleven around the circumference. The lubricant pressures at each node were determined by solving the Reynolds equation using a relaxation procedure, and integrating using Simpson's rule to determine the oil-film forces acting on the journal. The gravity load on each journal was that attributable to half the rotor mass, namely 4.37 kN. The starting position of the journal within the bearing was chosen arbitrarily at an eccentricity ratio of 0.7 and an attitude angle of 10°, and the subsequent journal motion examined at time intervals which enabled 50 points to be plotted per shaft revolution. Each of the Figures 6.5 corresponds to six shaft revolutions; the journal can be seen to orbit at approximately half the shaft rotational speed.

## 6.3 Resonant whip

Under certain operating conditions it is possible for a machine to exhibit the oil-whirl phenomenon and for the frequency of the subsequent vibrations to coincide with a system resonant frequency (usually the system fundamental frequency). In these circumstances very severe vibrations indeed are induced

INSTABILITY

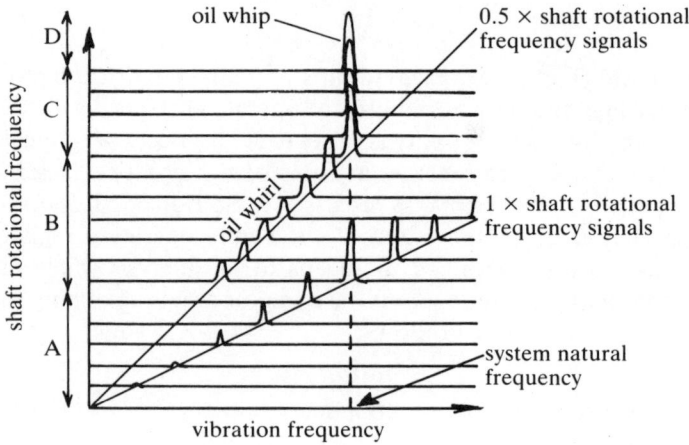

**Figure 6.6** Journal vibration frequency spectra showing oil whirl and oil whip. The signal level at a particular amplitude is indicated by peaks. The broken line indicates the system natural frequency.

and the situation is most grave. In effect the vibration associated with oil whirl combines with a system critical speed to produce most excessive vibrations at about half the shaft rotational frequency. Because this type of vibration is generally associated with the system fundamental frequency, it is likely to occur only when the shaft rotates at about twice the speed associated with the system fundamental frequency, when the bearings are functioning in an unstable operating regime.

The situation is represented by Figure 6.6, which indicates the frequency components of system vibration at various running speeds. In speed range A on the diagram there is no oil whirl present and the only significant vibration is that associated with out of balance, at $1 \times$ shaft rotational frequency. In speed range B, oil whirl is represented by the vibrations present at $\frac{1}{2} \times$ shaft rotational frequency. Speed range C represents the case when the oil-whirl vibrations correspond to the system resonant frequency and become of extremely high amplitude. At higher speeds still, range D on the diagram, the oil whirl may subside again leaving only vibrations associated with out-of-balance effects (including critical speeds).

## 6.4 Internal friction

Engineering materials show some resistance to their deformation which is a function of their rate of deformation. Such a material property may be represented as a damping force when modelling the material behaviour. Of greater significance than such material damping properties, however, is the

## INTERNAL FRICTION

damping effect produced by the friction forces between mating components on a shaft when the shaft deflects and the components move relative to each other. These forces are particularly significant where interference fit components are present. For example, Figure 6.7 indicates where compressive forces act on a shaft at the site of a component with an interference fit. As a consequence, when the shaft elongates at the location of the interface AA, friction forces will oppose the shaft deformation; these forces provide a hysteresis damping effect. A similar effect occurs as a result of friction forces opposing shaft material compression at interface BB.

The influence of the above-described friction forces on the behaviour of a rotating shaft has been studied by Tondl (1965). The analysis shown below is similar to Tondl's, where the friction damping effect is modelled as a force proportional to the rate of shaft deformation, as compared with the viscous damping present which is proportional to the absolute velocity of the rotor. Other contributions to the subject have been provided by Begg (1974) and by Zorzi and Nelson (1977). The system considered below is that shown in Figure 6.8, where the $u$ and $v$ axes are rotating. The equations of motion for the rotor of mass $m$ may then be written as

$$-ku - C_H \dot{u} - C_V(\dot{u} - \omega v) = m(\ddot{u} - 2\omega \dot{v} - \omega^2 u)$$
$$-kv - C_H \dot{v} - C_V(\dot{v} + \omega u) = m(\ddot{v} + 2\omega \dot{u} - \omega^2 v) \quad (6.14)$$

where $\omega$ is the whirl frequency, $k$ is the shaft stiffness, $C_V$ is the effective viscous damping coefficient and $C_H$ is the hysteresis damping coefficient. Equations (6.14) may be combined to give

$$-kw - C_H \dot{w} - C_V(\dot{w} + i\omega w) = m(\ddot{w} + 2i\omega \dot{w} - \omega^2 w) \quad (6.15)$$

where $w = u + iv$ is now a complex displacement. In general $w$ will vary with time according to

$$w = w_0 \exp(st) \quad (6.16)$$

which on differentiating and substituting into Equation (6.15) gives

$$ms^2 + (C_H + C_V)s + (k - m\omega^2) + 2i\omega ms + iC_V\omega = 0 \quad (6.17)$$

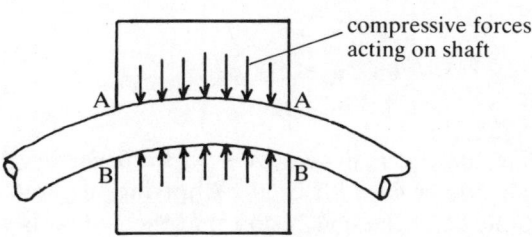

**Figure 6.7** Compressive forces acting on a built-up rotor indicating how hysteresis damping due to friction may occur.

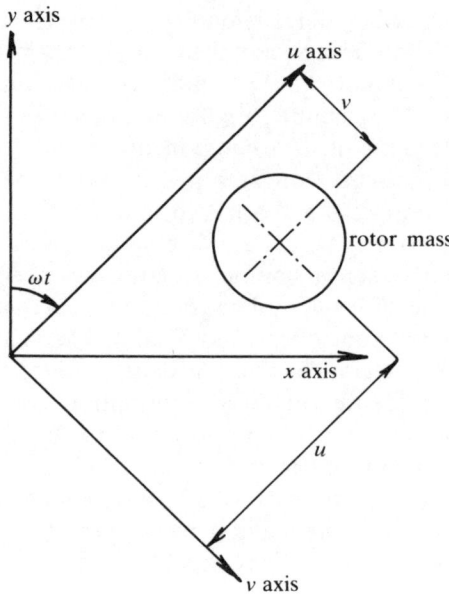

**Figure 6.8** General arrangement of a rotor described in terms of a rotating coordinate system. The $x$ and $y$ axes are stationary, the $u$ and $v$ axes are rotating.

which may be written in the form

$$ms^2 + (C_H + C_V + i2\omega m)s + (k - m\omega^2 + iC_V\omega) = 0 \qquad (6.18)$$

Equation (6.18) takes the form

$$(a_0 + ib_0)s^2 + (a_1 + ib_1)s + (a_2 + ib_2) = 0 \qquad (6.19)$$

for which the Routh–Hurwitz stability criteria, as used by Dimentberg (1961), give

$$-\begin{vmatrix} a_0 & a_1 \\ b_0 & b_1 \end{vmatrix} > 0$$

$$\begin{vmatrix} a_0 & a_1 & a_2 & 0 \\ b_0 & b_1 & b_2 & 0 \\ 0 & a_0 & a_1 & a_2 \\ 0 & b_0 & b_1 & b_2 \end{vmatrix} > 0 \qquad (6.20)$$

as the conditions for the real part of $s$ to be negative (that is, for the amplitude of $w$ to decrease with time). Applying the conditions given in Equations (6.20) leads to the conclusion that the system is stable provided that

$$C_H + C_V > 0, \qquad \omega < (1 + C_V/C_H)\omega_n \qquad (6.21)$$

where $\omega_n^2 = k/m$. The second of Equations (6.21) indicates that the system is always stable when operating below the critical speed $\omega_n$. If the hysteresis damping is large compared with the viscous damping then the system will be unstable at speeds greater than the critical speed. The presence of viscous damping, however, has the effect of raising the speed at which the system becomes unstable, so that if sufficient viscous damping is designed into the system then the instability threshold speed can be raised beyond the normal operating speed range of the machine.

EXAMPLE 6.3

A long rotor of mass 30 kg is mounted by means of a shrink fit onto a shaft whose lateral stiffness is 50 kN.m.$^{-1}$. When the shaft is stationary, an impulsive force is applied to the rotor and the subsequent transient vibrations indicate that the system has an effective damping ratio of $\zeta = 0.05$. The machine operates in an environment which is known to provide an effective viscous damping coefficient acting on the rotor of 100 N.s.m$^{-1}$; this value is independent of machine rotational speed. Calculate the speed at which the machine is likely to become unstable as a consequence of hysteresis damping.

SOLUTION

When the machine is stationary the overall effective damping coefficient is given by

$$C_{\text{eff}} = \zeta C_{\text{crit}} = \zeta 2(km)^{1/2} = 0.05 \times 2 \times (50 \times 10^3 \times 30)^{1/2} = 122 \text{ N.s.m}^{-1}$$

The transient vibrations recorded when the machine is stationary give rise to instantaneous absolute rotor velocities which are identical to the shaft deformation velocities. Therefore it follows that

$$C_H + C_V = 122 \text{ N.s.m}^{-1}$$

Substituting for $C_H$ into the second of Equations (6.21) gives the likely instability onset speed as

$$\omega = (1 + 100/22)(50 \times 10^3/30)^{1/2}$$
$$= 226 \text{ rad.s}^{-1}$$

## 6.5 Effect of rotor polar asymmetry

In many machines the lateral stiffness of the rotor is different in two orthogonal directions. For example, the rotor of an electrical motor or generator may have slots containing the electrical windings on some, but not all, parts of its surface, as shown in Figure 6.9. In this example the rotor stiffness for bending in the $y-y$ plane will be lower than that for bending in the $x-x$ plane because of the grouping of the winding slots from where material has been removed. In many cases this effect can be substantially offset by machining additional slots in the pole faces (in the case of electrical machinery) to compensate, thereby reducing the bending stiffness in the $x-x$ plane to a value similar to that for the $y-y$ plane. In some cases it may not be possible to guarantee that the rotor bending stiffness is constant for changes in the plane of bending, however, and in these instances it is important for the designer to be aware of the potential unstable machine operating regimes associated with the rotor polar asymmetry. Several researchers have contributed theoretical analyses of the subject; the analysis which follows is similar to those used by Tondl (1965) and by Rao (1983).

The equations of motion for the system under consideration may be developed as in section 6.4, with the exceptions that allowance must be made for different rotor bending stiffnesses in the $u$ and $v$ directions. If $k_u$ and $k_v$ are the rotor stiffnesses for bending in the $u$ and $v$ directions respectively (where $k_u < k_v$) then for the undamped system, if the effects of gravity and imbalance are ignored, the equations of motion may be written as

$$-k_u u = m(\ddot{u} - 2\omega \dot{v} - \omega^2 u), \qquad -k_v v = M(\ddot{v} + 2\omega \dot{u} - \omega^2 v) \quad (6.22)$$

The resulting motion of the system will be periodic and will take the form

$$u = u_0 \exp(st), \qquad v = v_0 \exp(st + \phi) \quad (6.23)$$

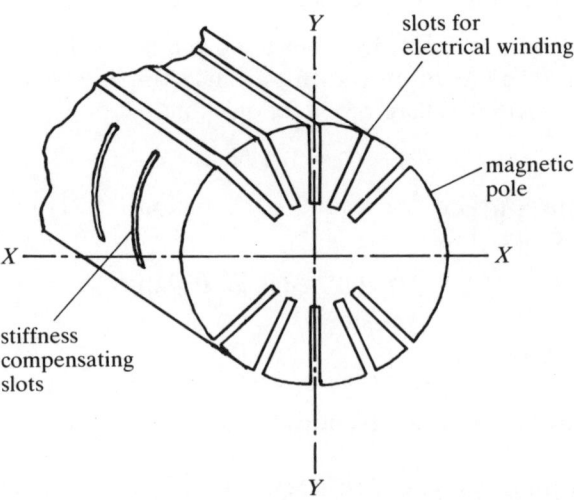

**Figure 6.9** Diagram showing how slots are machined into the rotor of an electrical machine to accommodate the electrical winding.

Dividing Equations (6.22) by $m$, and differentiating Equations (6.23) to obtain expressions for velocity and acceleration, substitution into Equations (6.22) gives

$$s^2 u_0 e^{st} - 2\omega s v_0 e^{st} e^{\phi} + (\omega_u^2 - \omega^2) u_0 e^{st} = 0$$
$$s^2 v_0 e^{st} e^{\phi} + 2\omega s u_0 e^{st} + (\omega_v^2 - \omega^2) v_0 e^{st} e^{\phi} = 0 \qquad (6.24)$$

where $\omega_u^2 = k_u/M$ and $\omega_v^2 = k_v/M$. Equations (6.24) may alternatively be written in matrix form as

$$\begin{bmatrix} (s^2 + \omega_u^2 - \omega^2) & (-2\omega s) \\ (2\omega s) & (s^2 + \omega_v^2 - \omega^2) \end{bmatrix} \begin{Bmatrix} u_0 e^{st} \\ v_0 e^{st} e^{\phi} \end{Bmatrix} = \begin{Bmatrix} 0 \\ 0 \end{Bmatrix} \qquad (6.25)$$

or more simply as

$$[C]\{X\} = \{0\} \qquad (6.26)$$

for which the values of $s$ for non-trivial solutions are given by equating the determinant of the matrix $[C]$ to zero. That is

$$s^4 + (\omega_u^2 + \omega_v^2 + 2\omega^2) s^2 + (\omega_u^2 - \omega^2)(\omega_v^2 - \omega^2) = 0 \qquad (6.27)$$

Equation (6.27) may be solved to determine the roots of $s$ for various running speeds $\omega$, whereupon stability will be indicated by the real part of $s$ being negative, thereby indicating decreasing values of $u$ and $v$ as time progresses, as discussed in the preceding sections. Alternatively, stability may be investigated using the Routh–Hurwitz criteria for a polynomial of degree 4. With either of these approaches it is found that there is a potential region of instability defined by

$$\omega_u < \omega < \omega_v \qquad (6.28)$$

In practice, there may be sufficient damping in the system to inhibit unstable vibrations.

The preceding analysis is based upon the assumption that the shaft vibration frequency $\omega$ corresponds to the machine running speed. This is a satisfactory assumption since in most cases the predominant vibration frequency component is that associated with machine imbalance. However, in the case of rotors with stiffness polar asymmetry which are mounted horizontally there is a component of vibration frequency at twice machine running speed. The reason for this may be understood by considering the cross-sections through the horizontal rotor shown in Figure 6.10. For the rotor position in Figure 6.10(a) the section second moment of area about the $x$–$x$ axis is greater than that when the rotor is in the position shown in Figure 6.10(b). For this reason there will be a greater sag of the rotor due to gravity for the latter rotor position as compared with the former. Since the major and minor axes of the rotor section change orientation twice per revolution, there will be a strong rotor vibration frequency component at twice machine running speed. For this reason there is an unstable machine

INSTABILITY

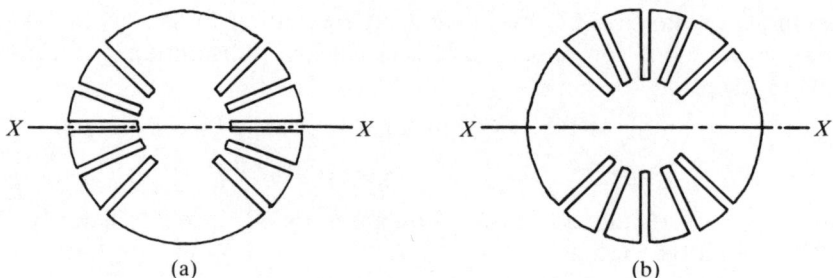

**Figure 6.10** Cross-sections of a rotor with winding slots, (a) initial position; (b) after rotation through 90°.

operating frequency range, for the horizontally mounted rotor, defined by

$$\omega_u < 2\omega < \omega_v \qquad (6.29)$$

Other publications dealing with instability caused by shaft and rotor polar asymmetry are those by Taylor (1940) and Yamamoto and Ota (1964).

EXAMPLE 6.4

Figure 6.11 shows the cross-section of a horizontal rotor of a turbo-generator which operates well above its fundamental critical speed. The machine fundamental frequency for shaft oscillations in the plane of least bending stiffness (plane $x-x$) is 12 Hz. It is proposed to machine slots in the rotor pole face in order to reduce the bending stiffness in plane $y-y$ to the same value as that for plane $x-x$. If the resultant bending stiffness in plane $y-y$ is approximately proportional to $d^4$, determine whether there is likely

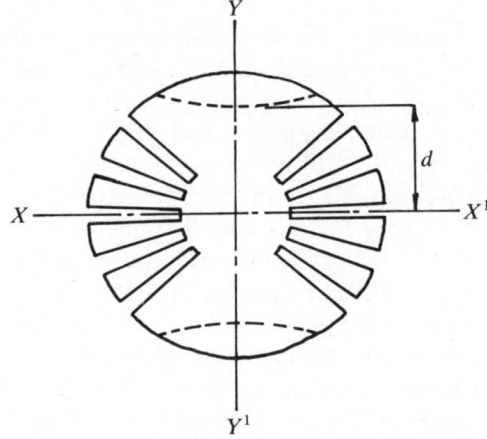

**Figure 6.11** Diagram showing how slits may be machined into the pole faces of the rotors of electrical machines to equalize the shaft bending stiffnesses in orthogonal directions.

to be an unstable operating speed range of the machine if there is an error of 0.5% in the machining of the distance $d$.

SOLUTION

From Equation (6.29) the unstable operating speed range is given by

$$\omega_{xx} < 2\omega < \omega_{yy}$$

Ideally $\omega_{xx} = \omega_{yy} = 12$ Hz. If the distance $d$ is machined oversize by 0.5%, then the corresponding bending stiffness is increased to $1.005^4 = 1.02$ times the required value. The corresponding natural frequency is increased to $1.02^{1/2} = 1.01$ times the required value (since the natural frequency is proportional to $k^{1/2}$). Such an error in the machining of distance $d$ would therefore give rise to an unstable operating speed range given by

$$12 < 2\omega < 12 \times 1.01 \text{ Hz}$$

that is

$$360 < \omega < 364 \text{ rev.min}^{-1}$$

Such an unstable operating speed range would be likely to go undetected since the machine would normally pass through such a speed band relatively quickly during run-up and run-down.

## 6.6 Instability due to aerodynamic forces

In Chapter 3 the reader was introduced to the concept of aerodynamic forces generated by shaft motion, which act upon the system and affect its dynamic behaviour. These forces are generally located at seals or at a turbine blade stage when present. One of the most important characteristics of such aerodynamic forces is that the direction in which the force is generated is frequently perpendicular to that of the shaft displacement, so that additional cross-coupling stiffness coefficients (similar to those used to describe hydrodynamic bearings) are in effect present, since the force magnitude is approximately proportional to the shaft displacement. These cross-coupling coefficients are generally of opposite sign (as is the case with $Kxy$ and $Kyx$ in the case of fluid film hydrodynamic bearings) and so may give rise to unstable whirling of the rotor.

Such aerodynamic effects have been noted with aeroengine compressor shafts and with steam turbines. In each case large vibrations of the shaft were recorded and were found to occur at frequencies corresponding to the fundamental critical speed of the system, rather than the machine running speed. Alford (1965) reports that the vibrations were found to occur at about $0.35$–$0.5 \times$ the machine running speed. The solution to the whirl

## INSTABILITY

problem related to aerodynamic effects is generally to stiffen the shaft and seal assembly, thereby reducing the likely shaft deflection and corresponding misalignment at the seals, and if necessary to change the bearings to ones which provide more viscous damping.

Theoretical stability analyses can be carried out which allow for aerodynamic cross-coupling provided that the effective cross-coupling coefficients are known. Such analyses normally take the form of those described in section 6.2, with the exception that the seal or turbine which has to be allowed for is represented as another bearing stage, with only cross-coupling stiffness coefficients, as described in section 3.8. The difficulty associated with such theoretical analyses is that very little is known

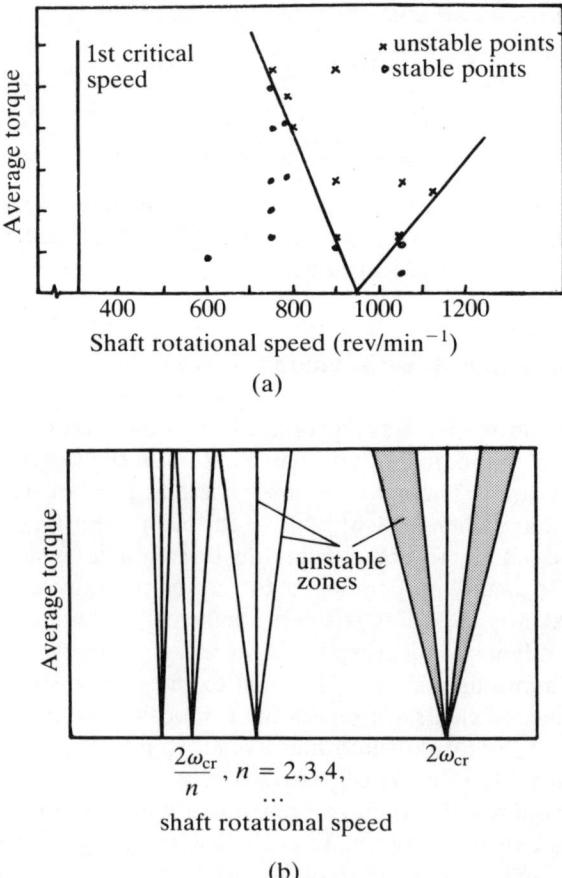

**Figure 6.12** Diagrams showing the forms of the unstable operating regimes on rotors subjected to pulsating torques, (a) experimental results (Eshleman and Eubanks, 1970); (b) theoretical results (Wehrli, 1963).

about the magnitude of the aerodynamic cross-coupling in real machines; however, some experimental measurements have been reported by Vance and Laudadio (1984).

## 6.7 Effects of a pulsating torque

Another unstable mode of operating rotating machinery, which relates to pulsating torques being applied to the shaft, has been reported by Wehrli (1963), and later by Eshleman and Eubanks (1970). The application of a torque to a shaft affects its natural frequency of lateral vibration, and when a pulsating torque is applied this has the effect of exciting the shaft nominal critical speed. The consequence is that for particular values of steady and pulsating torques the shaft vibrates at a frequency corresponding to its critical speed; this effect is independent of the torsional forcing frequency but occurs at particular ranges of the ratio of machine running speed/critical speed. The phenomenon is not completely understood as yet; some unstable zones of operation have been observed experimentally by Eshleman and Eubanks (1970) but, although these agree in general form with those predicted by Wehrli, the exact location of the unstable operating regime for their machine did not agree with the Wehrli prediction (see Figure 6.12).

## References

Akers. A., S. Michaelson & A. Cameron 1971. Stability contours for a whirling finite bearing. *Trans. ASME, J. Lub. Tech.*, **93**, January, 177–190.

Alford, J. S. 1965. Protecting turbomachinery from self-excited rotor whirl. *J. Eng. Power.* October, 333–344.

Bannister, R. H. & J. Makdissy 1980. The effects of unbalance on stability and its influence on non-synchronous whirling. *MechE. Conf. on Vibrations in Rotating Machinery*, Cambridge, Paper C310/80.

Barrett, L. E., E. J. Gunter & P. E. Allaire 1978. Optimum bearing and support damping for unbalance response and stability of rotating machinery. *Trans. ASME, J. Eng. Power*, January, 89–94.

Begg, I. C. 1974. Friction induced rotor whirl – a study in stability. *Trans. ASME, J. Eng. Indus.* 96B, **2**, 450–454.

Cameron, A. 1981. *Basic lubrication theory.* New York: Wiley.

Collar, A. R. & A. Simpson 1987. *Matrices and engineering dynamics.* New York: Wiley.

Dimentberg, F. M. 1961. *Flexural vibration of rotating shafts.* London: Butterworths.

Eshleman, R. L. & R. A. Eubanks 1970. Effects of axial torque on rotor response: an experimental investigation. *ASME Winter Annual Meeting*, New York. 29 Nov.–3 Dec. Paper no. 70-WA/DE-14.

Goodwin, M. J., M. P. Roach & J. E. T. Penny 1988. An analysis of combined squeeze-film and variable stiffness hydrostatic bearings, and their use in aircraft engine vibration control.

*Proc. IMechE. Int. Conf. Vibrations in rotating machinery.* Heriot-Watt Univ., Sept. Edinburgh. Paper C282/88 85–92.

Goodwin, M. J., M. P. Roach & J. E. T. Penny 1989. Stability of rotating machines running in variable stiffness bearings. *Proc. NATO/ASI Int. Conf. Vibration and wear damage in high speed rotating machinery*, Lisbon, April.

Kirk, R. G. & E. J. Gunter 1974. Transient response of rotor-bearing systems. *Trans. ASME, J. Eng. Ind.* **96** B, 682–693.

Lund, J. W. 1965. Computer programs for unbalance response and stability. *Rotor Bearing Dynamics Design Tech.* Part V, Mechanical Technology Inc., Report no. AFAPL-Tr-65-45.

Lund, J. W. 1974. Stability and damped critical speeds of a flexible rotor in fluid-film bearings. *Trans. ASME, J. Eng. Ind.* **96** B, 509–517.

Murphy, B. T. & J. M. Vance 1983. An improved method for calculating critical speeds and rotordynamic stability of turbomachinery. *Trans. ASME, J. Eng. Power* **105**, July, 591–595.

Newkirk, B. L. & H. D. Taylor 1925. Oil film whirl – an investigation of disturbances on oil films in journal bearings. *General Electric Review* **28**, 559–568.

Ogrodnik, P. J., Goodwin, M. J. & J. E. T. Penny 1985. The influence of design parameters on the occurrence of oil whirl in rotor-bearing systems. *Proc. Symp. Instability in Rotating Machinery*, NASA publication 2409, December, 135–144.

Ogrodnik, P. J., Goodwin, M. J. & J. E. T. Penny 1989. The effect of translational and conical bearing misalignment on the response and stability of a non-linear rotor bearing system. Poc. NATO/ASI Int. Conf. *Vibration and Wear Damage in High Speed Rotating Machinery.* Lisbon. April.

Rao, J. S. 1983. *Rotor dynamics.* New Delhi: Wiley Eastern.

Taylor, H. D. 1940. Critical speed behaviour of unsymmetrical shafts. *J. Appl. Mech.* June, 71–79.

Tondl, A. 1965. *Some problems in rotor dynamics.* London: Chapman and Hall.

Vance, J. M. & F. J. Laudadio 1984. Experimental measurement of Alford's force in axial flow turbomachinery. *J. Eng. Gas Turbines and Power* **106**, 585–590.

Wehrli, C. von 1963. Uber Kritische drehzahlhen unter pulsierender torsion. *Ingenieur Archiv.*, XXXIII band.

Yamamoto, T. & H. Ota 1964. On the unstable vibrations of a shaft carrying an unsymmetrical rotor. *J. Appl. Mech.* September, 515–522.

Zorzi, E. S. & H. D. Nelson 1977. Finite element simulation of rotor bearing systems with internal damping. *J. Eng. Ind.*m **99**, 71ff.

# 7 Balancing

## 7.1 Introduction

Imbalance is one of the most important disturbing forces in rotating machinery. When a machine is not properly balanced, high levels of vibration, noise and wear are generally evident; frequently there is also a reduction in the machine fatigue life. For these reasons most rotating machinery is balanced as part of the manufacturing process, before the machine is released to the customer.

Imbalance forces are set up by the rotor mass centre being eccentric to the shaft centre of rotation, so that centrifugal forces which act on the rotor mass are generated. For simple rotors with only one mass the correction procedure is straightforward, and is carried out by merely adding one balance weight to the rotor at the correct angular orientation. In the case of long flexible rotors, however, the procedure is more complicated because the distribution of the imbalance along the length of the shaft, and the shaft vibration mode shape, both affect the positions at which balance weights need to be added to achieve balance.

The aims of this chapter are to introduce the reader to the fundamental concepts of machine balancing, and to illustrate how they can be applied in practice. The first part of the chapter discusses the differences between static and dynamic imbalance, and shows how each can be corrected in the case of a rigid rotor system where the imbalance is known. The following section shows how to approach the balancing of a rigid shaft when the initial balance is not known. Two approaches for use when dealing with long flexible rotors, where the initial imbalance is not known, are subsequently described; these are the 'modal' and 'influence coefficient' balancing techniques. The last section of the chapter provides the reader with information on suitable balance tolerances applicable to different classes of machinery.

## 7.2 Rigid rotors whose initial imbalance is known

A rigid rotor is one which does not bend by an amount enough to cause a significant change in the centre of mass eccentricity from the axis of rotation during operation of the machine; it is, however, acceptable for the rotor mass to become eccentric from the axis of rotation as a consequence of deflection of the bearings and their foundations, and for the rotor still to be classified as rigid. In practice this means that rigid rotor machines are those which operate well below their pin–pin critical speed (the critical speed corresponding to the rotor mounted in bearings which behave as pinned supports). Rigid rotor machines can be balanced by adding balance weights to the machine at any two (or more) balance planes.

To balance a machine properly, the balancer must achieve both static and dynamic balance. Static balance of the rotor has been obtained if the rotor, when mounted on knife-edge supports, does not come to rest in any preferred angular orientation. This means that there must be no net moment, due to gravity forces on the masses, causing rotation about the shaft axis. For the system shown in Figure 7.1(a) this means that

$$m_1 g r_1 \cos \theta_1 + m_2 g r_2 \cos \theta_2 + m_3 g r_3 \cos \theta_3 + \cdots = 0 \quad (7.1)$$

or

$$\Sigma mr \cos \theta = 0 \quad (7.2)$$

If the rotor were now considered to be placed at an angular orientation at 90° to that shown in Figure 7.1(a) such that the imbalance masses were in the positions shown in Figure 7.1(b), then clearly for there to be no net moment about the shaft axis,

$$\Sigma mr \sin \theta = 0 \quad (7.3)$$

Since static balance is a condition which is independent of the assumed angular orientation of the rotor, Equations (7.2) and (7.3) must both be satisfied for the system to be in a state of static balance.

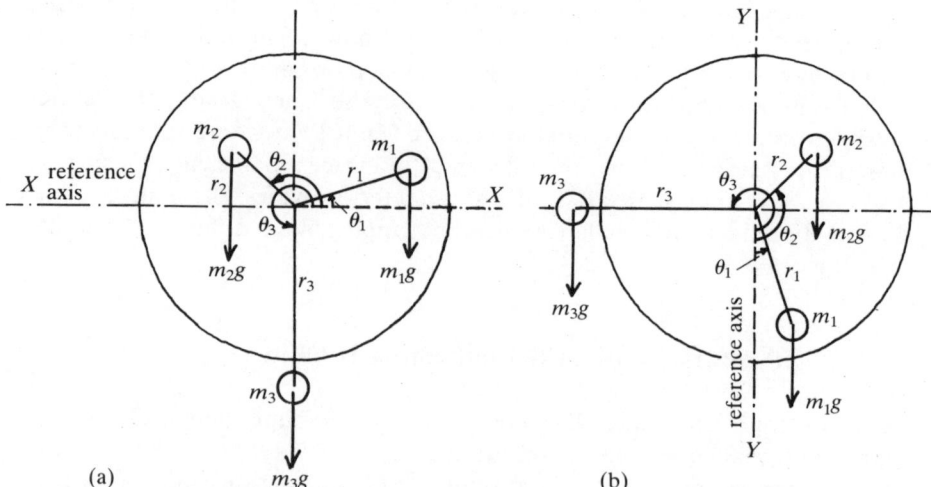

**Figure 7.1** Gravity forces acting on rotor masses, (a) application of gravity forces; (b) same forces acting when shaft is rotated through 90°.

## KNOWN IMBALANCE

Equations (7.2) and (7.3) are satisfied by closure of the '*mr* polygon', which is constructed by adding together vectors whose magnitude is given by the product of the imbalance mass which they represent and the eccentricity of the mass. Each vector is drawn in the direction indicated by the angular orientation of the imbalance mass on the system. For the system as shown in Figure 7.1(a) the *mr* polygon is constructed as shown in Figure 7.2, where the vectors are drawn to an arbitrary scale. It is noteworthy that had the polygon been drawn for the system orientation shown in Figure 7.1(b) then it would take the same form but would be rotated through $90°$. In Figure 7.2 it can be seen that when the *mr* polygon is drawn for the three masses considered, the polygon is not closed. The amount of residual static imbalance in the system is given by the vector $OA$ (broken line). To obtain static balance, an additional net balance weight needs to be added to the system at the angular position indicated by the vector $AO$, that is at an angle $\theta_4$ to the reference position. The magnitude of the balance weight to be added must be such that the product of its mass and eccentricity, that is its *mr* value, is represented by the length of the vector $AO$.

A system is said to be in a state of dynamic balance when the resultant of all centrifugal forces acting on the rotating system is zero *and* the centrifugal forces do not give rise to any couple acting on the machine. Centrifugal force is given by $mr\omega^2$ and acts radially outward in the direction

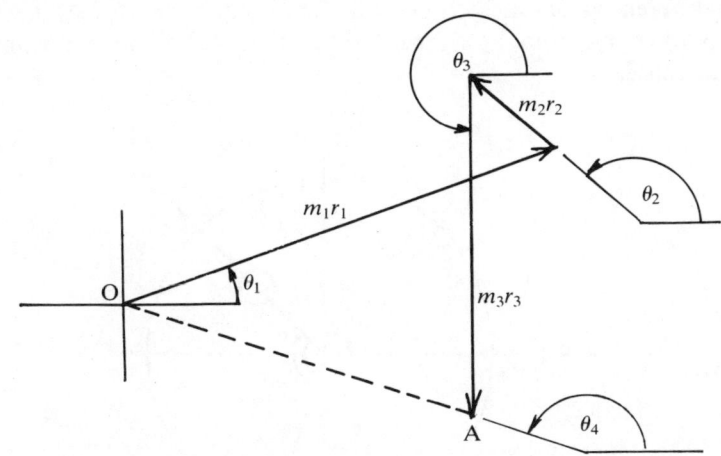

| mass | m/kg | r/m | mr/kg. m |
|------|------|-----|----------|
| 1 | 10 | 0.2 | 2 |
| 2 | 5 | 0.1 | 0.5 |
| 3 | 5 | 0.3 | 1.5 |

**Figure 7.2** Construction of an *mr* polygon. See text for details.

BALANCING

of the mass eccentricity (see Figure 7.3). In general there is no net centrifugal force acting on the rotor when the following hold:

$$\Sigma mr\omega^2 \sin \theta = 0, \qquad \Sigma mr\omega^2 \cos \theta = 0 \qquad (7.4)$$

Since $\omega^2$ is the same for all masses attached to the rotor, Equations (7.4) degenerate to Equations (7.2) and (7.3) and are again satisfied by the closure of the $mr$ polygon.

Centrifugal forces can give rise to couples acting on the rotor as indicated in Figure 7.4 where the rotor masses are mounted such that they have similar mass and eccentricity values but are eccentric in the opposite direction to each other. The system is therefore one in which the resultant centrifugal force acting on the rotor is zero (and so is in a state of static balance since the corresponding $mr$ vectors are equal and opposite) but is still not in a state of dynamic balance since, because centrifugal forces $mr\omega^2$ are set up as shown in the diagram when the system rotates, a couple $mr\omega^2 d$ acts on the rotor. If the machine is run in this condition, the rotor is not caused to 'overturn' by the imbalance couple because it is held in bearings which provide reaction forces $F_b$, as shown in the diagram. The bearing forces are such that they provide a couple $F_b l$ on the rotor, which is equal and opposite to that caused by the imbalance forces. Similar centrifugal and bearing reaction forces would, of course, be set up if the machine were also in a state of static imbalance, although in general the reaction forces would take on different values at each bearing. The couples and forces described above do, of course, rotate with the shaft, so that they do not act in any one particular sense.

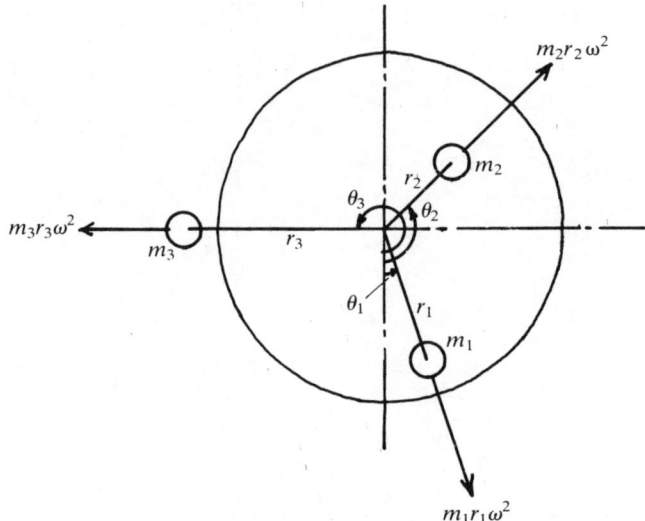

**Figure 7.3** Centrifugal forces acting on rotor masses.

## KNOWN IMBALANCE

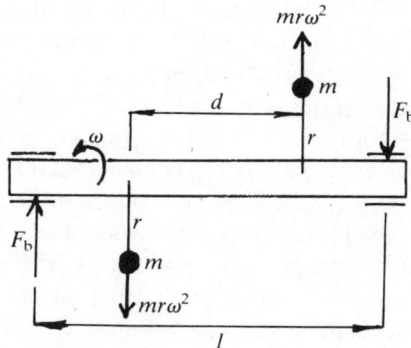

**Figure 7.4** Centrifugal forces generating an out-of-balance couple.

The magnitude of the couple produced by an imbalance force about some point is given by the product of the force and the distance between the line of action of the force and the point under consideration. In the case of the rotor shown in Figure 7.1a, shown from a different view in Figure 7.5, the couple produced in the $x$–$x$ plane about point $P$, by the centrifugal force acting on $m_1$, is $m_1 r_1 \omega^2 x_1 \cos \theta_1$. In general there is no net couple in the $x$–$x$ plane about point $P$ when

$$m_1 r_1 \omega^2 x_1 \cos \theta_1 + m_2 r_2 \omega^2 x_2 \cos \theta_2 + m_3 r_3 \omega^2 x_3 \cos \theta_3 + \cdots = 0 \quad (7.5)$$

that is, when

$$\Sigma mrx \cos \theta = 0 \quad (7.6)$$

Simiarly, it may be argued that there is no net couple in the $y$–$y$ plane about point $P$ provided that

$$\Sigma mrx \sin \theta = 0 \quad (7.7)$$

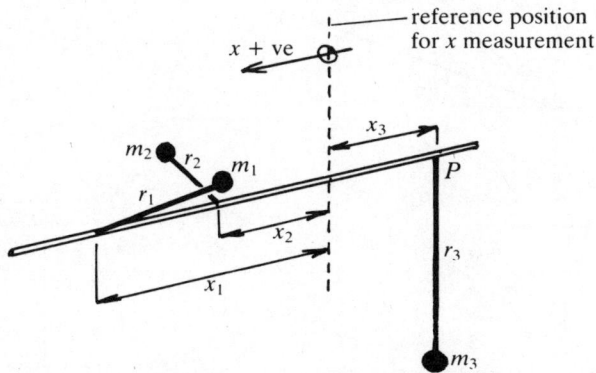

**Figure 7.5** A system with several rotor masses spaced axially.

Equations (7.6) and (7.7) are satisfied by closure of the *mrx* polygon for the system. The *mrx* polygon is constructed in the same way as the *mr* polygon, except that the magnitude of the vector representing a particular mass is given by the product *mrx* instead of simply *mr*. The angular orientation of the vectors, when drawn, is again given by the angular orientation of the imbalance mass being considered. The reference position from which $x$ values are measured may be chosen quite arbitrarily, as is the direction for positive $x$ values. When the measurement of $x$ is negative for an imbalance mass, the corresponding *mrx* vector is drawn in the opposite direction to that indicated by the angular orientation of the mass under consideration. For the system shown in Figure 7.5 the *mrx* polygon is constructed as shown in Figure 7.6; an arbitrary scale may again be chosen when drawing the vectors. It can be seen that the polygon does not close, the dynamic imbalance in the system being represented by the broken vector OB which when multiplied by $\omega^2$ gives the magnitude of the dynamic couple acting on the system. To obtain dynamic balance the *mrx* polygon must be closed by the addition of balance weights to the rotor which will be represented by the vector BO on the *mrx* polygon.

When balancing a machine, to obtain both static and dynamic balance, it is, of course, imperative that balance weights are added to the machine in such a way that when the *mrx* polygon is closed this does not upset closure

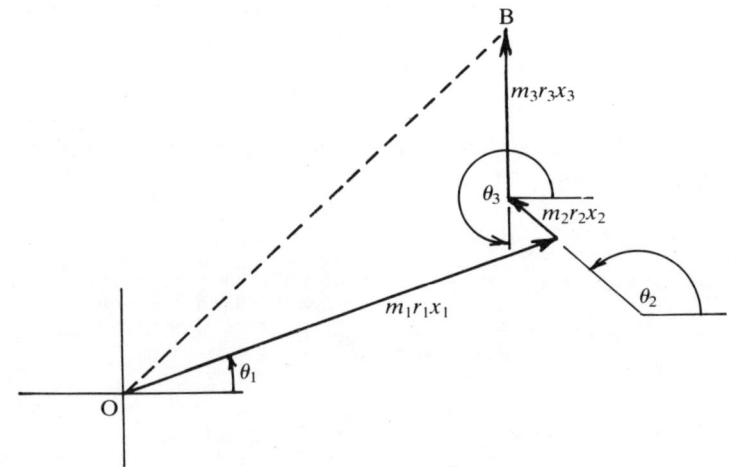

| mass | m/kg | r/m | x/m | mrx/kg. m² |
|------|------|-----|-----|------------|
| 1 | 10 | 0·2 | 2 | 4 |
| 2 | 5 | 0·1 | 1 | 0·5 |
| 3 | 5 | 0·3 | −1 | −1·5 |

**Figure 7.6** Construction of an *mrx* polygon. See text for details.

of the *mr* polygon, and vice versa. One way of ensuring that this is so is by adopting a procedure known as Dalby's method. The reference position, from which $x$ values are measured, is chosen to coincide with one of the planes at which balance weights are to be added (there must always be at least two planes at which balance weights are to be added if both static and dynamic balance are to be obtained). The *mrx* polygon is then the first to be closed, by the addition of balance weights at the other balancing plane. It is then possible to close the *mr* polygon by adding balance weights at the first balance plane (the reference position for $x$ measurement) and thereby achieve static balance of the machine; note that by adding balance weights at the reference plane to obtain static balance, closure of the *mrx* polygon is not upset because $x$ is zero for mass added at the reference plane, and so the *mrx* value for the mass is zero and the *mrx* polygon remains unchanged.

From the foregoing discussion the reader should understand that the presence of dynamic balance automatically implies static balance, but that static balance may be obtained without dynamic balance. (Dynamic balance requires closures of both the *mr* and *mrx* polygons, whilst static balance requires closure of only the *mr* polygon.)

The discussion so far has related to systems in which the initial distribution of imbalance is known. This might be the case for machines where the rotors are deliberately mounted eccentric to the shaft, or for example in the case of cranks on crankshafts. In these situations the methods described above may be used to obtain some nominal balance of the machine. Invariably, in all machines, there will remain some out-of-balance associated with irregular machining, anisotropic material properties, hand-built assemblies and other factors, which is not known at the design stage. To remove this out-of-balance it is necessary to resort to a different procedure; some possible alternatives are described in the following sections.

EXAMPLE 7.1

Determine the masses to be added at planes L and M at radii of 600 mm if the system shown in Figure 7.7(a) is to be dynamically balanced.

SOLUTION

Choosing the reference position for $x$ measurement at plane $L$, the *mrx* polygon may first be constructed as shown in Figure 7.7(b). The required *mrx* vector at plane M for balance is then found to be 165 kg.m$^2$. The balance mass required at M is therefore given by

$$m = \frac{mrx}{rx} = \frac{165}{0.6 \times 1.4} = 196 \text{ kg at } 47° \text{ from A (and } 2° \text{ from B)}$$

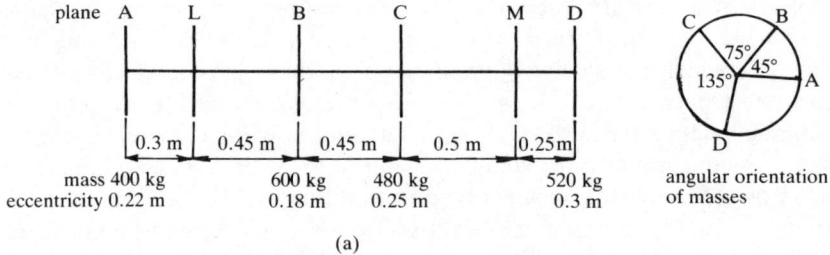

(a)

| plane | mass/kg | eccentricity/m | x/m | mr/kg. m | mrx/kg. m² |
|---|---|---|---|---|---|
| A | 400 | 0·22 | −0·3 | 88 | −26·4 |
| B | 600 | 0·18 | 0·45 | 108 | 48·6 |
| C | 480 | 0·25 | 0·9 | 120 | 108 |
| D | 520 | 0·3 | 1·65 | 156 | 257·4 |
| M | | | 1·4 | | |

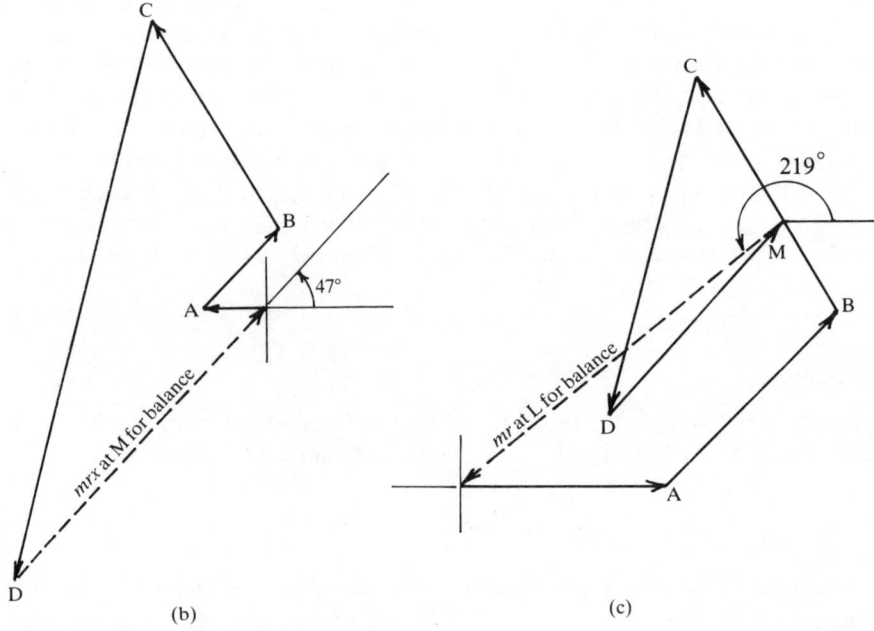

**Figure 7.7** Balancing a machine using two balance planes (Example 7.1), (a) unbalanced rotating mass system. The masses and their eccentricities are as shown in the table; (b) the *mrx* polygon; (c) the *mr* polygon.

The *mr* vector at plane M may now be evaluated as

$$mr = 196 \times 0.6 = 118 \text{ kg.m}$$

and so the *mr* polygon may be drawn as shown in Figure 7.7(c). The required vector for plane L, to produce balance, is then found to be 185 kg.m, so that the balance mass required at L is given by

$$m = \frac{mr}{r} = \frac{185}{0.6} = 308 \text{ kg at } 219° \text{ from A}$$

## 7.3 Rigid rotors with an unknown imbalance

In machines which have been manufactured with rotors mounted nominally concentric with their axis of rotation there will generally remain a residual amount of imbalance in the machine, after manufacture has been completed, caused by the factors discussed in the preceding section. This residual imbalance in the machine is generally present irrespective of the machine design features, and the designer will not know its magnitude or its location. Most machines, however, will incorporate in their design planes where balance weights may be added (or from where material may be removed) to effect a balance. The problem is to establish the correct magnitude and location of the balance weights when the imbalance is not known.

The approach to machine balancing in this situation is to run the machine and to take measurements of rotor vibration; normally measurements are taken at sites close to the bearings (alternatively, some special balancing machines make use of measurements of force transmitted through the bearings). Measurements must be taken of vibration amplitude and of phase angle relative to some reference position on the shaft (see also section 9.2). The measurements of vibration may be plotted as vectors using polar coordinates as shown in Figure 7.8, where phase lag behind the reference position on the shaft is plotted in the opposite direction to shaft rotation; with this approach each phase position on the diagram represents an actual angular position on the shaft. In the diagram the vectors OA and OB represent the vibrations measured at the two ends of the machine. These vibrations will, in general, be present as a result of the combined effects of both static and dynamic imbalance which normally will both be present. The effects of each of these types of imbalance are separated from each other, and then treated in isolation for the purposes of simplifying the balancing process. The method outlined below is similar to that described by Langlois and Rosecky (1968).

It will be understood by the reader, from the preceding section, that static imbalance forces the 'first mode' of vibration; that is, the type of vibration

BALANCING

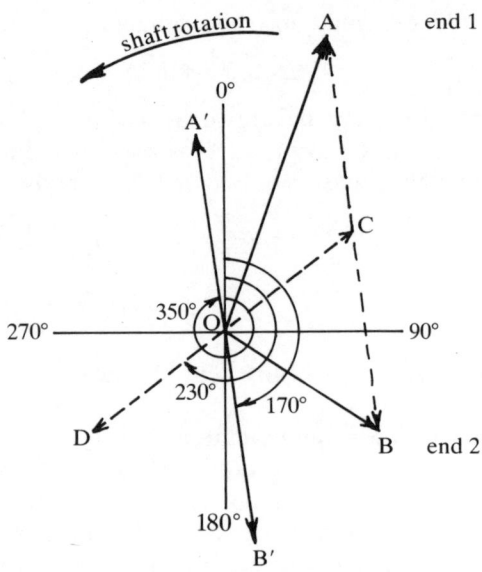

**Figure 7.8** Vibration vectors measured at each end of a shaft.

where both ends of the shaft tend to throw out in the same direction. Dynamic imbalance, on the other hand, forces the 'second mode' of vibration where the shaft ends tend to throw out in opposite directions to each other. The vibration vectors OA and OB represent the vector sum of the components of vibration associated with both static and dynamic imbalance, measured at the respective ends of the shaft. If, in Figure 7.8, C is now constructed as being the mid-point of AB, then CA and CB are the components of vibration related to dynamic imbalance at end 1 and end 2 respectively (equal in magnitude but opposite in direction), whilst *OC* is the vibration vector at each end of the shaft related to static imbalance (equal and in the same direction).

To remove the static imbalance from the machine, balance weights must be added to produce an additional vibration vector at each end of the machine which will cancel out the vector OC. This means that mass must be added which will force the rotor in the direction of the vector OD, equal and opposite to OC. On first considering the matter it might be assumed that this means that mass should be added to the rotor at an angle of 230° from the reference position. However, it should be remembered from elementary vibration studies that there is in general a phase lag angle of displacement behind force. For this reason the system phase lag angle for the first mode of vibration at the balancing speed under consideration must be known (if it is not known then it may be obtained from the Bode diagram for the system, which may be determined experimentally). If the machine is being run below its first critical speed (that corresponding to the first mode of vibration) the phase lag angle of shaft displacement behind imbalance force might be $30°$, for example. This would mean that balance weights must be added at both ends of the machine at an angle of $200°$ from the reference position. The amount of mass to be added will generally be determined by a combination of experience, and trial and error, and must be the same at each end of the machine to avoid introducing further dynamic imbalance. When the correct amount of mass has been added, point C in Figure 7.8 will become coincident with point O, and points A and B will move to positions A' and B' respectively.

The procedure for removing the dynamic imbalance in the machine is the same as that for removing the static imbalance, except that balance weights must be added $180°$ opposed to each other at the two ends of the machine. At end 1 mass must be added to produce a vibration vector in the direction of the vector OB' on the diagram, thereby cancelling out the dynamic imbalance represented by the vector OA'. The converse is equally applicable to remove the effects of dynamic imbalance as recorded at end 2. Once again the phase lag angle for the system must be accounted for; this time it is the phase lag angle corresponding to excitation of the second mode of vibration (once again, the phase lag angle may be found experimentally if necessary). If, for example, the phase lag angle was $10°$ for the second mode of vibration, then mass would have to be added at $160°$ at end 1, and an equal amount at $340°$ at end 2.

Sometimes an alternative approach to that described above is prefered in the case of rigid rotors. This approach, described by Rieger (1982) and by Rao (1983) involves the application of trial masses to the machine in order to determine the system influence coefficients experimentally. These can then be used to calculate the precise magnitudes and locations of balance weights to be added at each end of the machine when the initial imbalance vibration has been recorded. For example, in Figure 7.9 the vectors OA and OB represent the initial system vibration measurements, and the vectors OA' and OB' those measured when a trial mass $m_1$ has been added at end 1. It is then clear that the vectors AA' and BB' represent the vibrations introduced as a consequence of the trial mass addition. The influence coefficients, relating the vibrations introduced by the trial mass to the magnitude and location of the mass, are then given by

$$a_{11} = \text{AA}'/m_1$$
$$a_{21} = \text{BB}'/m_1$$
(7.8)

where all terms are written as complex numbers because they represent vectors (since both the vibration vectors and the trial mass position have specific angular orientations). Similarly, the influence coefficients relating

## BALANCING

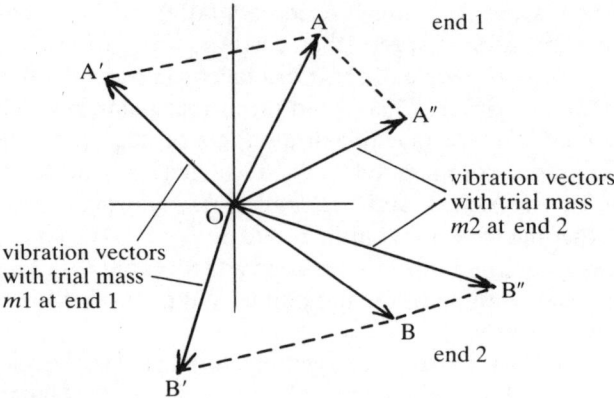

**Figure 7.9** Effect of trial mass addition on vibration vectors. **OA**, **OB** are the initial vibration vectors without any trial mass.

the addition of mass $m_2$ at end 2 to machine vibration are given by

$$a_{12} = AA''/m_2$$
$$a_{22} = BB''/m_2$$
(7.9)

where the terms are, again, complex. In general the vibration measured at end 1 (that is, in addition to the initial imbalance vibration OA) will be that caused by the addition of balance weights to end 1, plus that due to the addition of balance weights at end 2. If this vibration vector is $V_1$, due to the addition of masses $M_1$ and $M_2$ at end 1 and end 2 respectively, then

$$V_1 = a_{11}M_1 + a_{12}M_2 \tag{7.10}$$

where $V_1$, $M_1$ and $M_2$ are again complex terms. Similarly for end 2 we may write

$$V_2 = a_{21}M_1 + a_{22}M_2 \tag{7.11}$$

which may be combined with Equation (7.10) to give the matrix equation

$$\begin{Bmatrix} V_1 \\ V_2 \end{Bmatrix} = \begin{bmatrix} a_{11} & a_{12} \\ a_{21} & a_{22} \end{bmatrix} \begin{Bmatrix} M_1 \\ M_2 \end{Bmatrix} \tag{7.12}$$

which may be restated as

$$\begin{Bmatrix} M_1 \\ M_2 \end{Bmatrix} = \begin{bmatrix} a_{11} & a_{12} \\ a_{21} & a_{22} \end{bmatrix}^{-1} \begin{Bmatrix} V_1 \\ V_2 \end{Bmatrix} \tag{7.13}$$

Clearly, to balance the original system it is necessary to introduce additional vibration vectors AO and BO at end 1 and end 2 respectively. If these values of $V_1$ and $V_2$ are substituted into Equation (7.13), then the correct magnitudes and positions of the required balance weights $M_1$ and $M_2$ can be found.

UNKNOWN IMBALANCE

Further information concerning the balancing of rigid rotors may be obtained from Blake (1967), McQueary (1973) and Rieger (1981).

EXAMPLE 7.2

A large electric motor running at its normal operating speed, which is well below its first pin–pin critical speed, has the following shaft vibration measurements recorded at its bearings: 35 μm at 10° at bearing 1, and 50 μm at 100° at bearing 2. A trial balance mass of 0.05 kg is attached at end 1 of the rotor at the 0° (phase reference) position on the rotor, and new vibration measurements are found to be as follows: 68 μm at 11° at bearing 1, and 58 μm at 77° at bearing 2. Determine the additional balance masses to be added at each end of the machine to achieve both static and dynamic balance.

SOLUTION

The initial vibration vectors may be recorded as shown in Figure 7.10(a). Proceeding as described in section 7.3 enables the vibration associated with static imbalance to be calculated as 30 μm at 65°, in the first instance (vector OS). After addition of the trial mass the vibration associated with static imbalance is 53 μm at 41° (vector OS'). The change of static balance introduced by the addition of the trial balance mass is equivalent to a vibration vector of 28 μm at 15° (vector SS'), and so 15° is the static lag angle.

The addition of a 0.05 kg trial mass at 0° at end 1 only is equivalent to a 0.025 kg mass at 0° at both ends (static balance mass) plus a 0.025 kg mass at 0° at end 1 together with a 0.025 kg mass at 180° at end 2 (dynamic balance mass). To correct the static balance of the machine, additional mass must be added to the rotor so as to produce a vibration vector S'O which will cancel out OS'. The required magnitude of balance mass is therefore $0.025 \times 53/28$ kg at each end of the machine. These masses must be added at an angle position of $41° + 180° - 15° = 206°$ to obtain static balance.

Figure 7.10(b) shows the machine vibration vectors attributable to dynamic imbalance. The change in the dynamic imbalance vector brought about by the addition of the trial mass is given by the difference between vectors OB and OB' at end 2, and that between OA and OA' at end 1. That is, vectors of 5 μm at 188° at end 2 (vector $OD_2$) and 5 μm at 8° at end 1 (vector $OD_1$). From the diagram it can be seen that the dynamic lag angle is 8°. To correct the dynamic imbalance, balance masses must be added in order to produce additional vibration vectors which are equal in magnitude but opposite in direction to vectors S'A' (at end 1) and S'B' (at end 2). This requires additional balance masses of $0.025 \times 34/5$ kg at each end

BALANCING

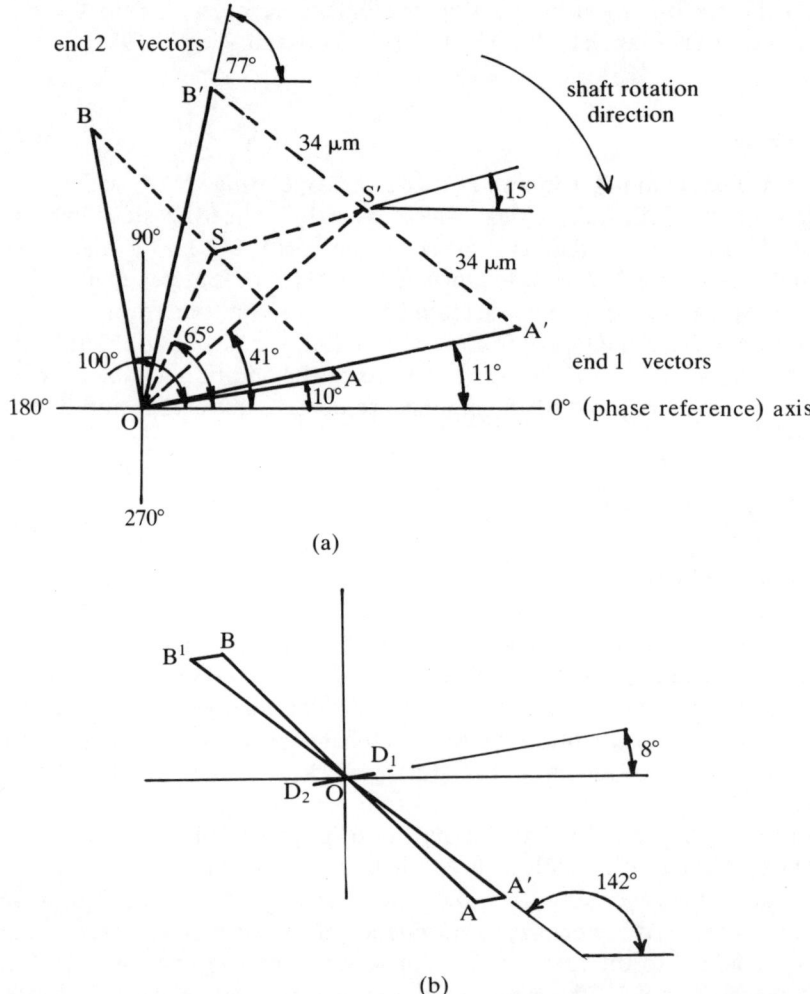

**Figure 7.10** Balancing a machine using plotted vibration vectors (Example 7.2), (a) measured vibration vectors; (b) machine vibration vectors attributable to dynamic imbalance.

of the machine, the required angular locations of these masses being $142° - 8° = 134°$ at end 1, and $-38° - 8° = -46°$ at end 2.

EXAMPLE 7.3

A rigid rotor machine is exhibiting vibration problems caused by imbalance. The machine is symmetrical about its centre-line. A trial balance mass of 0.5 kg is sited at end 1 at an angle of 20° relative to some reference position; this causes changes in vibration vectors of 30 $\mu$m at 31° at end 1 and 22 $\mu$m

## UNKNOWN IMBALANCE

at 40° at end 2. Determine the influence coefficients for use in balancing the machine, and calculate the balance mass required at each end of the machine if the measured imbalance vibrations are $-25$ μm at 210° at end 1 and $-60$ μm at 310° at end 2.

SOLUTION

The influence coefficients for the machine are given by

$$a_{11} = \frac{\text{change in vibration at end 1}}{\text{balance mass added at end 1}}$$

$$= \frac{30(\cos 31° + j \sin 31°)}{0.5(\cos 20° + j \sin 20°)}$$

$$= \frac{30(\cos 11° + j \sin 11°)}{0.5}$$

$$= (58.9 + j\, 11.4)\ \mu\text{m.kg}^{-1}$$

$$a_{21} = \frac{\text{change in vibration at end 2}}{\text{balance mass added at end 1}}$$

$$= \frac{22(\cos 40° + j \sin 40°)}{0.5(\cos 20° + j \sin 20°)}$$

$$= \frac{22(\cos 20° + j \sin 20°)}{0.5}$$

$$= (41.3 + j\, 15.0)\ \mu\text{m.kg}^{-1}$$

Also $a_{12} = a_{21}$ and $a_{22} = a_{11}$ because of the symmetry of the machine.

Using Equations (7.13), the balance masses required at each end of the machine are given by

$$\begin{Bmatrix} M_1 \\ M_2 \end{Bmatrix} = \begin{bmatrix} a_{11} & a_{12} \\ a_{21} & a_{22} \end{bmatrix}^{-1} \begin{Bmatrix} 25(\cos 210° + j \sin 210°) \\ 60(\cos 310° + j \sin 310°) \end{Bmatrix}$$

$$= \begin{bmatrix} (58.9 + j\, 11.4) & (41.3 + j\, 15.0) \\ (41.3 + j\, 15.0) & (58.9 + j\, 11.4) \end{bmatrix}^{-1} \begin{Bmatrix} (-21.7 - j\, 12.5) \\ (38.6 - j\, 46.0) \end{Bmatrix}$$

$$= \begin{bmatrix} (0.0319 + j\, 0.0044) & (0.0226 - j\, 0.0068) \\ (0.0026 - j\, 0.0068) & (0.0319 + j\, 0.0044) \end{bmatrix} \begin{Bmatrix} (-21.7 - j\, 12.5) \\ (38.6 - j\, 46.0) \end{Bmatrix}$$

$$= \begin{Bmatrix} (0.0777 - j\, 1.80) \\ (0.858 - j\, 1.43) \end{Bmatrix} \text{kg}$$

$$= \begin{Bmatrix} 1.8\ \text{kg at } -88° \\ 1.67\ \text{kg at } -59° \end{Bmatrix}$$

## 7.4 Modal balancing of long flexible rotors with an unknown imbalance

In the case of the rigid rotor systems discussed previously the rotor had only two possible modes of vibration, a lateral mode associated with static imbalance where both ends of the rotor throw out in the same direction, and a conical mode associated with dynamic imbalance where the rotor ends throw out in opposite directions to each other. In the case of long flexible rotors there are an infinite number of possible modes of vibration associated with the system, for example, the first three rotor mode shapes take the form shown in Figure 7.11. If the stationary rotor were given an impulsive load then the subsequent free vibrations of the system would be composed of vibrations associated with all of the system modes, all occurring at the same time. Each vibration mode of the rotor has a particular frequency associated with it, for example the first mode of vibration (known as the 'fundamental mode') always occurs at a frequency which is lower than that of the second mode; the second mode frequency is in turn lower than that of the third mode, and so on. Each of the system vibration mode frequencies corresponds to a system natural frequency, so that when the machine is forced close to one of the mode frequencies high levels of vibration set in; even when running at a speed which is remote from one of the system natural frequencies, the level of vibration is dependent on the magnitude of the forcing of the particular mode of vibration under consideration. For this reason it is necessary to balance flexible rotors, not only for the first two

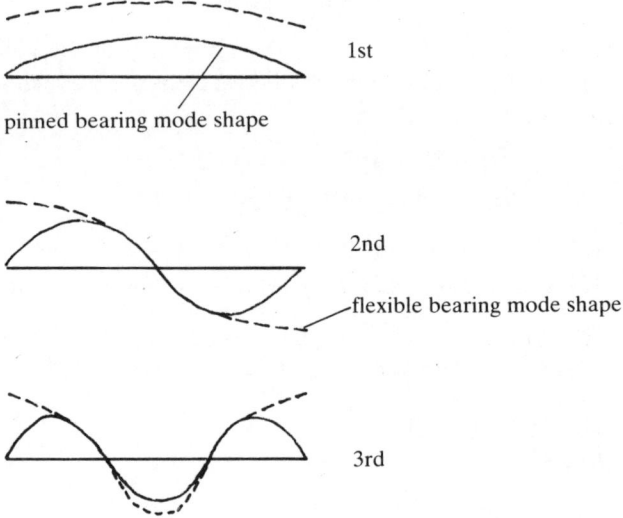

**Figure 7.11** The first three rotor mode shapes for a flexible rotor machine.

possible modes of vibration (the static and dynamic balance discussed in the previous sections), but for all modes of vibration up to the machine's maximum running speed. If there is a system natural frequency just beyond the normal maximum operating speed of the machine, the rotor will normally be balanced for that corresponding mode shape also, in case there is any significant excitation of that mode when operating at the maximum service speed. Evidently, it is important that, in balancing out one vibration mode, another mode is not excited further; for this reason it is important to be able to site balance weights at a number of axial locations on the rotor. In the case of the 'N-plane' balancing procedure described below, the minimum number of balance planes required is equal to the number of flexural vibration modes to be balanced.

In the first instance the balancing procedure for flexible rotors is similar to that described for rigid rotors in section 7.3, insofar as measurements of machine vibration are made, normally close to the bearing locations, and are used to determine the vibrations associated with static and dynamic imbalance as was shown in Figure 7.8. The measurements would normally be taken while the machine was operating at a speed close to the first critical speed of the machine, and balance weights would then be installed on the machine in order to eliminate the vibrations associated with static imbalance. The angular position and magnitude of the weights would be determined as before, taking due account of the appropriate phase lag angle and machine sensitivity to imbalance. The axial position of the balance weight is normally chosen to be close to the anti-node in the mode shape, since it is this region where the application of balance weight forces will have the most effect on straightening the shaft. Machine static imbalance is most accurately corrected at the first critical speed because it is at this speed that the machine is most sensitive to this type of imbalance.

The second mode of vibration is most accurately balanced when operating close to the machine second critical speed; again the machine is most sensitive to dynamic imbalance at this running speed. In some cases it may be necessary to make a balance attempt at a speed between the first and second criticals if the machine is so badly balanced that vibration levels prohibit its being run up to the second critical immediately (likewise, it is sometimes necessary to undertake a balance below the first critical). The required angular positions and magnitudes of the balance weights are determined as before from a knowledge of the phase lag angle for the machine second critical speed, and from experience of the sensitivity of the machine to balance weight addition. Once again the axial position of the balance weights is determined by the position of the anti-nodes in the mode shape under consideration.

Some flexible rotor machines operate close to, or even above, their third critical speed, so that the third mode of vibration must frequently be balanced also. Consideration of the system mode shapes discussed above enables one to conclude that when a machine exhibits apparent high levels of static imbalance, as recorded by transducers at the shaft ends, and the machine is running above its second critical speed, then if the vibration level at the first critical speed is low the machine must be operating close to the third critical speed. If balance weights are now added only at the anti-node at the centre of the rotor to correct these vibrations, then the level of vibration at the first critical speed will increase since the static balance of the machine will have been upset. Instead, it is necessary to add some mass at the centre of the rotor together with an equal amount 180° around the rotor, half at each end of the machine. In other words, three planes for balance weight addition are required for third critical balancing; in general these will be located close to the anti-nodes of the third mode of vibration. This process can, of course, be extended and applied to the balancing of machines with any number of critical speeds.

In machines which operate with very flexible bearings it is sometimes convenient to draw a distinction between rotor vibration as a rigid body, and vibration involving rotor flexure, as indicated in Figure 7.12. In these cases it may be preferable to balance the rigid body modes before attending to the flexural modes. Subsequent balancing of the flexural modes then requires balance mass addition at a number of balance planes equal to the number of the flexural mode under consideration plus 2, in order to ensure that the rigid body balance is not upset. This method of balancing is therefore called the 'N + 2 plane' method, as described by Kellenberger (1967, 1972) and by Kendig (1975).

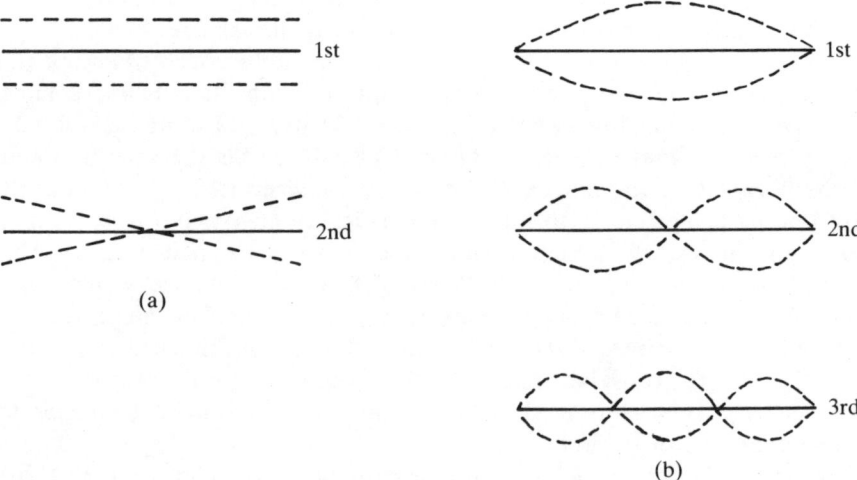

**Figure 7.12** (a) Rigid body modes and (b) flexural vibration modes for a shaft (showing only the first three flexural modes).

## MODAL BALANCING

In some types of machinery the balancing procedure may be complicated by the presence of rotors mounted on overhung shafts, or by the presence of two or more shafts coupled together. The general rule in these situations is to simplify the system as much as possible before the balancing process. In the case of overhung rotors, the shaft should be balanced in isolation before the rotor is assembled on the machine; a second balancing operation can then be carried out with attention focused on the rotor. In many instances it is convenient to introduce a balancing procedure as part of the manufacturing process, balancing separate components independently prior to assembly and final machine balance. In cases where machines consist of two or more shafts coupled together it is usually preferable to balance them as independent units.

Further details of the application of the modal balancing method may be obtained from Moore and Dodd (1964), Lindley and Bishop (1963), and Bishop and Gladwell (1959). Noteworthy contributions have also been provided by Parkinson and Bishop (1965), Parkinson (1965), Bishop and Mahalingham (1965) and Parkinson (1966). More recent reviews of the method have been given by Rieger (1982), Rao (1983), and Kellenberger and Rihak (1988).

EXAMPLE 7.4

The machine shown in Figure 7.13 is intended to operate above its third critical speed. It is balanced to within acceptable limits at its first and second critical speeds, and is run up to a speed just below its third critical where the vibrations measured at the bearings are:

$$\text{end 1: } 50 \ \mu\text{m at } 35°$$
$$\text{end 2: } 65 \ \mu\text{m at } 60°$$

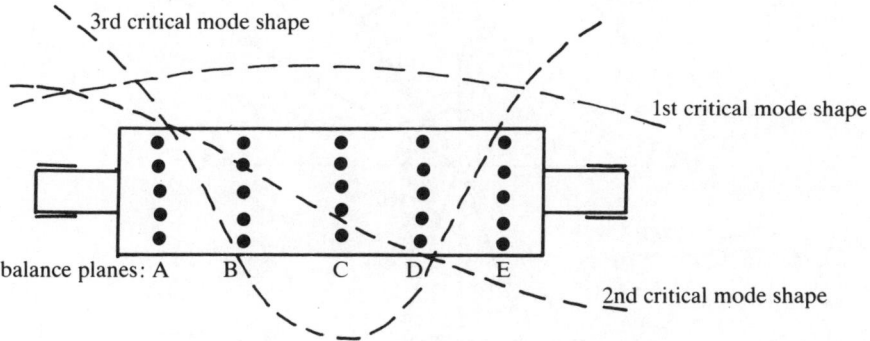

**Figure 7.13** A machine requiring balancing for operation above its third critical speed (Example 7.4).

# BALANCING

Similar machines have had an imbalance sensitivity at this speed of about 100 $\mu$m kg$^{-1}$.m$^{-1}$ for third-mode imbalance when the vibration is measured at the bearings. The machines' critical speed mode shapes have been determined theoretically and are shown by the broken lines in the diagram. For nulling purposes (see Section 9.10.2) the vibrations measured between the second and third critical speeds were:

$$\text{end 1: 20 } \mu\text{m at 200}°$$
$$\text{end 2: 30 } \mu\text{m at 160}°$$

Determine the balance masses and positions required to obtain 'static' balance at the third critical, and state which balance planes could be most readily omitted from the design and where additional balance planes might be useful.

Where applicable, use the most appropriate lag angles from the following list: 10°, 85°, 100° and 170°.

SOLUTION

The runout vectors are vectors $OR_1$ and $OR_2$ in Figure 7.14. The measured vibration vectors are vectors OA and OB which may be nulled to give actual vibration vectors OA' and OB'. The 'static' high side is then indicated by vector OC where C is the mid-point of A'B'. The low side, which must be

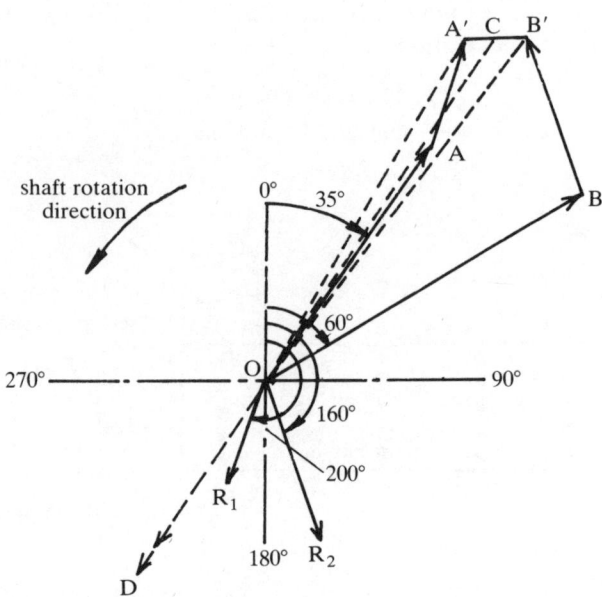

**Figure 7.14** Vibration vectors for the machine shown in Figure 7.13. (See Example 7.4).

forced out by the addition of balance mass, is 180° opposed to this and is in the direction of OD at 214°. Since vector OC is equivalent to 72 μm, 0.72 kg.m is required at each end of the machine to obtain third mode 'static' balance. Since vibration measurements are being taken at a speed just below the third critical, the phase lag angle of 85° is most appropriate, and the masses should be added at an angular position of 214° − 85° = 129°. These masses should be added close to the anti-nodes in the mode shape, planes A and E being most appropriate at their respective ends of the machine. A corresponding mass of 2 × 0.72 = 1.44 kg.m should be added at 129° + 180° = 309° on plane C so as not to upset the static balance of the machine.

Planes A, C and E are in suitable locations, relative to the anti-nodes in the machines' mode shapes, to use in balancing at the first and second critical speeds. Additional balance planes might be useful closer to the anti-nodes related to the third critical, that is outboard of planes A and E. Planes B and D are less useful.

## 7.5 Balancing long flexible rotors with an unknown initial imbalance using the method of influence coefficients

The influence coefficient method of balancing described in Section 7.3 for rigid rotors can also be used with flexible rotors (Tessarzik *et al.* 1972, Lund and Tonneson 1971). In this case more influence coefficients may be required, depending on the number of vibration modes to be balanced. The values of the system influence coefficients are found experimentally, as before, by measuring the effect on vibration of adding a trial mass at a number of locations. Collecting the system influence coefficients may be time-consuming, but the method may be advantageous, particularly when the machine design includes features which complicate the modal balancing procedure.

The first step in the process is to measure the effects of the imbalance in the machine by measuring the machine vibration at a number of locations $u$. In the case of a simple shaft with a bearing at each end $u = 2$ locations may be sufficient, but more may be chosen in the case of machines with many rotors coupled together or with overhung shaft sections. This process is then followed by installing a trial mass $m_1$ at balance plane 1 at some arbitrarily chosen angle, and measuring the new vibrations at the $u$ measuring positions. Both the vibrations and the trial mass have both magnitude and angular orientation and so may be represented as vectors; the subtraction of the initial vibration vectors from the new ones then gives the effective vibration vector $v$ measured at each of the $u$ positions as a consequence of the addition of the trial mass $m_1$. The influence coefficients relating the trial mass to its influence on machine vibration at the measuring positions are then given by

$$a_{11} = v_1/m_1$$
$$a_{21} = v_2/m_1$$
$$a_{31} = v_3/m_1 \qquad (7.14)$$
$$\vdots$$
$$a_{u1} = v_u/m_1$$

The procedure is then repeated to determine the influence coefficients relating to the addition of mass at the remainder of the $w$ balance planes to be used. Thus $a_{11}, a_{12}, \ldots, a_{u_1w-1}$ and $a_{u_1w}$ are determined. If we now consider adding masses $M_1, M_2, \ldots, M_w$ to all of the balance planes respectively simultaneously, then the vibrations $V$ introduced by these masses, as measured at the same locations will be given by

$$\begin{Bmatrix} V_1 \\ V_2 \\ V_3 \\ \vdots \\ V_u \end{Bmatrix} = \begin{bmatrix} a_{11} & a_{12} & a_{13} & \cdots & a_{1w} \\ a_{21} & a_{22} & a_{23} & \cdots & a_{2w} \\ a_{31} & a_{32} & a_{33} & \cdots & a_{3w} \\ & \cdots & \vdots & & \cdots \\ a_{u1} & a_{u2} & a_{u3} & \cdots & a_{uw} \end{bmatrix} \begin{Bmatrix} M_1 \\ M_2 \\ M_3 \\ \\ M_w \end{Bmatrix} \qquad (7.15)$$

which may be written more simply as

$$\{V\} = [A]\{M\} \qquad (7.16)$$

Equation (7.16) may then be rearranged to give

$$\{M\} = [A]^{-1}\{V\} \qquad (7.17)$$

If the values of the elements in the $\{V\}$ matrix are then chosen to be equal to the opposite (negative) of the imbalance vibration vectors as measured at the corresponding locations in the first instance, then Equation (7.17) gives the magnitude and angular orientation of the correct balance weights to be installed in the $w$ planes.

It is noteworthy that consideration must be given to the values of $u$ and $w$ to ensure that the influence coefficient matrix is square and can be inverted. Also the $u$ vibration measurements taken can in fact consist of $n < w$ measurements taken several times at different running speeds, so that $u = w$ readings in total are obtained; in this case some of the data collected may be discarded so that $u$ remains of the desired size. Alternatively, an optimization process may be used, allowing use of all of the data collected, to ensure that vibration is minimized (Goodman 1964, Lund and Tonnesen 1972, Little and Pilkey 1976).

Other publications relating to use of the influence coefficient method are Larsson (1976), Gu et al. (1988), and Sanderson (1988). A further publication by Darlow (1987) compares the influence coefficient method with modal balancing, and describes a unified approach where the advantages of each are capitalized on.

## 7.6 Balancing tolerances

The accuracy to which a machine should be balanced is determined by the type of machine and the duty it is expected to perform. Many manufacturers and operators of rotating machinery have their own standards to which they work. The balance quality of rigid rotors is frequently specified

Table 7.1 Balancing standards recommended by the International Standards Organization (1973, 1976) for rigid and flexible rotors. (a) Quality of balance required by machine type (rigid rotors)

| Quality group | | Machine type | Residual imbalance (kg.mm 100 kg$^{-1}$) | Balance ($\mu$m) |
|---|---|---|---|---|
| A | High-precision balancing | Gyro rotors<br>Centrifuges<br>Superfinish grinders | 0.0025–0.0120 | 0.2–1 |
| B | Precision balancing | Ultra high speed motors<br>Small gas turbines<br>Superchargers, precision grinding machines<br>Jet engine rotors<br>Small centrifuges | 0.005–0.025 | 0.5–1.75 |
| C | High-quality balancing | Small electric motors<br>Turbogenerators<br>Steam turbines<br>Gas turbines<br>Superchargers (medium speed)<br>Centrifuges | 0.025–0.12 | 1.75–10 |
| D | Good-quality balancing | Commercial electric motors<br>Fans, Blowers<br>Centrifugal pumps<br>Four-cylinder crankshafts<br>Flywheels<br>Torsional dampers | 0.05–0.25 | 5–25 |
| E | Average-quality balancing | Line shafts<br>Propeller shafts<br>Gear trains<br>One-to-three cylinder engines<br>Large flywheels<br>Farm machines<br>Paper machine cylinders | 0.15–0.8 | 20–100 |

*continued*

**Table 7.1(b)** Balance quality criteria for flexible rotors

| Rotor category | Ranges of effective pedestal vibration velocity at once-per-rev frequency $V_2$(mm.s$^{-1}$) r.m.s. | | | | | | | | | | | | | Correction factor† | | |
|---|---|---|---|---|---|---|---|---|---|---|---|---|---|---|---|---|
| | 0.28 | 0.45 | 0.71 | 1.12 | 1.8 | 2.8 | 4.5 | 7.1 | 11.2 | 18 | 28 | 45 | 71 | $C_1$ | $C_2$ | $C_3$ |
| I | \<--A--\> | | \<--B--\> | | \<--C--\> | | \<--D*--\> | | | | | | | | | |
| | Small electric motors up to 20 HP | | | | | | | | | | | | | 0.63 | | |
| | Superchargers | | | | | | | | | | | | | 0.63 | 2 | |
| | Gyroscopes | | | | | | | | | | | | | 0.63 | 2 | |
| II | | \<--A--\> | | \<--B--\> | | \<--C--\> | | \<--D--\> | | | | | | | | |
| | Paper-making machines | | | | | | | | | | | | | 0.63 | | |
| | Medium-size electric motors and generators, 20–100 HP on normal foundations | | | | | | | | | | | | | 0.63 | 4 | |
| | Electric motors and generators up to 400 HP on special foundations | | | | | | | | | | | | | 0.63 | 4 | 20 |
| | Pumps and compressors | | | | | | | | | | | | | 0.63 | 8 | 15 |
| | Small turbines | | | | | | | | | | | | | 0.63 | 4 | 8 |
| III | | | \<--A--\> | | \<--B--\> | | \<--C--\> | | \<--D--\> | | | | | | | |
| | Large electric motors | | | | | | | | | | | | | 0.63 | 5 | |
| | Turbines and generators on rigid and heavy foundations | | | | | | | | | | | | | 0.63 | 5 | 20 |
| IV | | | | \<--A--\> | | \<--B--\> | | \<--C--\> | | \<--D--\> | | | | | | |
| | Large electric motors, turbines and generators on lightweight foundations | | | | | | | | | | | | | 0.63 | 3 | 10 |
| | Small jet engines | | | | | | | | | | | | | 0.63 | | |
| V | | | | | \<--A--\> | | \<--B--\> | | \<--C--\> | | \<--D--\> | | | | | |
| | Jet engines larger than category IV | | | | | | | | | | | | | 0.63 | 2 | 10 |

*Quality bands.
A, precision quality; B, commercially acceptable; C, in need of attention at next overhaul; D, in need of immediate attention.

†$C_1$, balanced in a balancing machine; $C_2$, shaft vibrations measured at bearings; $C_3$ shaft vibrations measured at location of maximum lateral deflection.

as a balance quality grade number, for example G100. The number, in these cases, represents the maximum permissible tangential velocity of the rotor centre of gravity, in mm.s$^{-1}$. In the case of flexible rotors, balance quality is instead usually related to absolute vibration measurements on the machine.

One of the most common standards which is adhered to is that of the International Standards Organization (ISO 1973, 1976), whose suggested balance criteria are summarized in Table 7.1a and b.

EXAMPLE 7.5

A rigid rotor blower is being balanced. The initial vibration reading at the rotor is 70 $\mu$m at 225°, and that after the addition of a 0.003 kg balance

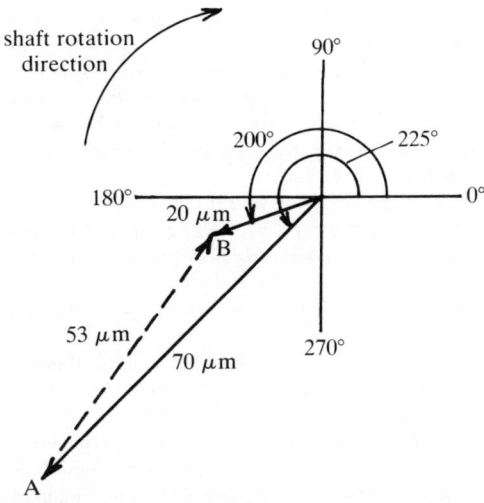

**Figure 7.15** Rotor vibration vectors recorded for a rigid rotor blower. (See Example 7.5).

mass at a radius of 200 mm is 20 μm at 200°. Determine the imbalance still remaining in the machine, and the requirement for further balance mass addition. The rotor mass is 150 kg.

SOLUTION

The difference in the measured vibration vectors represents the balance vector introduced by the addition of the balance mass. This is vector AB of magnitude 53 μm in Figure 7.15. It follows that the remaining imbalance in the blower is

$$mr = 0.003 \times 200 \times 20/53 = 0.226 \text{ kg.mm}$$

Referring to Table 7.1(a), the normal allowable imbalance range for good-quality balancing of a blower of mass 150 kg is 0.075–0.375 kg.mm. The blower in question is now balanced to within this range.

# References

Bishop, R. E. D. & G. M. L. Gladwell 1959. The vibration and balancing of an unbalanced flexible rotor. *J. Mech. Eng. Sci.* **1**, 66–77.

Bishop, R. E. D. & S. Mahalingham 1965. Some experiments in the vibration of a rotating shaft. *Proc. Roy. Soc. London A*, **259**.

Blake, M. P. 1967. Use phase measuring to balance rotors in place. *Hydrocarbon Processing*, August.

Darlow, M. S. 1987. Balancing of high speed machinery: theory, methods and experimental results. *Mechanical Systems and Signal Processing*, **1**, 1, 105–134.

Goodman, T. P. 1964. A least squares method for computing balance corrections. *Trans. ASME, J. Eng. Ind.* **86** B., 273–279.

Gu, H., H. Ren, & J. Yang 1988. An approach to improve least squares method for calculating balance correction masses by using successive iteration. *Proc. Second ASME. Int. Symp. Transport Phenomena, Dynamics and Design of Rotating Machinery*, Honolulu.

ISO 1973. *Balance quality of rotating rigid bodies.* International Standards Organization Geneva, (Document 1940).

ISO 1976. *The mechanical balancing of flexible rotors.* International Standards Organization, Geneva, Report no. ISO/TC. 108/SCI/WG2, Secretariat 7, Document 12.

Kellenberger, W. 1967. Balancing flexible rotors on two generally flexible bearings. *Brown Boveri Review.* **54**, No. 9.

Kellenberger, W. 1972. Should a flexible rotor be balanced in $N$ or $N+2$ planes? *J. Eng. Ind.* **94B**, 2, 548–560.

Kellenberger, W. & P. Rihak 1988. Bimodal (complex) balancing of large turbogenerator rotors having large or small unbalance. *Proc. IMechE. Conf. Vibrations in Rotating Machinery.* Heriot-Watt Univ. September, Paper C292/88 479–486.

Kendig, J. R. 1975. Current flexible rotor-bearing system balancing techniques using computer simulation. M.S. Thesis, Rochester Inst. Tech. Rochester, New York.

Langlois, A. B. & E. J. Rosecky 1968. *Field balancing.* Allis Chalmers.

Larsson, L. O. 1976. On the determination of the influence coefficients in rotor balancing, using linear regression analysis. *Proc. IMechE. Conf. Vibrations in Rotating Machinery*, Cambridge. 93–97.

Lindley, A. G. & R. E. D. Bishop 1963. Some recent research of the balancing of large flexible rotors. *Proc. Inst. Mech. Engrs.* **177**, 30, 811ff.

Little, R. M. & W. D. Pilkey 1976. A linear programming approach for balancing flexible rotors. *Trans. ASME, J. Eng. Ind.* **98**, 1030–1035.

Lund, J. W. & J. Tonnesen 1971. Analysis and experiments on multi-plane balancing of a flexible rotor. *Trans. ASME Third Vibn. Conf, Toronto.* Pap. 71 Vibr.

Lund, J. W. & J. Tonnesen 1972. Analysis and experiments on multi-plane balancing of a flexible rotor. *Trans. ASME, J. Eng. Ind.* **94B**, 233–242, appendix A.

McQueary, D. C. 1973. Understanding balancing machines. *American Machinist, Special Report* no. 656, June 11. Penton Publ. Co. Inc., Pittway Corp, Cleveland, Ohio.

Moore, L. S. and E. G. Dodd 1964. Mass balancing of large flexible rotors. *GEC Journal* **31**, 2, 74ff.

Parkinson, A. G. 1965. The vibration and balancing of shafts rotating in asymmetric bearings. *J. Sound and Vib.* **2**, 4, 477–501.

Parkinson, A. G. 1966. On the balancing of shafts with axial asymmetry. *Proc. Roy. Soc. London*, **292** A, 66ff.

Parkinson, A. G. & R. E. D. Bishop 1965. Residual vibration in modal balancing. *J. Mech. Engrg. Sci.* **7**, no. 1, 33–39.

Rao, J. S. 1983. *Rotor dynamics.* New Delhi: Wiley Eastern.

Rieger, N. 1981. *Balancing of rigid and flexible rotors.* Monograph 11, Shock and Vibration Information Center, Naval Research Lab., Washington D.C.

Rieger, N. 1982. *Vibrations of rotating machinery – part 1: rotor-bearing dynamics.* Clarendon Hills: Vibration Inst. Illinois. Feb.

Sanderson, A. F. P. 1988. Turbine generator trim balancing using optimized least squares methods. *Proc. IMechE. Int. Conf. Vibrations in Rotating Machinery*, Heriot-Watt Univ. September, Paper C308/88, 491–498.

Tessarzik, J. M., R. H. Badgley & W. J. Anderson 1972. Flexible rotor balancing by the exact point-speed influence coefficient method. *Trans. ASME, J. Eng. Ind.*, **94**, 148–158.

# 8 Measuring bearing impedances

## 8.1 Introduction

One of the most important factors governing the vibration characteristics of rotating machinery is the effective dynamic stiffness of the supports as seen by the rotor. In the case of rigid rotors it is the support dynamic stiffness, together with the rotor mass, which determines the critical speed and the vibration amplitude at running speeds remote from the critical speed. In the case of flexible rotors the support dynamic stiffness, together with the mass and flexibility of the rotor, determines these system characteristics. In both cases the dynamic stiffness of the supports is determined by the combined effects of flexibility of the bearing, the bearing pedestal assembly, and the foundations on which the pedestal is mounted. The system may be represented, in a simplified form, as shown in Figure 8.1(a); the analogy can be drawn with Figure 8.1(b), which has a single degree of freedom and whose vibration characteristics are very much influenced by the effective spring stiffness $K_e$. Of the three components of support stiffness described above, the one which most influences the value of $K_e$ for the equivalent system, and hence most influences the system vibration characteristics, is that which is the most flexible element of the supports since the three elements are effectively assembled in series. In the case of turbogenerator rotors mounted in oil-film journal bearings, the oil film itself might be of

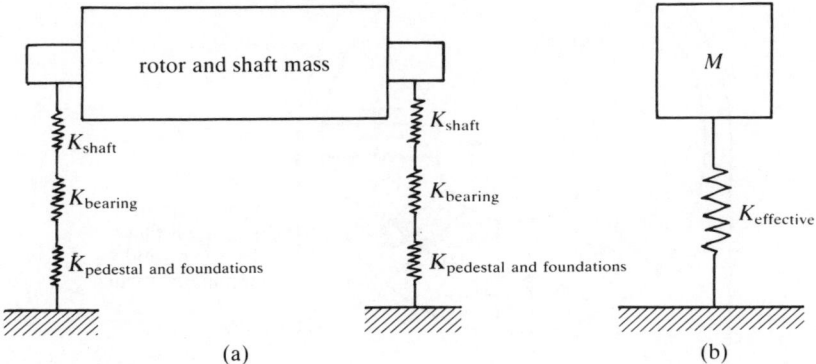

**Figure 8.1** (a) Simplified representation of a rotor-bearing system; (b) Single-degree-of-freedom analogue.

the order of three times more flexible than the pedestal and foundations, and thus it is the bearing oil film whose characteristics the designer needs to know. In other cases, for example that of aeroengine compressor shafts mounted in rolling element bearings, the bearing foundation is by far the most flexible element of the support and it is the foundation characteristics which are most important to the design engineer. .

Although the importance of rotor support dynamic stiffness is generally well recognized by the design engineer, it is often the case that theoretical models available for predicting it are insufficiently accurate, or are accurate only in very specific cases. It is for this reason that designers of high-speed rotating machinery must rely on empirically derived values for support stiffness and damping in their design calculations. The object of this chapter is to review some of the methods of measuring support dynamic stiffness used by previous investigators. The potential advantages and disadvantages associated with each method are also discussed. All of the procedures described allow for representation of the support dynamic stiffness in terms of eight coefficients of spring stiffness and damping as introduced in Chapter 2 and indicated in Figure 8.2. In the remainder of this chapter these measurement techniques are discussed with reference to bearing oil films since, in most cases, this is what the researcher will be investigating. In theory, however, the procedures could be applied to flexible elements composed of some other matter. It is impossible to say which of the methods discussed represents the 'best' approach because coefficients obtained using all methods show some general agreement with each other. There is, however, considerable scatter in the results, even in those associated with one measurement technique. Some of the more recently

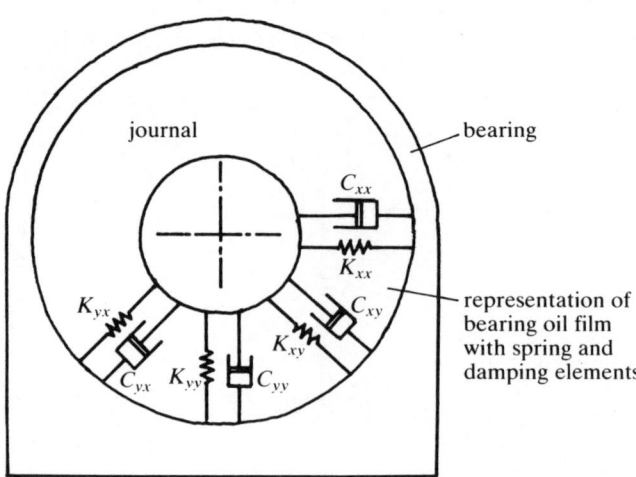

**Figure 8.2** Diagrammatic representation of rotor support stiffness and damping coefficients.

developed methods have shown a greater consistency and these hold much promise for the future.

## 8.2 Static force method

It is possible to determine all four stiffness coefficients of the bearing oil film by application of static loads only, as reported by Mitchell *et al.* (1966), Woodcock and Holmes (1970), Parkins (1979), and Tripp and Murphy (1984). Unfortunately this method of loading does not enable the oil-film damping coefficients to be determined. To find these another approach must be chosen; several alternatives are given in the remainder of this chapter. The system under consideration is that shown in Figure 8.2 where, because the application of static loads only is being considered, the oil-film damping has no effect on the system behaviour. The steady-state locus of the journal will be of the form shown in Figure 8.3, the exact operating position on this locus for any particular bearing being dependent on the Sommerfeld number. Because the bearing oil film coefficients are specific to particular locations on the static locus, a static load must first be applied in order to establish operation at the required point on the locus. The next step is to

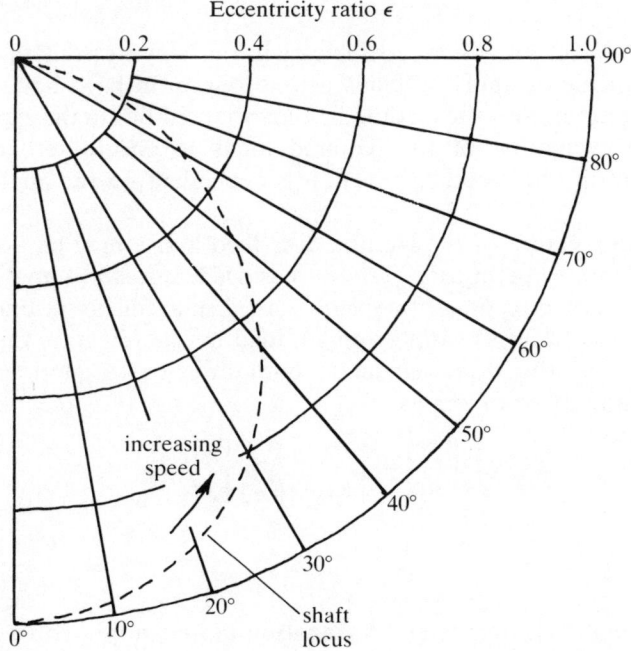

**Figure 8.3** Form of journal locus under steady operating conditions.

apply incremental loads in both horizontal and vertical directions which will cause changes in the journal horizontal and vertical displacement relative to the bearing bush. By relating the measured changes in displacements to the changes in static load it is possible to determine the four stiffness coefficients of the bearing oil film.

The increments in oil film forces, $F_x$ and $F_y$ in the $x$ and $y$ directions respectively, are related to the changes in journal displacement in these directions, $x$ and $y$, by the stiffness coefficients:

$$F_x = K_{xx}x + K_{xy}y, \qquad F_y = K_{yx}x + K_{yy}y \qquad (8.1)$$

At first sight these equations may seem impossible to solve since there are four unknowns and only two equations. However, if the journal can be loaded such that one of the displacements is zero, then the solution can be obtained. If the displacement in the $y$ direction is made to be zero by application of suitable loads $F_x$ and $F_y$ then

$$K_{xx} = \frac{F_x}{x}, \qquad K_{yx} = \frac{F_y}{x}$$

Similarly if the displacement in the $x$ direction is made zero then the values of $K_{xy}$ and $K_{yy}$ may be obtained, viz.

$$K_{xy} = \frac{F_x}{y}, \qquad K_{yy} = \frac{F_y}{y}$$

Determination of the oil-film coefficients in this way necessitates a test rig which is capable of applying loads to the journal in both horizontal and vertical directions; also the method is somewhat tedious in the experimental stage since evaluation of the required loads to ensure zero change in displacement in one or other direction is dependent on the application of trial loads.

The tedious nature of the method described above may be avoided if a slightly different experimental procedure is used. Instead of applying loads in both $x$ and $y$ directions, to ensure zero displacements in one of these directions, it is easier to simply apply a load in one direction only, and to measure the resulting displacements in both directions. Equations (8.1) are then written in matrix form as:

$$\begin{bmatrix} F_x \\ F_y \end{bmatrix} = \begin{bmatrix} K_{xx} & K_{xy} \\ K_{yx} & K_{yy} \end{bmatrix} \begin{bmatrix} x \\ y \end{bmatrix}$$

or

$$[F] = [K][d] \qquad (8.2)$$

If the $[K]$ matrix is inverted then Equation (8.2) can be written as

$$[d] = [K]^{-1}[F]$$

## STATIC FORCE METHOD

or more explicitly as

$$\begin{bmatrix} x \\ y \end{bmatrix} = \begin{bmatrix} a_{xx} & a_{xy} \\ a_{yx} & a_{yy} \end{bmatrix} \begin{bmatrix} F_x \\ F_y \end{bmatrix} \quad (8.3)$$

where the quantities $a_{xx}$, $a_{xy}$, etc. are called the oil-film influence coefficients. If the force in the $y$ direction is zero then

$$a_{xx} = \frac{x}{F_x}, \quad a_{yx} = \frac{y}{F_x}$$

Similarly, when the force in the $x$ direction is zero

$$a_{xy} = \frac{x}{F_y}, \quad a_{yy} = \frac{y}{F_y}$$

In this way it is possible to determine all of the elements in the matrix of influence coefficients, $[K]^{-1}$, which may then be re-inverted to determine the values of the oil-film stiffness coefficients. This method still requires a test rig which is capable of providing loads on the bearing in both $x$ and $y$ directions but is nevertheless more straightforward in the experimental stage.

If there is no facility on the test rig for applying loads transverse to the normal steady-state load direction of the bearing, it is still possible to obtain approximate values of the stiffness coefficients by consideration of the steady-state running position of the journal relative to the bearing centre. This approach was first demonstrated by Morrison (1962). In Figure 8.4 the steady running position of the journal, carrying a vertical load $W$, is point A. An imaginary force $F_x$ applied in the $x$ direction would change the steady state running position to point B, and the total load on the bearing would then be the vector sum of $F_x$ and $W$, and would act at an angle $d\theta$ to the vertical. The influence coefficient $a_{xx}$ is then given by

$$a_{xx} = \frac{x}{F_x} = \frac{(e + de)\sin(\phi + d\phi) - dx}{F_x}$$

$$= \frac{(e + de)(\sin\phi \cos d\phi + \cos\phi \sin d\phi) - dx}{F_x} \quad (8.4)$$

Setting $(e + de) \approx e$, $\cos d\phi = 1.0$, and $\sin d\phi = d\phi$ gives

$$a_{xx} \simeq \frac{e \, d\phi \cos \phi}{F_x} \quad (8.5)$$

A further simplification can be made if the resultant load $R$ is considered to be of virtually the same magnitude as the original load $W$ except that it has been turned through an angle $d\theta$. This approximation is valid provided that there is no difference between the form of the new active area of the bearing bush surface and that before the load $F_x$ was applied (for example in the

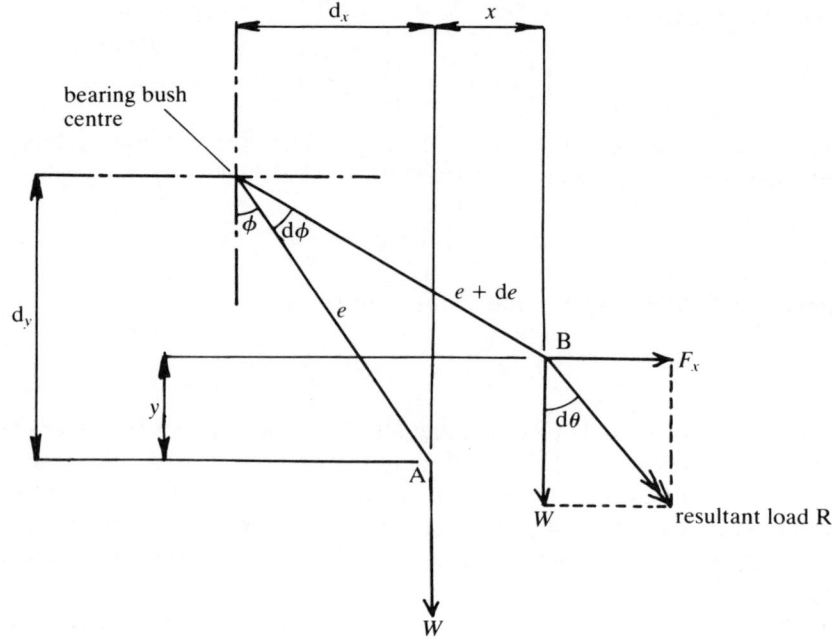

**Figure 8.4** Diagram showing the change in journal running position caused by the application of a horizontal force. See text for details.

case of a cylindrical bearing whose oil inlet ports have no effect on the bearing performance). Then we may write

$$d\phi \approx d\theta = \tan d\theta = F_x/W \qquad (8.6)$$

which on substitution into equation (8.5) gives

$$a_{xx} = \frac{e \sin \phi}{W} = \frac{d_y}{W}$$

Similarly it may be shown that

$$a_{yx} = -\frac{e \sin \phi}{W} = -\frac{d_x}{W}$$

The other influence coefficients $a_{yy}$ and $a_{xy}$ may be determined in the manner discussed earlier. Thus the matrix of influence coefficents, $[K]^{-1}$, can be inverted as before to determine the stiffness coefficients.

It is noteworthy that use of the static force method to obtain oil-film dynamic coefficients does imply a presumption that the oil-film 'static' stiffness is indeed the same as the oil-film 'dynamic' stiffness. There is some support for this assumption in that coefficients measured in this way show a general agreement with those obtained using other techniques. In cases

## STATIC FORCE METHOD

where the oil-film inertia forces are significant, this assumption is potentially invalid and instead a twelve-coefficient model which allows for film forces proportional to acceleration may be more accurate. However, in most cases it is acceptable to neglect oil-film inertia effects.

EXAMPLE 8.1

Under particular operating conditions, the theoretical values of the stiffness coefficients for a hydrodynamic bearing are found to be: $K_{xx} = 30$ MN.m$^{-1}$, $K_{xy} = 26.7$ MN.m$^{-1}$, $K_{yx} = -0.926$ MN.m$^{-1}$, $K_{yy} = 11.7$ MN.m$^{-1}$. A test rig is being designed so that these values can be confirmed experimentally. What increments in horizontal ($F_x$) and vertical ($F_y$) loads must the rig be capable of providing in order to provide

(a) a displacement increment of 12 $\mu$m in the horizontal direction whilst that in the vertical direction is maintained zero,
(b) a displacement increment of 12 $\mu$m in the vertical direction whilst that in the horizontal direction is maintained zero.

SOLUTION

(a) The relationship between the forces and displacements is given by Equation (8.2) as

$$\begin{Bmatrix} F_x \\ F_y \end{Bmatrix} = \begin{bmatrix} K_{xx} & K_{xy} \\ K_{yx} & K_{xy} \end{bmatrix} \begin{Bmatrix} x \\ y \end{Bmatrix}$$

so that if $y = 0$ and $x = 12$ $\mu$m then

$$F_x = K_{xx}x = 30 \times 10^6 \times 12 \times 10^{-6} = 360 \text{ N}$$
$$F_y = K_{yx}x = -0.926 \times 10^6 \times 12 \times 10^{-6} = -11.1 \text{ N}$$

(b) Similarly if $x = 0$ and $y = 12$ $\mu$m, then

$$F_x = K_{xy}y = 26.7 \times 10^6 \times 12 \times 10^{-6} = 320 \text{ N}$$
$$F_y = K_{yy}y = 11.7 \times 10^6 \times 12 \times 10^{-6} = 140 \text{ N}$$

EXAMPLE 8.2

The test rig described in Example 8.1 is used to measure the hydrodynamic bearing stiffness coefficients by applying first of all a horizontal load of 360 N, which is then removed and replaced by a vertical load of 320 N. The horizontal load produces displacements of 10.3 $\mu$m and 3.3 $\mu$m in the horizontal and vertical directions respectively, whilst the vertical load produces respective displacements of $-18.3$ $\mu$m and 19.7 $\mu$m. Calculate the value of the stiffness coefficients based on these measurements.

## MEASURING BEARING IMPEDANCES

SOLUTION

The relationship between the displacements and forces given by Equation 8.2 may be rewritten as

$$\begin{Bmatrix} x \\ y \end{Bmatrix} = \begin{bmatrix} K_{xx} & K_{xy} \\ K_{yx} & K_{yy} \end{bmatrix}^{-1} \begin{Bmatrix} F_x \\ F_y \end{Bmatrix}$$

$$= \begin{bmatrix} a_{xx} & a_{xy} \\ a_{yx} & a_{yy} \end{bmatrix} \begin{Bmatrix} F_x \\ F_y \end{Bmatrix}$$

In the first instance, when $F_x = 360$ N and $F_y = 0$ then

$$x = a_{xx} F_x + 0, \qquad y = a_{yx} F_x + 0$$

whereupon

$$a_{xx} = \frac{x}{F_x} = \frac{10.3 \times 10^6}{360} = 0.0286 \ \mu\text{m.N}^{-1}$$

$$a_{yx} = \frac{y}{F_x} = \frac{3.3 \times 10^{-6}}{360} = 0.00917 \ \mu\text{m.N}^{-1}$$

Similarly, when $F_x = 0$ and $F_y = 320$ N, the other influence coefficients may be evaluated as

$$a_{xy} = \frac{x}{F_y} = \frac{-18.3 \times 10^{-6}}{320} = -0.0572 \ \mu\text{m.N}^{-1}$$

$$a_{yy} = \frac{y}{F_y} = \frac{19.7 \times 10^{-6}}{320} = 0.0616 \ \mu\text{m.N}^{-1}$$

The stiffness coefficients are then obtained by re-inverting the influence coefficient matrix. That is,

$$\begin{bmatrix} K_{xx} & K_{xy} \\ K_{yx} & K_{yy} \end{bmatrix} = \begin{bmatrix} a_{xx} & a_{xy} \\ a_{yx} & a_{yy} \end{bmatrix}^{-1} = \begin{bmatrix} 0.0286 \times 10^{-6} & -0.0572 \times 10^{-6} \\ 0.00917 \times 10^{-6} & 0.0616 \times 10^{-6} \end{bmatrix}^{-1}$$

$$= \begin{bmatrix} 27 \times 10^6 & 25 \times 10^6 \\ -4 \times 10^6 & 12.5 \times 10^6 \end{bmatrix} \text{N.m}^{-1}$$

## 8.3 Use of an electromagnetic (or other) vibrator

In order to fully analyse the behaviour of a bearing under dynamic loading it is necessary to cause the journal to vibrate within the bearing bush under the action of a known exciting force. Alternatively, the bearing bush can be allowed to float freely on the journal, which is mounted in slave bearings, and the forcing is then applied to the bush. Experiments of this nature have been reported by Glienicke (1966), Morton (1971) and Parkins (1976) amongst others. By measuring the resulting system vibrations, and relating these to the force, it is possible to determine the effective oil-film stiffness and damping coefficients. One of the simplest ways of causing the journal to vibrate is to connect the output rod of an electromagnetic (or other type of) vibrator directly to the journal. By varying the amplitude, frequency and shape of the electrical signal input to the vibrator it is possible to exercise full control over the forcing applied to the system. The system under consideration may again be represented by Figure 8.2. Several variations of this approach are possible and these are described below.

### 8.3.1 Complex receptance method

This method involves applying a sinusoidally varying force to the journal in the horizontal direction, whilst the forcing in the vertical direction is zero, and measuring the resulting displacement amplitudes in the horizontal and vertical directions together with their respective phases relative to the exciting force. It is then necessary to repeat the procedure with the forcing applied only in the vertical direction. Knowledge of the applied force amplitude, and of the measured quantities, then enables the eight oil-film coefficients to be derived.

The forces transmitted across the oil film may be represented in the form

$$f_x = K_{xx}x + K_{xy}y + C_{xx}\dot{x} + C_{xy}\dot{y}$$
$$f_y = K_{yy}x + K_{yy}y + C_{yx}\dot{x} + C_{yy}\dot{y}$$
(8.7)

Assuming sinusoidal variations of $x$ and $y$, equations (8.7) can also be written as

$$f_x = (K_{xx} + j\omega C_{xx})x + (K_{xy} + j\omega C_{xy})y$$
$$f_y = (K_{yx} + j\omega C_{yx})x + (K_{yy} + j\omega C_{yy})y$$
(8.8)

which can be written in matrix form as

$$\begin{bmatrix} f_x \\ f_y \end{bmatrix} = \begin{bmatrix} Z_{xx} & Z_{xy} \\ Z_{yx} & Z_{yy} \end{bmatrix} \begin{bmatrix} x \\ y \end{bmatrix}$$
(8.9)

where $Z$ is a complex impedance coefficient given by $Z = K + j\omega C$. By inverting the complex impedance matrix, Equation (8.9) can also be written as

$$\begin{bmatrix} R_{xx} & R_{xy} \\ R_{yx} & R_{yy} \end{bmatrix} \begin{bmatrix} f_x \\ f_y \end{bmatrix} = \begin{bmatrix} x \\ y \end{bmatrix}$$
(8.10)

where the $[R]$ matrix is the inverted form of the $[Z]$ matrix, and is called the complex receptance matrix. Application of Equation (8.10) to the case when the forcing is only in the horizontal direction yields the complex terms

$$R_{xx} = \frac{x}{f_x}, \qquad R_{yx} = \frac{y}{f_x}$$

where $x$ and $y$ are the measured displacements in the horizontal and vertical directions at a particular instant in time, and $f_x$ is the force in the horizontal direction at that instant. In the example represented in Figure 8.5, the measured displacement in the $x$ direction lags behind the forcing in that direction by 20°. Thus the receptance $R_{xx}$ is given by

$$R_{xx} = \frac{X}{F_x \cos 20° + \mathrm{j} F_x \sin 20°}$$

where $F_x$ and $X$ are the force and displacement amplitudes in the horizontal direction. Similarly, when the forcing is only in the vertical direction the other receptance terms are derived, viz.

$$R_{xy} = \frac{x}{f_y}, \qquad R_{yy} = \frac{y}{f_y}$$

Re-inversion of the complex receptance matrix in Equation (8.10) again results in the complex impedance matrix whose elements are now known.

### 8.3.2 Direct complex impedance derivation

It is possible to determine the complex impedance coefficients $Z_{xx}$, $Z_{xy}$, etc. in Equation (8.9) directly without resorting to the use of receptances. This may be done provided that forcing in both the horizontal and vertical directions can be provided simultaneously, and with independent control over each input with respect to its amplitude and relative phase. The system force–displacement relationship is again given by Equation (8.9). With the complex receptance method the approach was to ensure that one of the force vectors was zero; the alternative is to ensure that one of the resulting system displacement vectors, say $Y$, is zero. This can be made to be the case

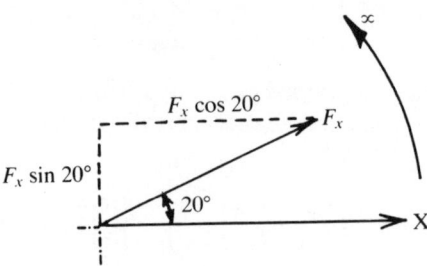

**Figure 8.5** Typical relationship between an applied force vector and journal displacement vector. See text for details.

by correctly adjusting the amplitude of the force in the $y$ direction and its phase relative to that in the $x$ direction. Suitable values for these quantities can be found relatively easily by trial and error. Thus the first line of Equation (8.9) could be written $f_x = Z_{xx}x$, which would allow the value of $Z_{xx}$ to be determined directly provided that the amplitude and phase of the horizontal displacement relative to the horizontal force had been measured. If the amplitude and phase of the force in the $y$ direction were also noted then the value of $Z_{yx}$ could also be determined as

$$Z_{yx} = \frac{f_y}{x}$$

In this case it is the phase of the $x$ direction displacement amplitude relative to the force in the $y$ direction that is significant. Similarly, if the forcing amplitudes and relative phases were adjusted so as to ensure a zero horizontal displacement, $x$, then the values of $Z_{xy}$ and $Z_{yy}$ could also be determined.

Although superficially this approach does appear to be simpler than that discussed in Section 8.3.1 it does suffer from the disadvantage of requiring a more complicated experimental procedure. It may also be more costly in terms of equipment, since two vibrators and additional control units are required.

With either of the above approaches to experimental measurement of the coefficients there are a number of further considerations to be made. The choice of forcing frequency, or forcing frequency range, will depend on the system resonances if these are known. If the system is excited close to its resonant frequency then a response of suitable magnitude may be obtained for a lower force amplitude input. This consideration may be important if the choice of vibrating equipment is limited. (It should, however, be remembered that, for example in the case of hydrodynamic oil-film bearings, the bearing impedance changes with journal vibration amplitude – the oil-film non-linearity effect – so that *too* large a vibration amplitude may result in measured stiffness and damping coefficients which are only applicable for cases when amplitude is large). There is also another potential advantage of exciting the system at a frequency in the region of its resonant frequency, that is that phase lag angles will be generally greater than zero with the result that small inaccuracies in their measurement are less likely to substantially alter the magnitude of the coefficients which are derived. This is not the case when the lag angle is very small or when it is close to $90°$. In these cases 'ill conditioning' of the equations of motion results in significant changes in the magnitude of the derived coefficients for even a change of only 3–4 degrees of phase (which may be about the accuracy to which phase is measured). Fortunately, although coefficients derived using data generated well away from the critical speed may well be considerably inaccurate (perhaps even of the order of 100% in some cases), this is not what is usually of most importance. The paramount consideration is usually the resulting whirl orbit of the rotor that is implied by the coefficients, and under stable running conditions such substantial changes in the values of the measured coefficients are likely only to result in a few degrees of phase change in the whirl orbit. If, on the other hand, the stability of a system is to be questioned, and perhaps measured bearing coefficients are to be used in the design office to predict the onset of instability of the fully assembled machine, then the result of this calculation is likely to be very much influenced by particular values of the coefficients and it will be important to use measured coefficient values which are very accurate.

EXAMPLE 8.3

A bearing is forced in the horizontal direction by a force $F_x = 200 \sin 150t$ N. The resulting journal vibrations are $x = 12 \times 10^{-6} \sin(150t - 0.35)$ m (in the horizontal direction) and $y = 20 \times 10^{-6} \sin(150t - 0.4)$ m in the vertical direction. When the same forcing is applied in the vertical direction the horizontal and vertical displacements take the respective forms $x = 13 \times 10^{-6} \sin(150t + 0.3)$ and $y = 25 \times 10^{-6} \sin(150t - 0.38)$. Determine the elements of the complex receptance matrix for the bearing.

SOLUTION

When the forcing is in the horizontal direction only, application of Equations (8.10) yields

$$R_{xx} = \frac{x}{f_x} = \frac{12}{200 \cos 0.35 + j\, 200 \sin 0.35}$$

$$= \frac{12}{243 + j\, 71.5}$$

$$= \frac{12(243 - j\, 71.5)}{(243^2 + 71.5^2)}$$

$$= (0.0456 - j\, 0.0134)\ \mu\text{m.N}^{-1}$$

and, similarly

$$R_{yx} = \frac{y}{F_x} = \frac{20}{(200 \cos 0.4 + j\, 200 \sin 0.4)}$$

$$= (0.0921 - j\, 0.0389)\ \mu\text{m.N}^{-1}$$

The other elements are found similarly to be

$$R_{xy} = (0.0621 + j\, 0.0192)\ \mu\text{m.N}^{-1}$$
$$R_{yy} = (0.116 - j\, 0.0463)\ \mu\text{m.N}^{-1}$$

## 8.3.3 Multifrequency testing

Some of the disadvantages associated with the other methods of coefficient measurement described in this chapter can be overcome by the use of multifrequency testing as discussed by Burrows and Sahinkaya (1982), Stanway *et al.* (1979) and Burrows *et al.* (1981). This is a relatively new technique which has yet to find widespread acceptance but it offers several advantages which include certainty of exciting all system modes within the prescribed frequency range, and inherent high noise rejection. The method involves forcing the system in both $x$ and $y$ directions, at all frequencies within the range of interest, simultaneously.

The aim is to arrive at more accurate values of the coefficients, which are assumed to be independent of frequency, by means of some averaging procedure. Although, in theory, excitation at two frequencies alone would enable the coefficients to be derived, knowledge of the bearing behaviour at many different frequencies should enable more accurate results to be obtained. The fact that all frequencies are excited simultaneously allows a saving in laboratory time. Fourier analysis can be used in the processing to convert measured input and output signals from the time domain to the frequency domain. Recent advances in laboratory instrumentation, for example, the emergence of spectrum analysers capable of carrying out the Fourier transformation, have helped the technique to evolve.

In theory, any shape of input signal with a multifrequency content can be used to force the system. For example an impulse signal is actually composed of signals at all frequencies in coexistence. Because of the likely

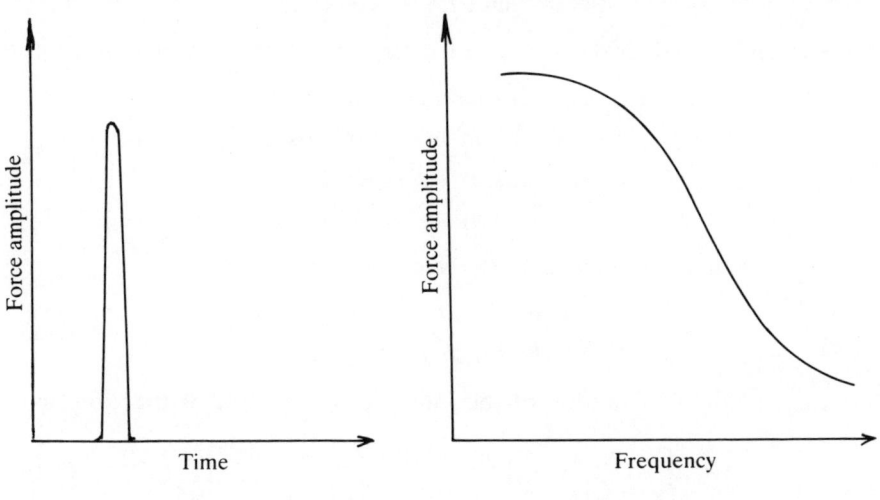

(a)  in the time domain  (b)  in the frequency domain

**Figure 8.6** An impulse signal represented, (a) in the time domain; (b) in the frequency domain.

concentration of the signal at the low-frequency end of the spectrum however (see Figure 8.6), an impulse in practice provides useful signals over only a relatively small frequency range because for the higher frequencies the signal/noise ratio becomes too low. An alternative is a white noise signal which contains all frequencies within its spectrum. Band-limited white noise, sometimes referred to as coloured noise, contains all frequencies within a prescribed range; one way of producing such a signal is with 'pseudo random binary sequences' (PRBS) where the frequency range that is present is chosen to excite appropriate modes in the system under test (Burrows and Stanway 1977). Unfortunately, both with impulse and with PRBS signals there is a danger of saturating the system so that amplitudes at some frequencies are so large that non-linearities are encountered and the test becomes invalid. Also whilst the PRBS signal overcomes one problem with impulse testing in that it excites all frequencies persistently, it is still not possible to suppress particular harmonics to prevent exciting particular modes in a composite structure. These disadvantages can be overcome by using a signal made up of equal-amplitude sinusoidal signals whose frequencies are those which one wishes to excite within a particular frequency range. One signal of this type is 'Schroeder phased harmonics' (Burrows *et al.* 1981).

If the system response to a multifrequency signal is recorded, the bearing properties may be obtained as follows. The displacements in the $x$ and $y$ directions occurring at a frequency $\omega$ are written in the form

$$x = X^s \sin \omega t + X^c \cos \omega t$$
$$y = Y^s \sin \omega t + Y^c \cos \omega t \tag{8.11}$$

respectively, then their derivatives with respect to time may be defined as

$$\dot{x} = \omega X^s \cos \omega t - \omega X^c \sin \omega t$$
$$\ddot{x} = -\omega^2 X^s \sin \omega t - \omega^2 X^c \cos \omega t$$
$$\dot{y} = \omega Y^s \cos \omega t - \omega Y^c \sin \omega t$$
$$\ddot{y} = -\omega^2 Y^s \sin \omega t - \omega^2 Y^c \cos \omega t \tag{8.12}$$

The forcing function may similarly be defined as

$$F_x = F_x^s \sin \omega t + F_x^c \cos \omega t$$
$$F_y = F_y^s \sin \omega t + F_y^c \cos \omega t \tag{8.13}$$

and the equations of motion of the journal, in the $x$ and $y$ directions as

$$\left. \begin{array}{l} F_x - K_{xx}x - K_{xy}y - C_{xx}\dot{x} - C_{xy}\dot{y} = M\ddot{x} \\ F_y - K_{yx}x - K_{yy}y - C_{yx}\dot{x} - C_{yy}\dot{y} = M\ddot{y} \end{array} \right\} \tag{8.14}$$

where $M$ is the mass of the journal and $K_{xx}$, $K_{yy}$ ... etc. are the oil film stiffness and damping coefficients. Substituting of Equations (8.11), (8.12), and (8.13) into (8.14) and comparing coefficients of sin $\omega t$ and cos $\omega t$ leads to the matrix equation

$$\begin{bmatrix} X^c & \omega X^s & Y^c & \omega Y^s & -F_x^c & -F_y^c \\ X^s & -\omega X^c & Y^s & -\omega Y^c & -F_x^s & -F_y^s \end{bmatrix} \begin{bmatrix} K_{xx} & K_{yx} \\ C_{xx} & C_{yx} \\ K_{xy} & K_{yy} \\ C_{xy} & C_{yy} \\ 1 & 0 \\ 0 & 1 \end{bmatrix} = \begin{bmatrix} M\omega^2 X^c & M\omega^2 Y^c \\ M\omega^2 X^s & M\omega^2 Y^s \end{bmatrix}$$

(8.15)

Equation (8.15) may be written for $\omega = \omega_0, 2\omega_0, 3\omega_0, \ldots, n\omega_0$, i.e. a total of $n$ times in all where the values of $\omega$ and the quantities in the first and last matrix of Equation (8.15) are determined by the Fourier transformation. All of these Equations (8.15) may be grouped as a single matrix equation (8.16):

$$[D]_{2n \times 6}[Z]_{6 \times 2} = [A]_{2n \times 2} \qquad (8.16)$$

Work by Burrows and Sahinkaya (1982) has suggested that the contents of the $[Z]$ matrix might be best obtained by means of a least squares estimator. This involves recognizing that measurements obtained in the laboratory will be inaccurate and so there are no values of the coefficients in the $[Z]$ matrix which will satisfy all lines of Equation (8.16). Instead a residual matrix is established which defines the 'errors' between the left-hand and right-hand sides of Equation (8.16) i.e. $[E] = [A] - [D][Z]$, where $[E]$ is a $2n \times 2$ matrix. The contents of the $[Z]$ matrix are defined as being those values which ensure that the sums of the squares of the elements in the $[E]$ matrix are minimized. It can be shown that such a value of $[Z]$ is given by

$$[Z] = [[D]^T[D]]^{-1}[D]^T[A]$$

Since the measured terms used to make up the $[D]$ and $[A]$ matrices are obtained via Fourier transformation of the input and output signals, noise occurring at a frequency greater than $n\omega_0$ is automatically filtered out of the analysis.

## 8.4 Use of centrifugal forces

One of the simplest ways of exciting a journal in a sinusoidal manner is by means of centrifugal forcing, simply by attaching imbalance masses of known magnitude to a rotating shaft. Clearly, one of the advantages of this method is that the capital outlay associated with electromagnetic and other forms of vibrating equipment is avoided. In this section three methods of coefficient measurement which use imbalance to excite the system are discussed, all of them involving processing of data measured in the time domain. Two of these approaches make use of an imbalance of known magnitude attached to the journal in question (and are based on the assumption that inherent rotor imbalance is insignificantly small). The other method involves use of a separate imbalance mass shaft, which can rotate at frequencies independent of the journal rotational frequency. Each of these methods is described.

### 8.4.1 Imbalance mass attached to the journal

This means of exciting the journal can be used to determine the bearing oil-film damping coefficients when the bearing stiffness coefficients are already known, for example as a result of applying the methods described in section 8.2. The experimental method involves measurement of the horizontal and vertical displacement amplitudes of the journal relative to the bearing, and of the bearing or pedestal itself relative to space; in addition, measurements of the corresponding phase lag angles of each of these displacements behind the imbalance force vector are also made.

It is not possible to use these measured system responses to synchronous imbalance to determine all eight bearing oil-film coefficients because there is insufficient information available to form more than four simultaneous equations in the eight coefficients. Consider the rotor–bearing–pedestal system shown in Figure 8.7. The system is considered to be symmetrical, with the central rotor of mass $2M$ mounted on a light rigid shaft which is supported in two identical journal bearings. The rotor is subjected to a known imbalance force of magnitude $2F$, which in the case of an experimental test rig would be provided by attaching an imbalance mass of known magnitude to the rotor. The imbalance mass must be sufficiently large to ensure that any other residual imbalance in the system is insignificantly small in comparison. Because the excitation is provided by imbalance, the corresponding forces on the rotor in the horizontal and vertical directions are constrained to be one quarter of a revolution, or $90°$, out of

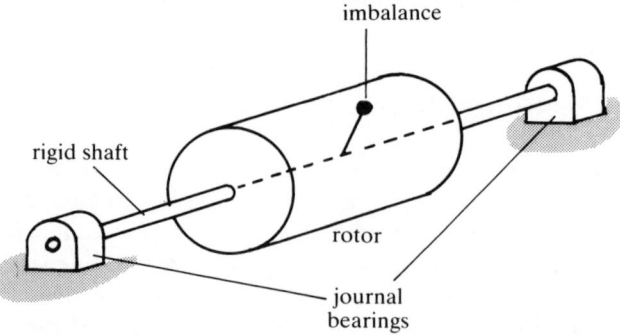

**Figure 8.7** Diagram of an unbalanced rotor-bearing system.

phase with each other, thus they can be written as

$F_x = 2F \sin \omega t$ (in the horizontal direction)

$F_y = 2F \cos \omega t$ (in the vertical direction).

In addition to the imbalance forces the rotor also has the oil film forces acting on it, these being transmitted to the rotor by the shaft. These forces act to oppose motion of the shaft away from its steady state running position. Thus the equations of motion for the rotor mass may be written as

$$2F_x \sin \omega t - 2K_{xx}x - 2K_{xy}y - C_{xx}\dot{x} - C_{xy}\dot{y} = 2M(\ddot{x} + \ddot{x}_p) \quad (8.17)$$

in the horizontal direction, and

$$2F_y \cos \omega t - 2K_{yx}x - 2K_{yy}y - 2C_{yx}\dot{x} - 2C_{yy}\dot{y} = 2M(\ddot{y} + \ddot{y}_p) \quad (8.18)$$

in the vertical direction, where the stiffness and damping coefficients $K_{xx}$, $K_{xy}$ ... etc., are those of one bearing and where the subscript p indicates reference to the bearing pedestal assembly. The displacements of the journal relative to the bearing or pedestal take the form

$x = X \sin(\omega t - \alpha_1)$ (in the horizontal direction) (8.19)

$y = Y \cos(\omega t - \alpha_2)$ (in the vertical direction). (8.20)

Similarly, the displacements of the bearing pedestal in the horizontal and vertical directions are non-zero, since the pedestal itself is flexible, and these may be written as

$x_p = X_p \sin(\omega t - \gamma_1)$ (in the horizontal direction) (8.21)

$y_p = Y_p \cos(\omega t - \gamma_2)$ (in the vertical direction). (8.22)

Equations (8.19)–(8.22) may be differentiated with respect to time to obtain expressions for $\dot{x}$, $\dot{y}$, $\ddot{x}$, $\ddot{y}$, $\ddot{x}_p$ and $\ddot{y}_p$ which may be substituted back into Equations (8.17) and (8.18) to give the following matrix equation:

$$\begin{bmatrix} M\omega^2 - K_{xx} & K_{xy} & \omega C_{xx} & \omega C_{xy} \\ -K_{yx} & K_{yy} - M\omega^2 & \omega C_{yx} & \omega C_{yy} \\ \omega C_{xx} & -\omega C_{xy} & K_{xx} - M\omega^2 & K_{xy} \\ \omega C_{yx} & -\omega C_{yy} & K_{yx} & K_{yy} - M\omega^2 \end{bmatrix} \begin{bmatrix} X \sin \alpha_1 \\ Y \cos \alpha_2 \\ X \cos \alpha_1 \\ Y \sin \alpha_2 \end{bmatrix}$$

$$= \begin{bmatrix} -M\omega^2 X_p \sin \gamma_1 \\ F_y + M\omega^2 Y_p \cos \gamma_2 \\ F_x + M\omega^2 X_p \cos \gamma_1 \\ M\omega^2 Y_p \sin \gamma_2 \end{bmatrix} \quad (8.23)$$

All of the quantities in Equation (8.23) are either known or are measured during the course of the experiment, other than the oil-film coefficients.

Thus the four equations which go to make up the matrix equation (8.23) can be solved simultaneously to yield the bearing coefficients.

Unfortunately, the fact that there are only four equations means that only four of the unknown coefficients can be solved for, and these 'solutions' will be in terms of the other unknowns if there are any. Thus it is convenient to obtain the four stiffness coefficients by the method described in section 8.2 and to use this technique to obtain the damping coefficients.

An alternative approach to the treatment of experimental data captured from test rigs similar to that shown in Figure 8.7 has been developed by Sahinkaya and Burrows (1984). The method utilizes the Fourier transform of the measured shaft displacements in the $x$ and $y$ directions, defined as

$$x = X^s \sin \omega t + X^c \cos \omega t \quad (8.24)$$

$$y = Y^s \sin \omega t + Y^c \cos \omega t \quad (8.25)$$

These expressions can also be differentiated with respect to time to give expressions for $\dot{x}$, $\dot{y}$, $\ddot{x}$, and $\ddot{y}$ also in terms of $X^s$, $X^c$, $Y^s$, $Y^c$ and $\omega$. Such Fourier transforms are obtained for many measurements, each taken at a discrete time interval. These values are then substituted into the system equations of motion (8.17) and (8.18) which are written for a system with rigid pedestals in the following matrix form.

$$[x \ y \ \dot{x} \ \dot{y} \ Fx \ Fy] \cdot \begin{bmatrix} K_{xx} & K_{yx} \\ K_{xy} & K_{yy} \\ C_{xx} & C_{yx} \\ C_{xy} & C_{yy} \\ -1 & 0 \\ 0 & -1 \end{bmatrix} = [-M\ddot{x} \ -M\ddot{y}] \quad (8.26)$$

or simply

$$[D][Z_1 \ Z_2] = [A_1 \ A_2]$$

Estimation of the oil-film coefficients in the $[Z_1 \ Z_2]$ matrix is then based on minimizing the sum of the squares of the errors between the left-hand and right-hand sides of Equation (8.26), recognizing that if $n$ measurements are taken these will not all be identical. The errors may be defined as

$$[e_1 \ e_2] = [A_1 \ A_2] - [D][Z_1 \ Z_2]$$

and the cost function to be minimized as the integral over a vibration cycle of the sum of the errors $e_1$ and $e_2$ squared, i.e.

$$C = \int_0^{2\pi/\omega} [e_1^2 + e_2^2] \, dt$$

On differentiation and equating to zero this gives

$$\left[\int_0^{2\pi/\omega} [D]^T[D] \, dt\right] [Z_1 \quad Z_2] = \left[\int_0^{2\pi/\omega} [D]^T[A_1] \, dt \quad \int_0^{2\pi/\omega} [D]^T[A_2] \, dt\right]$$

or simply

$$[D]_{6\times 6}[Z_1 \quad Z_2]_{6\times 2} = [M]_{6\times 2}$$

from which the contents of the $[Z_1 \quad Z_2]$ matrix can be obtained as

$$[Z_1 \quad Z_2] = [D]^{-1}[M]$$

In practice the form of the $[D]$ matrix is such that it can only be inverted provided its order is no more than $2 \times 2$. Thus again only four coefficient estimates may be obtained.

If all eight coefficients are to be obtained then some other means has to be found to obtain additional information about the system behaviour under different test rig operating conditions. For example, if in Equation (8.23) either the journal mass or the pedestal stiffness could be changed so as to significantly affect the values of $M$, or of $X$ and $Y$, then a further set of equations describing the system behaviour could be obtained. These changes to the test rig would probably not be easy to provide, however, and the author is not aware of any previous attempts to measure bearing coefficients in this way. Changing the magnitude of the imbalance is not an acceptable alternative since this would simply result in a proportional change in the measured amplitudes, so long as these were maintained small enough for the oil film to still behave linearly. Thus the same set of four equations (8.23) would be repeated. Alternatively if a large change in imbalance were made, so as not to obtain proportional changes in measured amplitudes, then the oil-film non-linearity would be affecting the results and a different set of oil-film coefficients would be necessary to describe the oil film properly for the second phase of tests. It is equally unacceptable to change the journal rotational frequency, and thereby change $\omega$ in Equation (8.23), because a change in journal rotational frequency upsets the bearing Sommerfeld number and steady-state operating position described by the eccentricity ratio and attitude angle. Thus under these circumstances we would no longer expect the oil film to behave as though the same set of bearing coefficients were appropriate anyway. This problem has, however, been overcome by using the experimental approach described below.

### 8.4.2 *Imbalance mass attached to an independent vibrator shaft*

In the preceding section 8.4.1 it was shown that the matrix equation (8.23) would describe the behaviour of an experimental test rig under particular operating conditions, and it was shown that a further set of four simultaneous equations are necessary in order to solve for the eight oil-film coefficients. This problem has been overcome (Bannister 1972, Goodwin 1981), by using a test rig capable of providing excitation by means of imbalance forcing, where the forcing frequency could be varied without upsetting the journal rotational frequency. Thus a second equation (8.23) could be obtained, resulting in eight simultaneous equations in all, by using a different value of $\omega$ without upsetting the bearing Sommerfeld number, etc.

Previous investigations using this approach have involved the use of a rigid rotor test rig similar to that shown in Figure 8.8. The test journal itself has contained within it a second shaft on which imbalance masses are sited, the second shaft being driven from the end of the rig opposite to the main journal drive. There are auxiliary bearings between the main journal and the imbalance mass shaft, such that the two can be driven independently at different speeds. In this way it is possible to excite the test bearing by means of a rotating imbalance, and to vary the rotational frequency of the imbalance mass shaft without changing that of the test journal itself. Furthermore the second value of exciting frequency $\omega$ can be chosen to be the same speed as that of the first, but in the opposite direction, thereby enabling the test to be completed with a constant value of exciting force (thereby ensuring that oil-film non-linearity effects are kept to a minimum), without having to change the imbalance mass.

Experimental measurements of vibration amplitude should be made with instrumentation which incorporates a tracking filter to ensure that the vibrations measured are those caused by the known rotating imbalance and not by test journal imbalance. In this way the effect of noise at frequencies not close to the excitation frequency, and of journal imbalance, and run out, is filtered out of the measurements.

The advantages of using imbalance force to excite the journal are that the method is very simple to use and does not necessitate the purchase of electromagnetic or other types of shaking equipment which are relatively

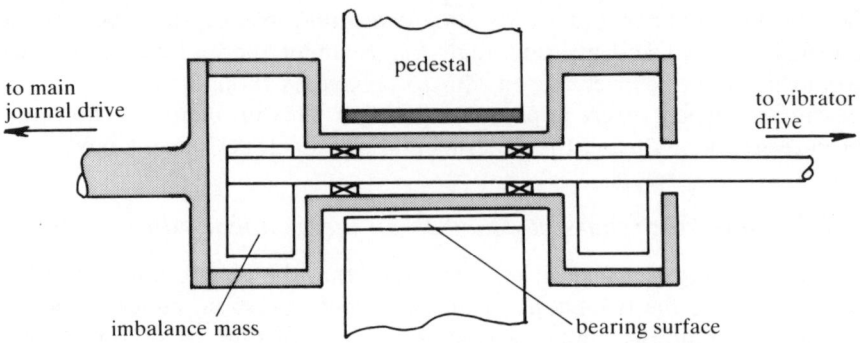

**Figure 8.8** A method of forcing the journal using an independent unbalanced vibrator shaft.

expensive. Furthermore, the experimental data may be processed relatively easily to determine the oil-film coefficients. On the other hand, this method of coefficient derivation utilizes Equations (8.23) which are frequently ill-conditioned so that only small inaccuracies in the measured values of phase lag angle can result in significant changes in derived coefficient value. In addition, changing the imbalance force or frequency may necessitate stopping the rig completely to increase or decrease the imbalance which is present. Unlike the methods involving system excitation using another form of vibrator, the use of centrifugal forcing necessitates system excitation at only one frequency at a time, namely the imbalance shaft rotational frequency, and thus if system response at several different frequencies is to be taken into account it is not possible to save laboratory time by investigating all of these at once.

EXAMPLE 8.4

The eight bearing stiffness and damping coefficients are to be determined by using the method described in section 8.4.2. Experimental measurements of journal vibration amplitude and phase lag angle are given in Table 8.1; pedestal vibrations are found to be negligible. Determine the values of the oil-film coefficients implied by these measurements, and the maximum change in the direct and cross-coupling terms introduced by an error of $+4°$ in the measurement of phase recorded as $42.5°$.

SOLUTION

Substitution of the data from Table 8.1 into Equations (8.23) for both forward and reverse excitation (negative values are assigned to phase lag angles measured for reverse excitation) results eight simultaneous equations (p. 236):

**Table 8.1** Some test data used to calculate bearing stiffness and damping coefficients

|  | Forward excitation | Reverse excitation |
|---|---|---|
| Horizontal vibration amplitude | 66.4 $\mu$m | 46.6 $\mu$m |
| Horizontal phase lag | 42.5° | 20.9° |
| Vertical vibration amplitude | 55.5 $\mu$m | 38.4 $\mu$m |
| Vertical phase lag | 9.9° | 111° |
| Force amplitude | 1.0 kN | 1.0 kN |
| Forcing frequency | 12.6 Hz | 12.6 Hz |
| Journal mass | 150 kg | 150 kg |

$$\begin{bmatrix} -44.89 \times 10^{-6} & 54.7 \times 10^{-6} & 0 & 0 & 3.91 \times 10^{-3} & 0 & 7.61 \times 10^{-4} & 0 \\ 0 & 0 & -44.89 \times 10^{-6} & 54.7 \times 10^{-6} & 0 & 3.91 \times 10^{-3} & 0 & 7.61 \times 10^{-4} \\ -48.92 \times 10^{-6} & 9.52 \times 10^{-6} & 0 & 0 & 3.59 \times 10^{-3} & 0 & -4.38 \times 10^{-3} & 0 \\ 0 & 0 & 48.92 \times 10^{-6} & 9.52 \times 10^{-6} & 0 & 3.59 \times 10^{-3} & 0 & -4.38 \times 10^{-3} \\ 16.65 \times 10^{-6} & -13.7 \times 10^{-6} & 0 & 0 & -3.48 \times 10^{-3} & 0 & 2.87 \times 10^{-3} & 0 \\ 0 & 0 & 16.65 \times 10^{-6} & -13.7 \times 10^{-6} & 0 & -3.48 \times 10^{-3} & 0 & 2.87 \times 10^{-3} \\ 43.48 \times 10^{-6} & -35.88 \times 10^{-6} & 0 & 0 & 1.33 \times 10^{-3} & 0 & -1.1 \times 10^{-3} & 0 \\ 0 & 0 & 43.48 \times 10^{-6} & -35.88 \times 10^{-6} & 0 & 1.33 \times 10^{-3} & 0 & -1.1 \times 10^{-3} \end{bmatrix} \times \begin{Bmatrix} K_{xx} \\ K_{xy} \\ K_{yx} \\ K_{yy} \\ C_{xx} \\ C_{xy} \\ C_{yx} \\ C_{yy} \end{Bmatrix} = \begin{Bmatrix} -4.49 \times 10^{-5} \\ 1.0 \times 10^{3} \\ 1.0 \times 10^{3} \\ 9.52 \times 10^{-6} \\ 1.67 \times 10^{-5} \\ 1.0 \times 10^{3} \\ 1.0 \times 10^{3} \\ -3.59 \times 10^{-5} \end{Bmatrix}$$

These may be solved to give the following values of bearing coefficients:

$K_{xx} = 18.0$ MN.m$^{-1}$     $C_{xx} = 213.8$ kN.s.m$^{-1}$
$K_{xy} = -2.49$ MN.m$^{-1}$    $C_{xy} = 142.5$ kN.s.m$^{-1}$
$K_{yx} = 27.6$ MN.m$^{-1}$     $C_{yx} = 140.0$ kN.s.m$^{-1}$
$K_{yy} = 24.1$ MN.m$^{-1}$     $C_{yy} = 476.3$ kN.s.m$^{-1}$

If the analysis of the data is repeated, but with the angle recorded as 42.5° assigned, instead of a value of 46.5°, then the coefficient values obtained are

$K_{xx} = 18.6$ MN.m$^{-1}$     $C_{xx} = 236.3$ kN.s.m$^{-1}$
$K_{xy} = -1.77$ MN.m$^{-1}$    $C_{xy} = 170.0$ kN.s.m$^{-1}$
$K_{yx} = 31.2$ MN.m$^{-1}$     $C_{yx} = 133.8$ kN.s.m$^{-1}$
$K_{yy} = 27.4$ MN.m$^{-1}$     $C_{yy} = 503.8$ kN.s.m$^{-1}$

This represents a change of $-29\%$ in $K_{xy}$ and of $+14\%$ in $K_{yy}$.

## 8.5 Transient methods – measurement on a running machine

In previous sections the methods used to measure oil-film impedance have involved the use of a vibration source which would operate continuously at a specified frequency (or frequencies). Only in the case of the multi-frequency method was it possible to investigate the oil-film behaviour at several vibration frequencies at the same time, and hence to save time in the laboratory. This section describes two alternative techniques used to measure oil-film impedance which make use of transient shaft vibrations. In each case the transient vibrations themselves are composed of shaft displacements occurring at many different frequencies simultaneously. The use of transients enables many frequencies to be investigated, thereby enabling more accurate values of oil-film impedance to be obtained, and allows these investigations to take place simultaneously, thereby saving laboratory time (as in the case of multifrequency testing). The methods described do not require special signal generating equipment, as in the case of some of the multifrequency methods, and the necessary data analysis is no more complex. The first method described has been used by Nordmann and Schöllhorn (1980) with a model rotor-bearing test rig, and with such a system test repetition is relatively easy. The second method described has been used by Morton (1974, 1975) to investigate the oil-film characteristics of journal bearings on a running machine comprising bearing, pedestals, and a massive flexible rotor. In this instance the sheer size of the equipment means that repetition of the test is relatively more involved. In both cases it is important that instrumentation capable of recording all of the appropriate test signals throughout the duration of the test (which may last for only a few seconds) should be available.

The method used by Nordman makes use of a system consisting of a symmetrical rigid rotor mounted in two identical journal bearings similar to that shown in Figure 8.9. Transient vibrations of the rotor in the bearings are caused by applying a force impulse to the rotor centre of gravity. In practice this is provided by striking the rotor with a 'calibrated' hammer whose head mass is known. This means of excitation results in an impulse which lasts for a finite period of time (typically a fraction of a second) and, provided that an accelerometer can be mounted in the hammer head, it is possible to determine the instantaneous force which is applied to the rotor. The electrical output from the hammer will then indicate the variation of the applied force with time, and hence describes the impulse that has been applied. The shape of the resulting force time graph, as shown in Figure 8.10(a) (as for Figure 8.6(a) and (b)), can actually be considered to be made up of a number of sine waves of different frequencies, all occurring simultaneously. The same signal is shown in the frequency domain in Figure 8.10(b). By varying the hammer head mass, the stiffness of the hammer impact face, and the initial hammer head velocity, it is possible to vary the amplitude, frequency content, and duration of the applied impulse. Thus it is possible to ensure that the bearing response to signals occurring at particular frequencies is taken into consideration. For example, in the case of steam turbine bearings imbalance forcing occurs at 50 Hz, so it would be logical for this to be close to the centre frequency of the frequency band in Figure 8.10(b) during testing of such a bearing. Shaft displacements, relative to the bearing bush, and the force impulse, are measured and the resulting electrical signals can be processed immediately to yield oil-film coefficients, provided that suitable instrumentation is available. Alternatively, all of the signals can be stored on magnetic tape (which must be

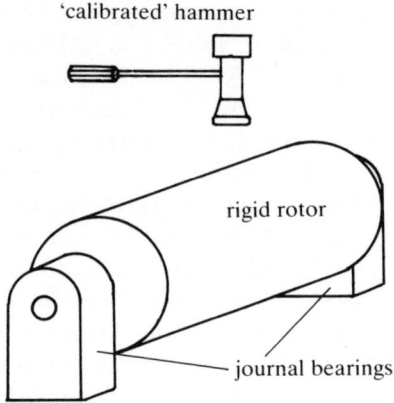

**Figure 8.9** Impulse testing to obtain bearing coefficients.

controlled manner. The alternative approach described makes use of a form of step input to the system, as opposed to an impulse. Morton suggested that a suitable input could be provided by providing a slowly increasing 'static' load on the rotor in one direction, for example with a crane, via a calibrated link which would fail at a known load. Such a method of loading would provide a force function similar to that indicated in Figure 8.13(a) with a resulting system response of the form indicated in Figure 8.13(b). In Morton's experiment the loading strap is connected to the machine rotor by means of an oil-lubricated 'foil' bearing, a system which itself has an oil-film impedance and a mass associated with it. Because of the inertia associated with the foil bearing, when the loading link fails the force applied to the rotor does not decay immediately to zero; instead the forcing function in Figure 8.13(a) may be considered to be modified as shown in Figure 8.13(c). A receptance may be obtained as described above by dividing the Fourier transform of the displacement by that of the force input, which for a force function as shown in Figure 8.13(c) is given by

$$R(\omega) = \left[ \frac{X_0}{F_0} - \frac{j\omega}{F_0} \int_0^\infty (x)t \exp(-j\omega t)\, dt \right] M \qquad (8.35)$$

where

$$M = \left[ 1 - \frac{j\omega(j\omega + \omega_1 \cdot \exp(-j\omega\, \Delta t))}{\omega_1^2 - \omega^2} \right]^{-1}$$

In Equation (8.35), the term $M$ is a factor which modifies the right-hand side of the equation to allow for the forcing function being modified as shown in Figure 8.13(c).

The flexibility of the shaft may be allowed for in the analysis as follows. In Figure 8.14 the shaft AC is mounted in a bearing at C (and at any other number of axial locations along the shaft). The shaft has a force F applied to it at B, which would cause some displacement $_cX_1$ at the location of C if no bearing were sited there. The displacement $_cX_1$ is related to the force $F$

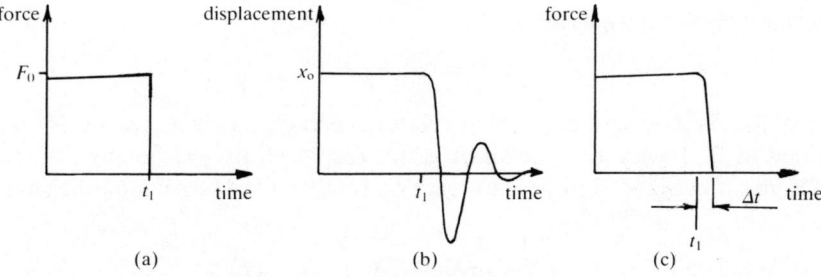

**Figure 8.13** Step input forcing function and displacement response, (a) ideal step input force; (b) system displacement response; (c) modification of input force by loading system inertia.

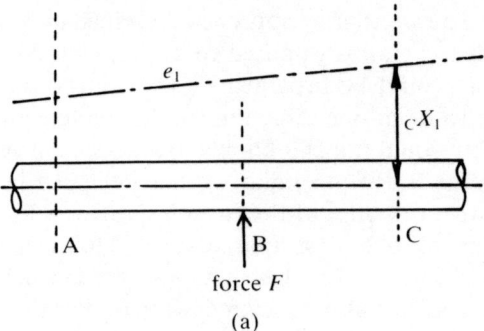

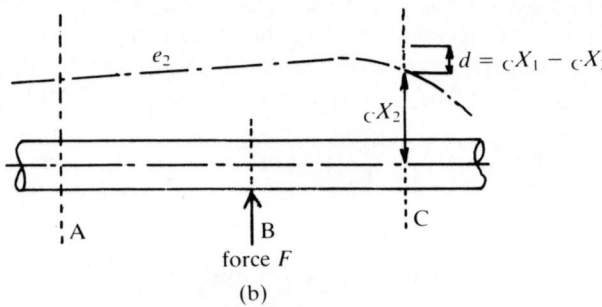

**Figure 8.14** Shaft displacements caused by an external force, (a) force F at B, no bearing reaction; (b) force F at B, bearing reaction at C.

by the mass/elastic properties of the shaft, these are described by an effective shaft impedance relating the forcing at B to the displacement at C, i.e.

$$_B F_x = {_x Z_{BC}} {_C X_1}$$

where $Z$ is the shaft impedance relating a forcing at B to the response at C. Alternatively we may write

$$_C X_1 = {_x R_{BC}} {_B F_x}$$

where $R_{BC}$ is a receptance which relates the response at C to the forcing applied at B. In practice the shaft is not displaced by $_C x_1$ but by $_C x_2$, the difference $d$ being attributable to reaction forces of the bearing on the shaft, where

$$d = {_C X_2} - {_C X_1} = {_C X_2} - {_x R_{BC}} {_B F_x}$$

This response of the shaft $d$ is related to the force provided by the bearing on the shaft at C by a shaft impedance term $_xZ_{CC}$ where

$$_{CC}F_x = {_xZ_{CC}}[{_cX_2} - {_xR_{BC}}{_BF_x}]$$

Similarly the shaft displacement at any other bearing location N may offer a contribution to the total force on the shaft at C given by

$$_{CN}F_x = {_xZ_{CN}}[{_NX_2} - {_xR_{BN}}{_BF_x}]$$

Thus, for a system with bearings located at A and C, the total force at bearing C would be given by

$$_cF_x = [{_xZ_{CA}}\ {_xZ_{CC}}]\begin{bmatrix} {_AX_2} - {_xR_{BA}}{_BF_x} \\ {_cX_2} - {_xR_{BC}}{_BF_x} \end{bmatrix}$$

A similar expression can then be developed which describes the total force on the shaft at C in the $y$ direction, i.e.

$$_cF_y = [{_yZ_{CA}}\ {_yZ_{CC}}]\begin{bmatrix} {_AY_2} - {_yR_{BA}}{_BF_y} \\ {_cY_2} - {_yR_{BC}}{_BF_y} \end{bmatrix}$$

and these two equations may be combined into the single matrix equation

$$\begin{bmatrix} {_cF_x} \\ {_cF_y} \end{bmatrix} = \begin{bmatrix} {_xZ_{CA}} & {_xZ_{CC}} & 0 & 0 \\ 0 & 0 & {_yZ_{CA}} & {_yZ_{CC}} \end{bmatrix} \begin{bmatrix} {_AX_2} - {_xR_{BA}}{_BF_x} \\ {_cX_2} - {_xR_{BA}}{_BF_x'} \\ {_AY_2} - {_yR_{BA}}{_BF_y} \\ {_cY_2} - {_yR_{BC}}{_BF_y} \end{bmatrix} \quad (8.36)$$

Equation (8.36) may be written for two specific cases: when the forcing at B is in the $x$ direction only, and also when it is in the $y$ direction only. Combining the two resulting matrix equations gives

$$\begin{bmatrix} {_{xc}F_x} & {_{yc}F_x} \\ {_{xc}F_y} & {_{yc}D_y} \end{bmatrix} = \begin{bmatrix} {_xZ_{CA}} & {_xZ_{CC}} & 0 & 0 \\ 0 & 0 & {_yZ_{CA}} & {_yZ_{CC}} \end{bmatrix}$$

$$\cdot \begin{bmatrix} {_AX_2} - {_xR_{BA}}{_BF_x} & {_AX_2} \\ {_cX_2} - {_xR_{BA}}{_BF_x} & {_cX_2} \\ {_AY_2} & {_AY_2} - {_yR_{BA}}{_BF_y} \\ {_cY_2} & {_cY_2} - {_yR_{BC}}{_BF_y} \end{bmatrix} \quad (8.37)$$

But the forces described by Equation (8.37) are related to the oil-film stiffness and damping of the bearing at C by the matrix equation

$$-\begin{bmatrix} {_{xc}F_x} & {_{yc}F_x} \\ {_{xc}F_y} & {_{yc}F_y} \end{bmatrix} = \begin{bmatrix} K_{xx} + j\omega C_{xx} & K_{xy} + j\omega C_{xy} \\ K_{yx} + j\omega C_{yx} & K_{yy} + j\omega C_{yy} \end{bmatrix} \begin{bmatrix} {_{xc}X_2} & {_{yc}X_2} \\ {_{xc}Y_2} & {_{yc}Y_2} \end{bmatrix} \quad (8.38)$$

Dividing throughout by $_BF_x$ and $_BF_y$ in the first and second columns respectively of Equations (8.37) and (8.38), and then combining the two, leads to the following expression describing the bearing oil-film coefficients:

$$\begin{bmatrix} K_{xx} + j\omega C_{xx} & K_{xy} + j\omega C_{xy} \\ K_{yx} + j\omega C_{yx} & K_{yy} + j\omega C_{yy} \end{bmatrix} = \begin{bmatrix} {}_xZ_{CA} & {}_xZ_{CC} & 0 & 0 \\ 0 & 0 & {}_yZ_{CA} & {}_yZ_{CC} \end{bmatrix}$$

$$\begin{bmatrix} \left[ {}_xR_{BA} - \frac{{}_AX_2}{{}_BF_x} \right] & -\frac{{}_AX_2}{{}_BF_y} \\ \left[ {}_xR_{BC} - \frac{{}_CX_2}{{}_BF_x} \right] & -\frac{{}_CX_2}{{}_BF_y} \\ -\frac{{}_AY_2}{{}_BF_x} & \left[ {}_yR_{BA} - \frac{{}_AY_2}{{}_BF_y} \right] \\ -\frac{{}_CY_2}{{}_BF_x} & \left[ {}_yR_{BC} - \frac{{}_CY_2}{{}_BF_y} \right] \end{bmatrix} \begin{bmatrix} \frac{{}_{xC}X_2}{{}_BF_x} & \frac{{}_{yC}X_2}{{}_BF_y} \\ \frac{{}_{xC}Y_2}{{}_BF_y} & \frac{{}_{yC}Y_2}{{}_BF_y} \end{bmatrix} \quad (8.39)$$

The individual values of the oil-film stiffness and damping coefficients may then be obtained, from the real and imaginary components, respectively, of corresponding elements in the resulting matrix on the right-hand side of Equation (8.39). If care is taken in the choice of loading position, such that B may be considered to be coincident with C, and if the distances between bearings is large, such that resulting shaft displacements at other bearings are negligible, then Equation (8.39) reduces to

$$\begin{bmatrix} K_{xx} + j\omega C_{xx} & K_{xy} + j\omega C_{xy} \\ K_{yx} + j\omega C_{yx} & K_{yy} + j\omega C_{yy} \end{bmatrix} =$$

$$\begin{bmatrix} \left[ 1 - {}_xZ_{CC} \frac{{}_CX_z}{{}_BF_x} \right] & -{}_xZ_{CC} \cdot \frac{{}_CX_2}{{}_BF_y} \\ -{}_yZ_{CC} \cdot \frac{{}_CY_z}{{}_BF_x} & \left[ 1 - {}_yZ_{CC} \cdot \frac{{}_CY_2}{{}_BF_y} \right] \end{bmatrix} \begin{bmatrix} \frac{{}_{xC}X_2}{{}_BF_x} & \frac{{}_{yC}X_2}{{}_BF_y} \\ \frac{{}_{xC}Y_2}{{}_BF_x} & \frac{{}_{yC}Y_2}{{}_BF_y} \end{bmatrix} \quad (8.40)$$

The terms $X/F$ and $Y/F$ in Equations (8.39) and (8.40) are then determined according to Equation (8.35), whilst the shaft impedance and receptance terms, $Z$ and $R$ respectively, are best found by experimental measurement, the shaft being supported so that its motion is completely unrestricted ('free–free' support conditions). In practice, these terms may be found to be complex depending on the significance of shaft internal damping.

# References

Bannister, R. H. 1972. Non-linear oil film force coefficients for a journal bearing operating under aligned and misaligned conditions. Ph.D. thesis, Aston University.

Burrows, C. R. & M. N. Sahinkaya 1982. Frequency domain estimation of linearised oil film coefficients. *Trans. ASME, J. Lub. Tech.* **104**, April, 210–215.

Burrows, C. R. & R. Stanway 1977. Identification of journal bearing characteristics. *J. Dynam. Systems Measurements and Control*. September, 167–173.

Burrows, C. R., R. Stanway & R. Sayed-Esfahani 1981. A comparison of multifrequency techniques for measuring the dynamics of squeeze-film bearings. *Trans. ASME, J. Lub. Tech.* **103**, January, 137–143.

Glienicke, J. 1966. Experimental investigaton of the stiffness and damping coefficients of turbine bearings and their application to instability prediction. *Proc. Inst. Mech. Engrs.* **181**, 3b, 116–129.

Goodwin, M. J. 1981. Variable impedance bearings for large rotating machinery. Ph.D. thesis, Aston University.

Mitchell, J. R., R. Holmes & H. Van Ballegooyen 1966. Experimental determination of a bearing oil film stiffness. *Proc. Inst. Mech. Engrs.* **180**, 3k, 90–96.

Morrison, D. 1962. Influence of plain journal bearings on the whirling action of an elastic rotor. *Proc. Inst. Mech. Engrs.* **176**, 22, 542–553.

Morton, P. G. 1971. Measurement of the dynamic characteristics of a large sleeve bearing. *Trans. ASME, J. Lub. Tech.* January, 143–150.

Morton, P. G. 1974. The derivaton of bearing characteristics by means of transient excitation applied directly to a rotating shaft. *Proc. IUTAM Symp. Dynamics of Rotors*, Copenhagen.

Morton, P. G. 1975. Dynamic characteristics of bearings – measurement under operating conditions. *GEC J. Science and Technology* **42**, 1, 37–47.

Nordmann, R. & K. Schöllhorn. 1980. Identification of stiffness and damping coefficients of journal bearings by means of the impact method. *IMechE Conf. Vibrations in Rotating Machinery*, Cambridge. Paper C285/80, 231–238.

Parkins, D. W. 1976. Static and dynamic characteristics of a hydrodynamic journal bearing. Ph.D. thesis, Cranfield Institute of Technology.

Parkins, D. W. 1979. Theoretical and experimental determination of the dynamic characteristics of a hydrodynamic journal bearing. *Trans. ASME, J. Lub. Tech.* **101**, 129–139.

Sahinkaya, M. N. & C. R. Burrows 1984. Estimation of linearised oil film parameters from the out-of-balance response. *Proc. Inst. Mech. Engrs.* **198c**, 8, 131–135.

Stanway, R., C. R. Burrows & R. Holmes 1979. Discrete time modelling of a squeeze film bearing. *J. Mech. Eng. Sci. Inst. Mech. Engrs.* **21**, 6, 419–427.

Tripp, H. & B. T. Murphy 1984. Eccentricity measurements on a tilting pad bearing. *Trans. ASLE*, **28**, 2, 217–224.

Woodcock, J. S. & R. Holmes 1970. Determination and application of the dynamic properties of a turbo-rotor bearing oil film. *Proc. Inst. Mech. Engrs.* **184**, 31, 111–119.

# 9 Measurements and diagnostics

## 9.1 Introduction

On most types of rotating machinery the system vibration is measured regularly in order to monitor the condition of the machine, and to enable machine faults to be detected and corrected as soon as they arise. High levels of vibration are indicative of high levels of component stresses, high noise levels, and reduced machine fatigue life. Measurements are usually taken of the system vibration amplitude, its phase, and its frequency; in many instances the vibration may be composed of several sinusoidal signals all at different frequencies and it is necessary to distinguish the component signals from each other. In the case of flexible bearings the steady running position of the journal in the bearing is also usually monitored. These measurements can be processed and displayed in such a way as to enable judgements to be made about the condition of the machine in question, and will usually help in the diagnosis of some fault conditions. The aim of this chapter is to describe some of the means of measuring and recording data, and to discuss the diagnosis of the more common fault conditions. Monitoring strategies and standards are also described.

## 9.2 Signal measurement and display

Any vibration signal collected has ultimately to be measured and displayed. The quantities that have to be measured are usually the vibration amplitude, its frequency, and its phase. The transducer from where the signal originates will usually be either

- a non-contacting proximity transducer providing a voltage signal proportional to the gap between the probe tip and the shaft,
- a velocity transducer mounted on the machine, providing a signal proportional to its velocity, or
- an accelerometer mounted on the vibrating machine, providing a signal proportional to its acceleration.

In applications where phase is to be measured, a reference signal indicating a particular location on the shaft will be required in addition to the vibration signal which is to be measured; this 'key phasor' signal might be obtained by using, say, a proximity transducer to detect the passing of a bolt head on the shaft, or by an optical transducer detecting the passing of light-reflecting tape adhered to the shaft; alternatively some other signal which produces a well-defined peak can be used. The key phasor signal is used in conjunction with the vibration signal, the phase angle recorded being that between the peaks of the two signals. The arrangement is shown in Figure 9.1.

Signals received from the transducers have to be measured and displayed in convenient forms. The instruments most frequently used for signal measuring in monitoring of rotating machine vibration are oscilloscopes, tracking filters and spectrum analysers. The choice of measuring instrument also affects the way in which the information collected is displayed, as explained below.

An oscilloscope provides an immediate visual representation of the vibration signal, enabling the engineer to observe the form of the signal (sine wave or otherwise), its frequency and its amplitude. This is known as 'time domain' data presentation, since the value indicated by the measuring instrument is a function of time. Most oscilloscopes also provide an '$x$ on $y$' facility which allows two vibration signals from, say, two proximity transducers mounted perpendicular to each other, to be input to the oscilloscope. These signals can then be displayed in the form of a shaft whirl orbit instead of the normal time-base representation. Some oscilloscopes enable a phase-indicating pulse of extra-bright beam to be displayed on the orbit once per revolution of the shaft. The shape of the whirl orbit, or Lissajous figure, can itself be a useful tool in monitoring machine health. A typical whirl orbit obtained for a machine subject to a small amount of

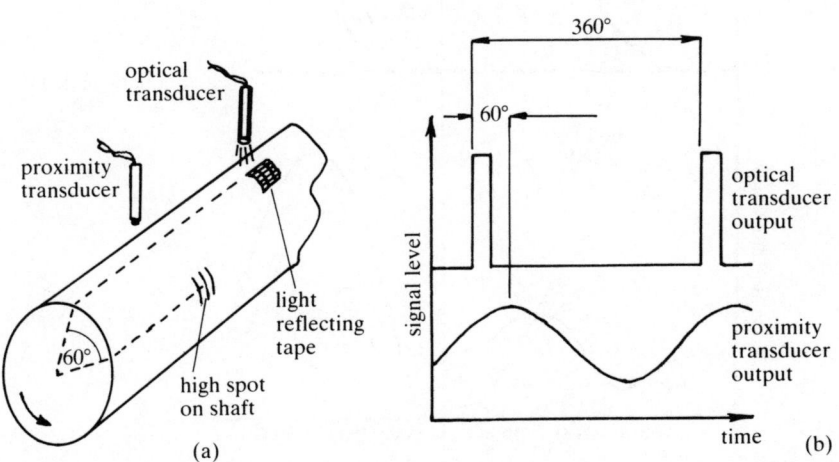

**Figure 9.1** Phase measurement using an optical transducer, (a) shaft and transducers; (b) transducer output.

imbalance is shown in Figure 9.2; some further examples of shaft whirl orbits obtained with different machine faults are shown later in this chapter.

A tracking filter is a device which accepts two input signals, one being the vibration signal under consideration and the other being a phase reference signal as discussed above. The tracking filter removes from the vibration signal any components which are not of the same frequency as the reference signal. The amplitude and phase (relative to the reference signal) of the

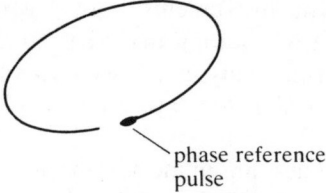

**Figure 9.2** Shaft whirl orbit as displayed on an oscilloscope.

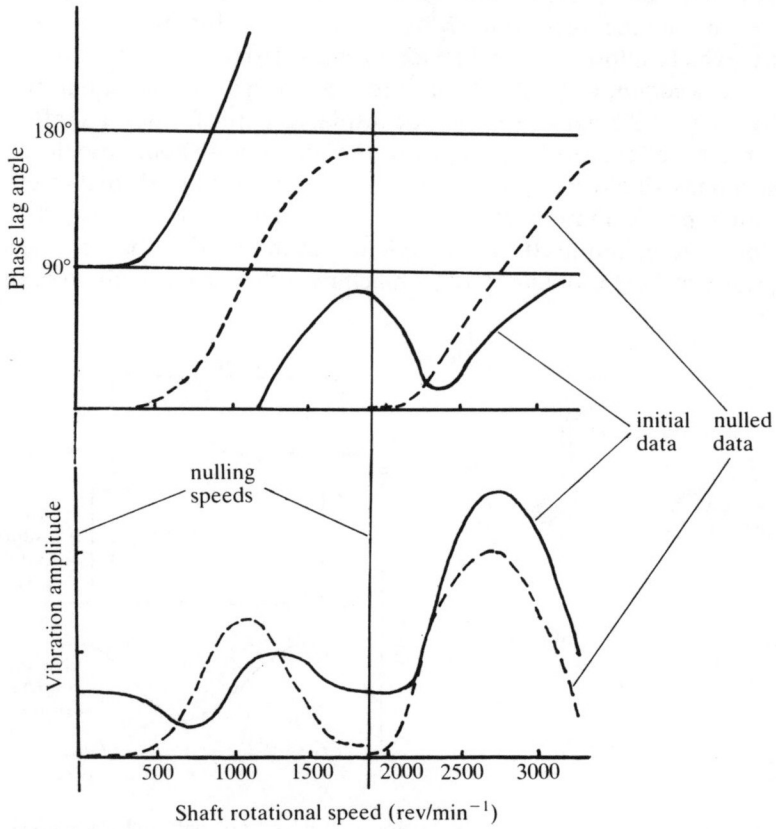

(a)

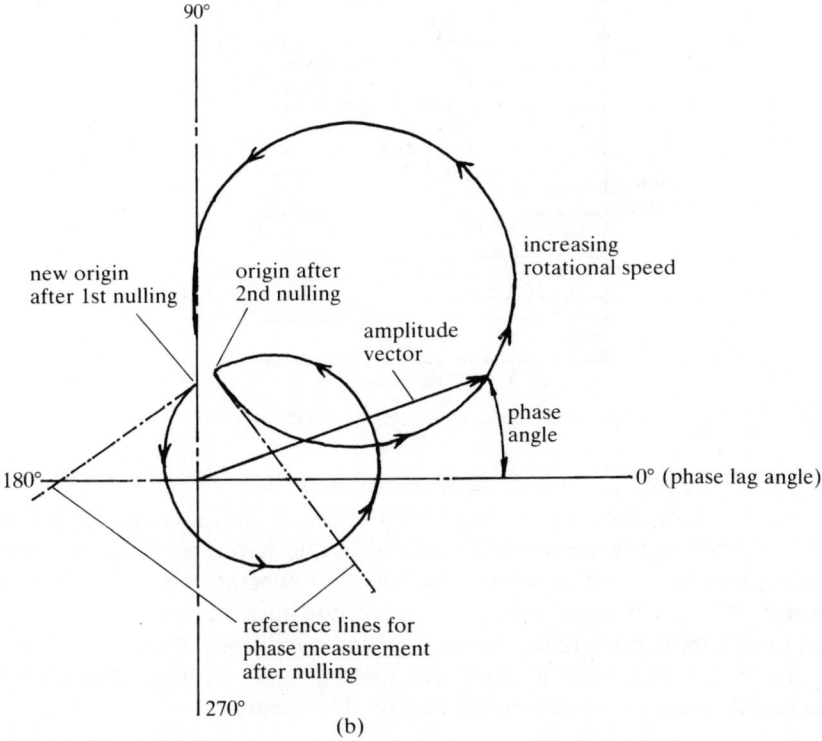

**Figure 9.3** (a) A form of Bode plot used in analysis of rotating machinery; (b) A form of Nyquist diagram used in analysis of rotating machinery.

remaining signal are then displayed, usually in digital form. The output from a tracking filter can be used to construct Bode plots and Nyquist diagrams used, for example, during balancing operations. Examples of a Bode plot and a Nyquist diagram are shown in Figures 9.3. Both diagrams should be plotted using readings which have been 'nulled' at slow speed and between critical speeds, thereby removing the effects of runout and the effects of vibration modes which do not relate to the critical speed under consideration. 'Nulling' is described in Section 9.10.2.

The spectrum analyser is used to separate out the incoming vibration signal into all of the frequencies from which the total signal is composed. The amplitude and phase, relative to some reference signal, of all of the frequency components is displayed in the form of a graph of amplitude (or phase) against frequency. The data is thus said to be displayed in the frequency domain. Many spectrum analysers have the facility to plot several such graphs 'in cascade', as shown in Figure 9.4. Such diagrams help to determine the relationship between vibration signal and machine running speed, which in turn helps in the monitoring of the machine condition.

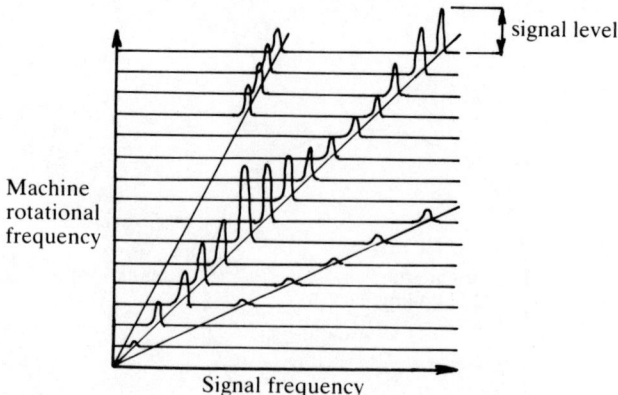

**Figure 9.4** Cascade plot of frequency spectrum.

Once the vibration signals have been collected and measured, they are used to judge whether or not the machine in question is operating properly. These judgements are made on the basis of whether there are unusual features of the vibration signal which are not normally present, or which may usually be present under particular fault conditions. Examples of some of the better-documented fault conditions, and of how they can be diagnosed, are given in the remainder of this chapter.

## 9.3 Shaft imbalance

This is one of the most common machine faults, and its effect is to cause vibration which acts mainly in the radial direction. The vibration occurs at $1 \times$ machine rotational frequency in general, but sometimes higher harmonics of rotational speed are excited (Jackson 1975, Haddad and Corless 1978). Machine vibration caused by imbalance can most easily be detected by monitoring shaft displacement amplitude and phase as the machine is run through its critical speed, filtering out non-rotational speed frequencies. The amplitude peaks at the critical speed, and phase changes by about 180° passing through about 90° at the critical, as indicated in the Bode plot of Figure 9.5(a). The shaft whirl orbit generally takes on an elliptical shape (unless the shaft support impedance is isotropic, in which case it is circular), the orientation of the ellipse changing as the critical speed is passed through.

In some instances the magnitude and phase of the imbalance vibration vector might change with time. Such behaviour of the machine is frequently attributable to some form of temporary imbalance, as discussed by Shatoff (1976). When the symptoms of imbalance exhibit this feature the correct

## MISALIGNMENT

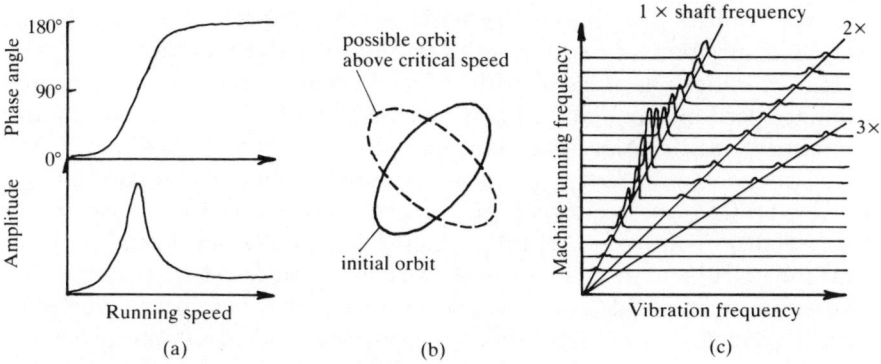

**Figure 9.5** Some ways of presenting vibration data symptomatic of imbalance, (a) Bode diagram; (b) shaft orbit; (c) waterfall plot of vibration spectrum.

treatment is not normally simply to re-balance the machine. Examples of faults in this category are

- hot spots on the shaft causing a thermal bow,
- components mounted loosely on the shaft, and
- moisture entering a hollow shaft.

## 9.4 Misalignment, pre-loaded shaft

Shafts with a heavy pre-load carried by the bearings, as distinct from the out-of-balance load, can show vibration characteristics similar to those caused by bearing misalignment. This feature of pre-loads is discussed by Collacott (1979). Many other investigations of the effects of bearing misalignment on machine vibration have also been reported, for example those by Saugerud (1977), Spettel (1985) and Haddad and Corless (1978). This category of fault is probably the second most common cause of machine vibration, after imbalance.

A pre-load might be applied at a bearing as a consequence of gear mesh forces, aerodynamic forces, and hydrodynamic forces. (Lobed bearings also pre-load the journal, but this is not a fault condition.) Misalignment may be present because of improper machine assembly or as a consequence of thermal distortion, and itself results in additional loads being applied to the bearings.

Misalignment of adjoining shafts, or of the bearings of one shaft, causes abnormal loads to be transmitted through the bearings, and imposes additional bending stresses on the shaft thereby reducing its fatigue life. In some cases misalignment may cause one bearing to be unloaded (if the pre-load so applied is in the opposite direction to the normal gravity load);

this can result in lowering of the machine critical speed. These symptoms are sometimes present in addition to that of excessive vibration.

The vibration associated with misalignment occurs at $1 \times$ machine running speed but, unlike the imbalance case, there is usually a substantial component in the axial direction which may be greater than that in the radial direction. Substantial amounts of misalignment (or pre-load) can also cause vibrations at frequencies of $2 \times$ machine running speed, and sometimes higher multiples. Vibrations caused by misalignment tend to peak near the machine critical speed, but away from the critical may either stay about constant or tend to increase; in many cases they have been reported to appear suddenly as machine speed is increased, and to disappear equally quickly for a particular operating speed range. The 'cascade' plot of the frequency spectra for a machine with misaligned bearings is shown in Figure 9.6. Also shown in Figure 9.6 are shaft whirl orbits which are typical when misalignment is present. For moderate misalignment the shaft

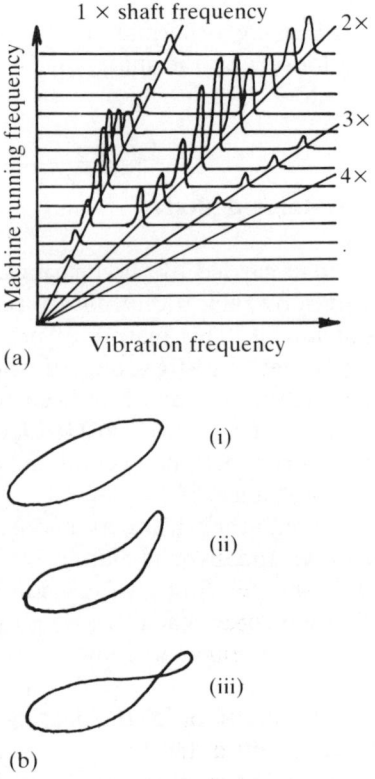

**Figure 9.6** Frequency spectra and whirl orbits typical of a machine with misaligned bearings, (a) waterfall plot of vibration spectrum; (b) typical shaft whirl orbits, showing (i) moderate, (ii) heavier and (iii) severe misalignment.

amplitude is reduced in the sense of the applied pre-load, so that if it were originally circular it might become elliptical; this is because there is more resistance to motion in the direction opposing the pre-load and so the rotor 'feels' a higher support stiffness in this direction. For more severe cases the orbit may become banana-shaped or even figure-8 shaped.

If the misalignment has been introduced when two shafts were improperly coupled, some reduction of the undesirable symptoms may be obtained by using a flexible coupling; the amount of improvement obtained will depend upon the coupling stiffness. Coupling alignment may be checked by taking dial gauge readings and gap measurements, as shown in Figure 9.7. The variations in these measurements should be recorded as both shafts are rotated simultaneously, and averaged values used to calculate the misalignment present. Alignment of bearings may be checked using optical methods (one bearing with another), or using feeler gauges in the clearance between journal and bush; also proximity transducer measurements may be used to record the gap between the shaft and each end of a bearing and so indicate alignment. Other indicators of alignment in fluid film bearings are measurements of the bearing temperature and lubricant pressure distributions.

EXAMPLE 9.1

A large commercial electric motor, normally operating well above its fundamental critical speed, is found to exhibit effective pedestal vibration velocities of 45 mm.s$^{-1}$ in the radial direction and 15 mm.s$^{-1}$ in the axial direction. The vibration frequency is always at $1 \times$ shaft rotational speed, and its amplitude and phase are unchanging with time. Comment on these measurements.

SOLUTION

The $1 \times$ rotational speed frequency of the vibrations would suggest shaft imbalance as the most likely cause of the vibrations. If this was found to be

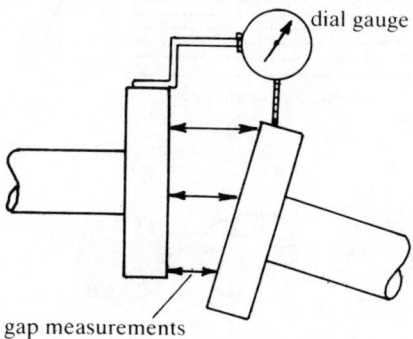

**Figure 9.7** Dial gauge and gap measurements taken to check shaft alignment at a coupling.

the case then it is likely that rebalancing would enable the problem to be corrected since the symptomatic vibrations are unchanging with time. The class of machine described would fall into the flexible rotor group III in Table 7.1, and the vibration levels recorded suggest that immediate attention to the machine is advisable. The presence of some axial vibration suggests that misalignment might also be suspected as a possible cause of the problem; shaft alignment should be examined closely.

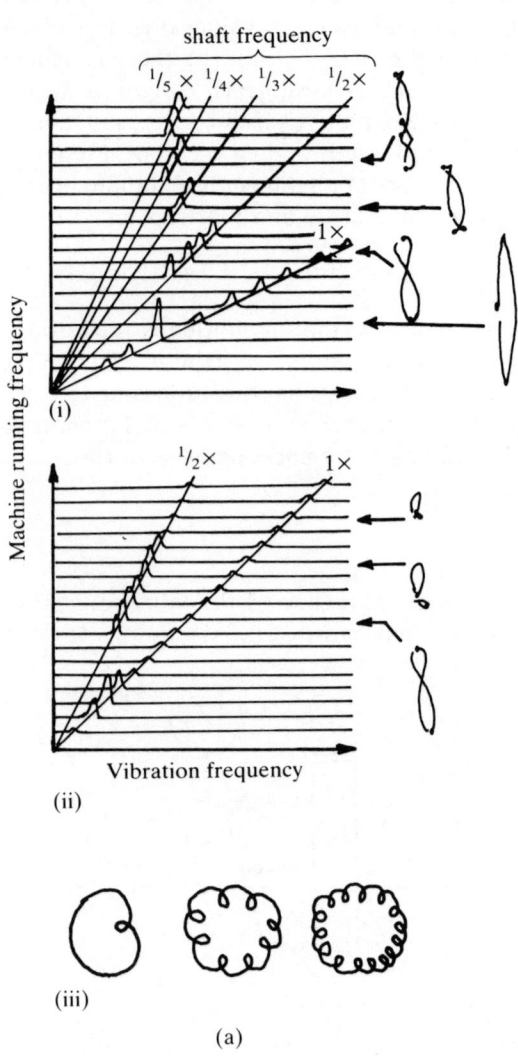

(a)

RUBS

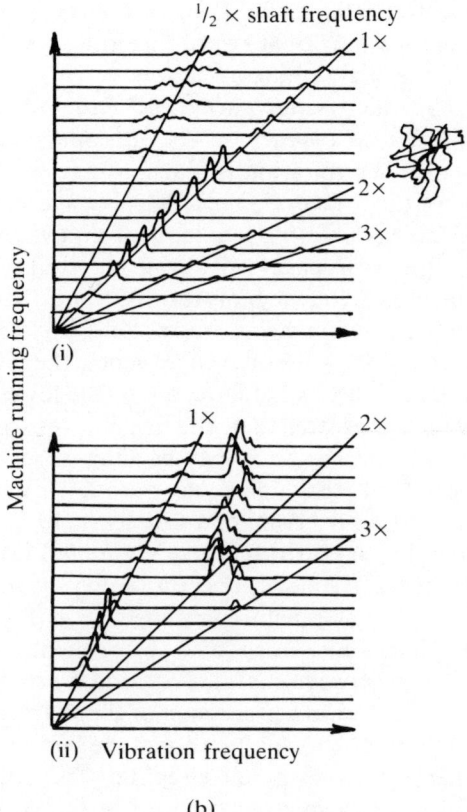

**Figure 9.8** Frequency spectra and whirl orbits typical of a machine suffering partial or full rub, (a) (left) partial rub: Waterfall plots for (i) light rub and (ii) heavier rub, obtained by Bently (1983) and (iii) typical shaft orbits described by Collacot (1979) (b) full rub: (i) with synchronous precession, (ii) with reverse precession during machine run up in speed.

## 9.5 Rubs

Rubs are produced when the rotating shaft comes into contact with stationary components of the machine. They may be broadly classified as partial rubs or full rubs, or may take the form of a stick-slip action.

A partial rub is that where contact between the shaft and stationary component exists only for part of the cycle time, for example when a high spot on a shaft rubs against part of a seal. In studies by Bently (1983) and Collacott (1979) partial rubs were found to take the form of a 'hit and bounce' action of the shaft against the stationary component resulting in shaft orbits and frequency spectra 'waterfall' plots as shown in Figures 9.8. The frequency spectra for a partial rub always show some synchronous vibration (1 × shaft rotational speed) together with some subharmonics which are related to free lateral vibrations of the rotor. The subharmonics may be in either the forward or backward directions (the forward direction being that in which the shaft rotates).

If the machine is operated with a partial rub acting for a sufficient period of time, then the point of contact on the stationary member may wear sufficiently to lengthen the duration of each contact so that it becomes a significant fraction of the total cycle time. The average support stiffness felt by the shaft then increases with a consequent equivalent increase in the machine natural frequency. A case such as this is described by Spettel (1986) where the fundamental frequency of the machine then coincided with $0.5 \times$ machine running speed and was substantially excited. If the rotational speed is more than $2 \times$ normal resonant frequency, the increase in resonant frequency will often result in its 'latching on' to the lowest subharmonic of rotational speed which is greater than the original resonant frequency.

In the case of a full rub, the shaft may no longer make contact with the stationary member only at local high spots, but may instead bounce its way all around its orbit. In Bently (1983) and Collacott (1979) the orbit itself is, in this case, shown to be very distorted as in Figure 9.8b(i). The corresponding 'waterfall' plot of the frequency spectra is also shown in Figure 9.8b(i) where it can be seen that the main component is at $1 \times$ rotational speed; harmonics and subharmonics may also be present. Alternatively, the tangential friction force between shaft and stationary component may be so large that it results in a backward precession of the shaft; the motion may be considered as a rolling of the shaft around the inside of the stationary component together with substantial slipping. The corresponding whirl orbit and 'waterfall' frequency spectra plot are shown in Figure 9.8b(ii). Such backward precession of the shaft is an extremely dangerous situation which may result in destruction of the machine. In Bently (1983) the backward precession is reported to have commenced at a rotational speed just below the first balance resonance during run-up, but to have been present even at very low speeds during run-down. This type of rub has also been discussed by den Hartog (1956), Collacott (1979) and Ehrich (1969).

A rub may also give rise to stick-slip action of the shaft against the stationary component. This type of rub has been discussed by Sadowy (1959), Ehrich (1972) and Cameron (1981). It is a vibration set up as a consequence of the friction coefficient between shaft and stationary component changing with relative velocity. In Figure 9.9 the rotor rotation initially causes it to roll over the stationary component so that its centre moves from left to right; when the spring force in the shaft is equal to the friction force which initiates the motion then the rotor starts to slip over the stationary component instead of rolling over it. At this moment there is a change in friction coefficient which leads to a smaller friction force and the rotor moves now from right to left, overshooting the point where the new friction force is equal to the shaft spring force because of the inertia of the

# RUBS

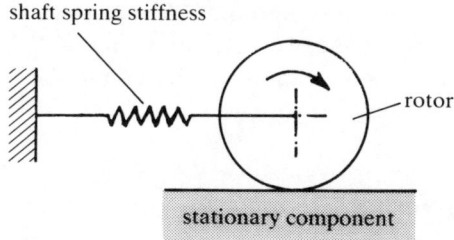

**Figure 9.9** Diagram of a type of mechanism likely to generate a 'stick-slip' rub.

rotor. At some point the rotor stops moving from right to left and instead begins to roll on the stationary component once more, and the cycle of events starts again. The vibration frequency is usually much higher than the shaft rotational frequency, and the fault frequently shows up as a torsional vibration as well as a lateral one.

If a rub is suspected on a machine then the machine should be shut down and inspected for damage. In some instances it may be helpful to remove some of the close-fitting components during a trial run to help diagnose the problem. In cases where subsequent rubs cannot be avoided by increasing clearance then either the rotor surface or that of the stationary component should be manufactured from a soft material which will easily wear away; this will help to avoid backward precession. The situation may also be helped by lubricating the area of contact. In an earlier study by Ehrich (1969) it was found that reverse precession could also be avoided by designing the machine with a very stiff rotor and a flexibly mounted stator.

EXAMPLE 9.2

A rotor of mass 50 kg is mounted on a shaft of stiffness $0.5 \text{ MN.m}^{-1}$. The rotor runs in bearings which behave as pinned supports, and has been operating for some time with high vibration levels. The machine has a number of components fitted close to the shaft, and it is suspected that a partial rub may be the cause of the vibrations. As a consequence of this situation the presence of subharmonics in the vibration frequency spectrum is to be examined. At what machine running speeds might the subharmonics of $\frac{1}{2} \times$ and $\frac{1}{3} \times$ machine running speed be present?

SOLUTION

The subharmonics are related to machine running speed, and tend to occur when the subharmonic frequency coincides with the machine fundamental frequency. The machine fundamental frequency is

$$\omega_n = (k/m)^{1/2} = (0.5 \times 10^6/50)^{1/2} = 100 \text{ rad.s}^{-1}$$

This coincides with a rotational speed of 955 rev. min$^{-1}$. The $\frac{1}{2}\times$ rotational speed subharmonic might therefore set in at about 1910 rev. min$^{-1}$, and might cessate at about 2865 rev. min$^{-1}$, whereupon the $\frac{1}{3}\times$ rotational speed subharmonic might set in.

## 9.6 Loose components

In some machines vibration levels may be excessive as a consequence of components being assembled too loosely, for example in the case of a bearing which is not properly secured. The vibration signal detected from a transducer is nominally a sine wave, but is truncated at one extremity as shown in Figure 9.10(c). The truncation of the sine wave occurs because of the non-linearity of, for example, the bearing mount which suddenly becomes very stiff when the loose component reaches the limit of its allowable travel. This is tantamount to inputting a series of displacement impulses which are superimposed on the normal vibration signal; these impulses show up as a series of harmonics of rotational frequency when viewed in the frequency domain. The $2\times$ machine running speed component is usually the harmonic most easily detected, and it is this that has

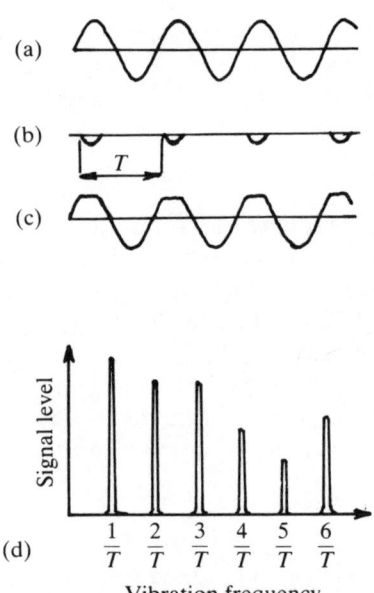

**Figure 9.10** Vibration signals relating to a machine with a loosely mounted component, (a) nominal imbalance vibration signal; (b) effective addition to signal due to a loose component; (c) truncated form of resultant signal; (d) resultant signal in frequency domain.

been reported by most researchers as the symptom of a loose component; see for example Carmody (1972), Colby (1978).

In cases where it is the bearing which is not properly secured, the shaft average support stiffness over one cycle is effectively reduced. When the shaft speed is less than $2 \times$ normal resonant frequency, the reduction in shaft support stiffness acts to lower the resonant frequency to $\frac{1}{2} \times$ shaft rotational speed so that vibration at this frequency becomes significant. A similar effect may occur with other subharmonics of rotational speed. Examples of instances where mechanical looseness has caused vibrations at $\frac{1}{2} \times$ shaft rotational frequency are reported by Jackson (1975) and Shatoff (1976).

## 9.7 Shaft cracks

Shaft cracks are potentially the cause of catastrophic machine failures. They are particularly likely to occur in instances where shaft stresses are high and where the machine has endured many operation cycles throughout its life, so that material failure occurs as a consequence of fatigue. If shaft cracks can be recognized before catastropic failure occurs then the machine can be temporarily taken out of service and repaired before the situation gets out of hand.

The presence of a transverse shaft crack can sometimes be detected by monitoring changes in vibration characteristics of the machine. The shaft stiffness at the location of the crack is reduced, by an amount depending on the crack size. This in turn affects the machine natural frequencies, so that changes in natural frequency may be symptomatic of a shaft crack. Theoretical studies, for example those by Dimarogonas and Paipetis (1983) and by Henry and Okah-Avae (1976), have confirmed this, but unfortunately have also shown that substantial changes in natural frequency may not occur until the crack has reached a dangerously large size. For this reason most users of rotating machinery depend upon changes in vibration amplitude, phase, and frequency spectrum to detect shaft cracks, rather than on changes in natural frequency. Some typical examples have been discussed by Bently (1986), where a transverse crack results in significant changes in both the $1 \times$ and $2 \times$ rotational speed vibration frequencies. The $1 \times$ rotational speed component may change in both amplitude (which may either grow or decay) and phase, as a consequence of the change in rotor bending stiffness. A transverse crack also results in a bigger rotor bow due to a steady load (for example gravity) which may be detected under 'slow roll' conditions. The classical symptom of a cracked shaft is the occurrence of a vibration component at $2 \times$ shaft rotational frequency as a consequence of the shaft asymmetry at the crack location in the presence of a steady load (see also section 6.5). The $2 \times$ rotational frequency component

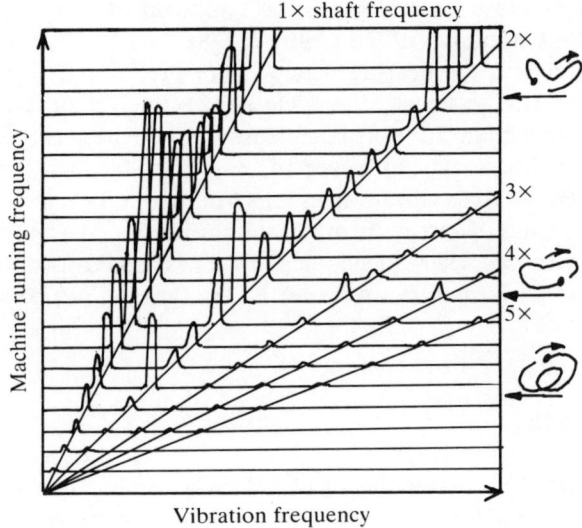

**Figure 9.11** Frequency spectra and whirl orbits for a shaft with a transverse crack.

is usually particularly prevalent when the machine is running at a critical speed or at $\frac{1}{2} \times$ any critical speed. Typical examples of the frequency spectra and shaft whirl orbits are shown in Figure 9.11.

Other types of crack, in particular those which originate from the centre of the shaft, may not cause a change in shaft bending stiffness which is significant enough to enable the crack to be detected by $2 \times$ rotational speed components in the vibration spectrum. In these cases the machine operator must rely on non-destructive test techniques, such as ultrasonic or X-ray inspection, to detect the crack during a routine servicing.

## 9.8 Rolling element bearing faults

Faults in rolling element bearings may occur prematurely as a consequence of operating the bearing under inappropriate loading conditions (including misalignment) and at excessive speeds, or alternatively they may be produced simply as part of the normal wear process during the life of the bearing. Traditionally machines with rolling element bearings would have their bearings renewed regularly, as part of the normal maintenance schedule, irrespective of their wear. This would be done in an effort to avoid a bearing failure at a later time which would necessitate machine stoppage at an inconvenient (and more costly) moment. The growing trend, however, is to monitor the condition of rolling element bearings continually so that bearing wear may be detected at an early stage, and enable the engineer to ensure that the bearing is replaced at a convenient time before the bearing fails completely. In this way bearings are replaced only when they are worn out, not as a matter of routine, and at the same time the machine overhaul can be well ordered and planned in advance. Condition monitoring of rolling element bearings has enabled cost savings of over 50%; as compared with traditional maintenance methods; a typical example has been reported by Sturrock and Hoelscher (1985).

The most common method of monitoring the condition of rolling element bearings is to measure the vibration of the machine at the bearing at regular intervals using a velocity transducer or an accelerometer mounted on the machine casing. More recently, observation of the bearing outer-race deformations using fibre optics (Hirschfeld 1978) and high-sensitivity proximity transducers (Bently 1982) has also been used to monitor bearing condition. Defects in the bearing may develop on either raceway, on the rolling elements themselves, or on the cage; subsequent vibrations are forced as a consequence of impact between the fault and other bearing components, so that the frequency of the resulting vibrations is largely dependent on the frequency of impacting. For example, a defect in a bearing outer raceway would set up vibrations corresponding to the frequency with which rolling elements passed over the defect. Consideration of the bearing geometry enables calculation of the frequencies associated with defects in different bearing components, as indicated in Figure 9.12.

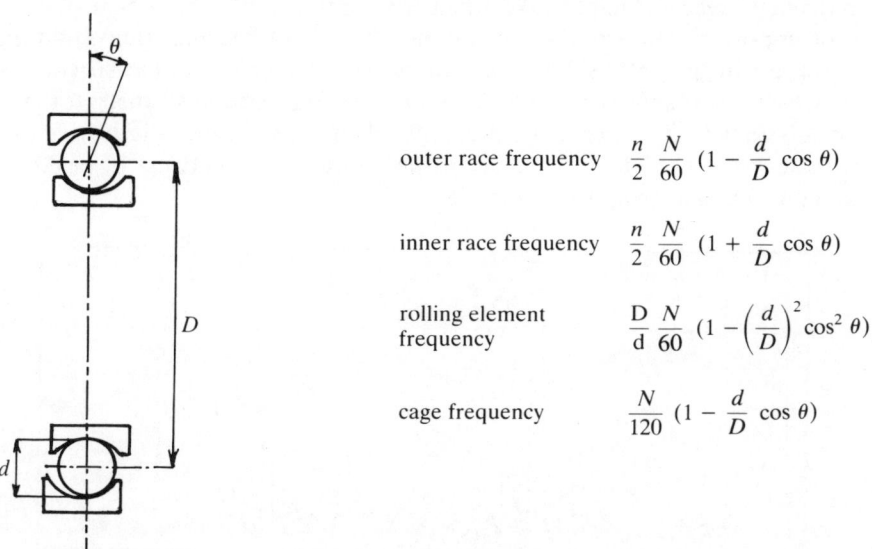

| | |
|---|---|
| outer race frequency | $\dfrac{n}{2} \dfrac{N}{60} \left(1 - \dfrac{d}{D} \cos \theta\right)$ |
| inner race frequency | $\dfrac{n}{2} \dfrac{N}{60} \left(1 + \dfrac{d}{D} \cos \theta\right)$ |
| rolling element frequency | $\dfrac{D}{d} \dfrac{N}{60} \left(1 - \left(\dfrac{d}{D}\right)^2 \cos^2 \theta\right)$ |
| cage frequency | $\dfrac{N}{120} \left(1 - \dfrac{d}{D} \cos \theta\right)$ |

**Figure 9.12** Calculation formulae for rolling element bearing faults. $d$, rolling element diameter; $D$, Bearing pitch diameter; $\theta$, contact angle; $n$, number of rolling elements; $N$, shaft frequency in rev.min$^{-1}$.

Whilst the characteristic frequencies can be easily calculated, the process of diagnosing a fault can be complicated by a number of factors. Some of the characteristic frequencies may be very close to harmonics of rotational speed; this means that a narrow-band spectrum analyser is required in order to distinguish vibration components caused by bearing failure from those caused by, for example, imbalance. A guide to the use of such signal analysers has been published by Hewlett Packard (1983). Even when the instrumentation provides a suitable resolution in the frequency domain, additional 'sum and difference' frequencies may be apparent as a result of interaction between two or more characteristic frequencies (Eshelman 1980). Such sum and difference frequencies appear as sidebands, that is frequency peaks either side of the higher main frequency component of the signal. For example, in Figure 9.13 the frequency component $f_1$ is caused by imbalance while frequency $f_2$ might be characteristic of a raceway defect; the spectrum also shows sidebands related to the sum and difference of these frequencies. As wear of the bearing progresses the frequency spectrum changes further. Sometimes higher-order harmonics of the defect frequency become present (Downham 1976), sometimes with their own sidebands, and can dominate the spectrum; in addition, wear particles are transported around the bearing and accelerate the development of further defects at other locations, leading to high levels of vibration at many frequencies so that peaks which are characteristic of particular defects become difficult to distinguish. Figure 9.14 shows the general forms that the frequency spectrum might take when the bearing fault becomes severe. A most important feature of condition monitoring of bearings (and rotating machinery in general), is the collection of 'baseline' reference measurements of vibration taken when the machine is first commissioned (or re-commissioned after overhaul). It is only when the engineer is in possession of these that a confident diagnosis of the significance and cause of peaks in the vibration spectrum can be made.

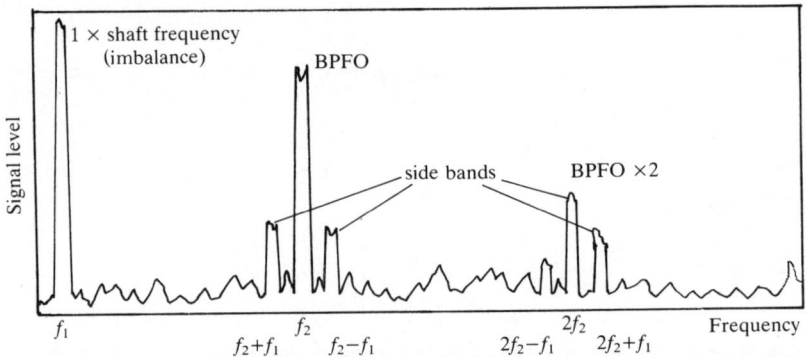

**Figure 9.13** An example of the occurrence of sidebands in the frequency spectrum. BPFO, outer race ball pass frequency.

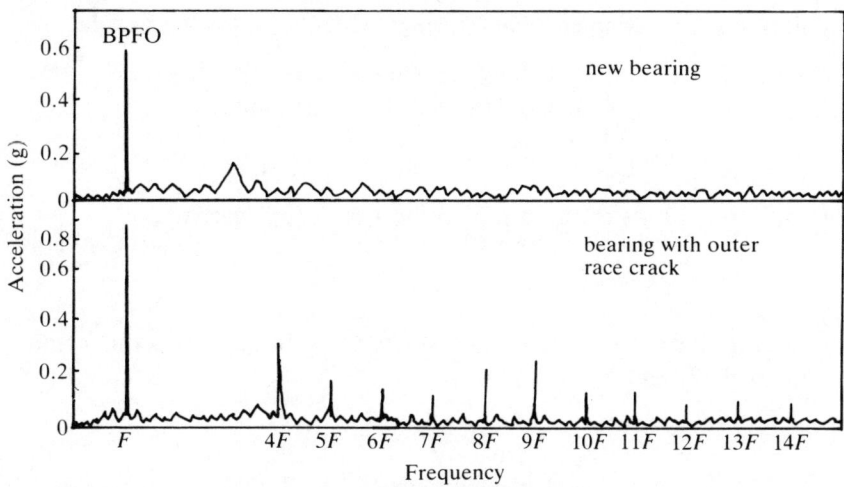

**Figure 9.14** Changes in the vibration spectrum typical of a deteriorating fault condition in a rolling element bearing.

An alternative method of monitoring rolling element bearing condition is the 'shock pulse method' (SPM) described by Brown (1977). This method depends upon the signal input to the measuring instrument being passed through a narrow-band filter whose centre frequency corresponds to the accelerometer natural frequency. As part of the bearing collides with a defect, a shock wave is transmitted through the bearing and machine casing thus exciting the accelerometer by means of a pulse input. The accelerometer output is a damped transient waveform whose frequency is much higher than the frequencies characteristic of specific bearing faults, and whose magnitude is dependent upon the magnitude of the defect in the bearing. Again, this method relies heavily upon establishing reliable baseline data and on monitoring changes in output level rather than on absolute values (though the latter can give a vague impression of bearing condition). Unlike the method described in the preceding paragraph, the SPM cannot indicate the cause of the vibration levels, but it is certainly a less expensive technique.

EXAMPLE 9.3

Monitoring of rolling element bearing condition is carried out as part of a routine planned maintenance schedule at a manufacturing plant, using narrow-band spectrum analysis. On one particular machine, where the shaft rotates at 1000 rev.min$^{-1}$, the vibration spectrum shows peaks at the following frequencies: 17 Hz, 95 Hz, 97 Hz, 114 Hz, and 132 Hz. The machine runs in rolling element bearings with the following details:

>    no. of rolling elements      10
>    rolling element diameter     16 mm
>    contact angle                0°
>    bearing pitch diameter       112 mm

Suggest likely causes of the peaks in the frequency spectrum.

SOLUTION

The bearing characteristic frequencies may be calculated using the formulae in Figures 9.12, and are as follows:

>    outer race frequency         71 Hz
>    inner race frequency         95 Hz
>    rolling element frequency    114 Hz
>    cage frequency               7.1 Hz

Machine running frequency is 17 Hz. It seems likely, therefore, that the 17 Hz peak is caused by imbalance, and the 114 Hz peak by a fault on one of the rolling elements. The strong 97 Hz and 132 Hz signals are sidebands to the rolling element fault frequency ('sum and difference' interaction with the 17 Hz signal). The peak at 95 Hz indicates the presence of inner race wear.

Because of the absence of absolute values of the vibration data, and of trend data accumulated over a period of time, it is not possible from these data to say whether the bearing should be replaced.

## 9.9 Faults in gears

Vibration measurements taken at the bearings can also indicate the condition of the gearbox. A narrow-band spectrum analyser is usually a necessary piece of equipment because the monitoring process involves the engineer in detecting discrete frequency components which must be distinguished from frequencies generated by other vibration mechanisms, as in the monitoring of rolling element bearing condition. The characteristic frequency of a particular gear set is the gear mesh frequency (the product of the number of teeth on the gear and the shaft rotational frequency) which will be evident in the spectrum relating to any gearbox, in good condition or otherwise. When one gear becomes damaged the gear mesh frequency component of vibration may increase substantially, as compared to baseline vibration measurements (Borhaug and Mitchell 1973), but this is not always the case. Harmonics of gear mesh frequency may also become more apparent. Another frequency which is often excited by gear defects is the

# SPURIOUS SIGNALS

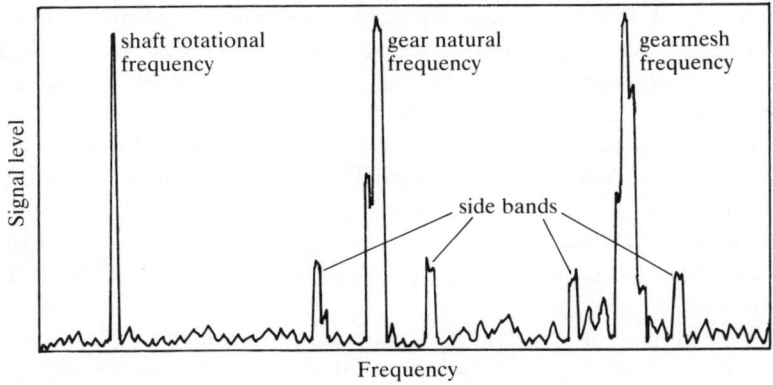

**Figure 9.15** Effect of a gear defect on the vibration frequency spectrum.

resonant frequency of the geared shaft itself; this frequency can usually be measured by impulse testing if it is not already known, so that it may readily be identified in the vibration spectrum. Both the natural frequency of the geared shaft and the gear mesh frequency may have accompanying sidebands; in some cases the sidebands themselves may be the main indicator of a defective gearbox. Examples of a vibration spectrum related to gear defects is shown in Figure 9.15. A more detailed account of gear defect identification by vibration analysis is given by Taylor (1979), and Hewlett Packard (1983).

## 9.10 Protection against spurious signals

When studying vibration signals it is important to be sure that the signal under consideration represents vibration which is actually present; in some situations it is possible for the electrical signal received by the measuring system to indicate the presence of vibration which does not in fact exist. Alternatively, a spurious signal may be superimposed on a genuine vibration signal and thereby upset its measurement. The two most common causes of spurious signals are 'noise' and 'runout'.

### 9.10.1 Electrical noise

Electrical noise may arise from a number of different sources; some common sources include random electron motion, local magnetic fields, arcing, and earth loop faults.

Noise created by the random motion of electrons is known as 'thermal' noise. It is generally only of the order of $\mu$V, however, and is not normally significant in measurements of rotating machinery vibrations.

The noise set up by local magnetic fields is usually of the order of mV, and is more significant. It is set up as a consequence of magnetic fields in nearby electrical apparatus inducing noise signals in electrical leads conveying the signal which is to be measured. Some protection against this type of noise-generating mechanism can be obtained by screening the leads with a high-conductivity material which is earthed. A similar effect can also arise when one piece of instrumentation is sited close to another, because the magnetic fields inside one can induce noise signals in the circuit of the other. In this instance the entire instrument should be screened by placing it inside a metal case.

Another source of electromagnetic radiation which can induce noise signals is that emitted as a consequence of electrical arcing in switches and commutators. This type of noise-generating mechanism can also be protected against by screening as described above.

Earth loop noise can occur when there are too many earth connections in the instrumentation. An example is shown in Figure 9.16 where a coaxial cable is used to connect a transducer to an amplifier. The screen of the cable should be connected to earth to protect against local magnetic effects, but in this example it is earthed at both ends. Such earth connections may be made inside the transducer and amplifier, and may not be obvious at first sight. Electrical fields set up between different points on the Earth's surface by large electrical machinery, usually at 50 Hz, are then able to generate earth loop currents as shown in the diagram. The earth loop current $i_l$ results in a voltage drop $i_l R_s$ in the screen which is in series with the transducer output and so upsets the measurement. The current $i_h$ is relatively small because of the high amplifier input impedance. The effects of earth loop currents can be reduced by decreasing the value of the significant lead resistance, that of the screen $R_s$ in this instance, which means reducing the lead length; alternatively the impedance between the amplifier low terminal and earth can be increased. Another alternative with instruments which have a guard

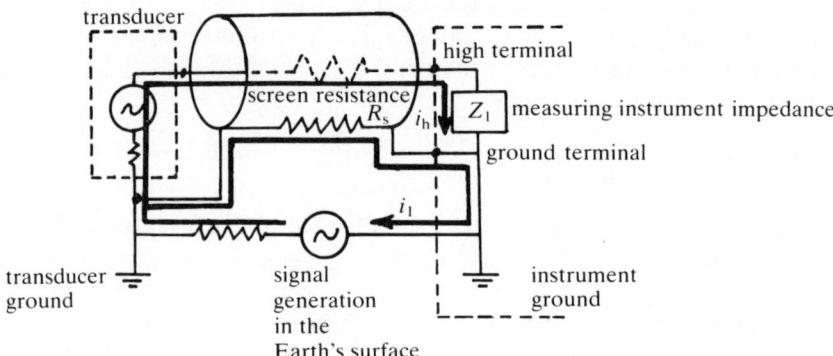

**Figure 9.16** Earth loop currents in an unguarded measuring system.

# SPURIOUS SIGNALS

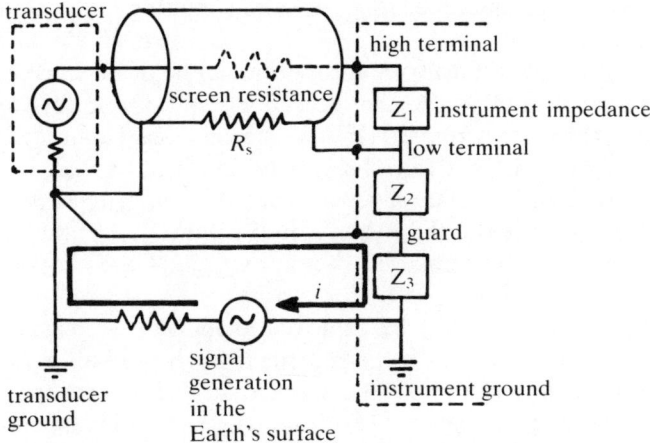

**Figure 9.17** Measuring system protected by a guard terminal.

terminal is to connect this terminal so that earth loop current is directed away from the signal leads, as shown in Figure 9.17.

Further discussion of noise-generating mechanisms and methods of protection can be found in Probert *et al.* (1971) and Hewlett Packard (1970).

## 9.10.2 Runout

Proximity transducers used to monitor rotating machinery vibrations depend for their operation on a change in transducer reactance. Changes in transducer reactance may be present as a consequence of either mechanical or electrical runout.

Mechanical runout is present when the shaft section being monitored by the transducer is eccentric to the axis of rotation, or has significant surface undulations; alternatively, the shaft may be bent. Each of these conditions results in motion of the shaft surface towards or away from the transducer tip when the shaft is rotated, motion which is not caused by shaft vibration. To remove mechanical runout the shaft must be re-machined; if the shaft is bent it must be straightened prior to re-machining.

Electrical runout is present when, around the shaft circumference, there are variations in the permeability of the shaft material to the electrical field set up by the transducer. This might be caused by residual magnetic fields in the shaft surface (due for example to particular methods of manufacture or non-destructive testing), by material inhomogeneity, or by local residual stresses. Residual magnetic fields can be removed by degaussing; if residual magnetism is not the cause of the electrical runout then special treatments

## MEASUREMENTS AND DIAGNOSTICS

such as burnishing, micropeening, or electroplating may be necessary (Bently Nevada 1978).

The amount of mechanical runout present can be determined by mounting a dial test indicator next to the transducer, and noting the variation in reading when the shaft is rotated slowly. The electrical runout present may be determined by noting the transducer reading when the shaft is rotated at low speed (sufficient to obtain a reading on the measuring instrument but not so high as to cause the shaft to vibrate); subtraction of the mechanical runout vector from the measured vector then gives the electrical runout vector.

All runout vectors should be removed from signal measurements to enable true vibration vectors to be recorded. Such 'nulling' is carried out simply by subtracting the total residual runout vector from the measured vector, as indicated in Figure 9.18. Some types of tracking filter enable runout vectors to be automatically removed from the incoming signal, so that the true vibration vector is displayed immediately. This facility can also be programmed into data-logging equipment. Nulling is also carried out between critical speeds when producing a Bode plot, as shown in Figure 9.3(a), to obtain information about phase for use in balancing. In this case it is not only the runout that is removed from the incoming signal, but also components of vibration associated mainly with modes whose characteristic frequencies are lower than the frequency at which the nulling is being carried out.

### 9.10.3 *Electronic differentiation and integration*

In many instances the vibration signal is to be differentiated or integrated prior to being measured. For example, many signal conditioning devices enable accelerometer output signals to be integrated twice to provide a signal which is proportional to displacement, or alternatively they might differentiate a velocity transducer output signal to provide a measurement of acceleration. Generally, passive electronic differentiators and integrators result in a loss of signal accuracy of the order of several per cent because

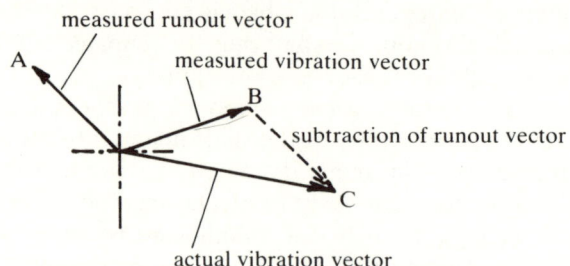

**Figure 9.18** Removing runout from a vibration signal.

their mathematical foundation involves an approximation. Active devices, requiring a power source for internal amplifiers, are more accurate but are also more expensive.

It is particularly important to guard against spurious signals when differentiation or integration is necessary. This is because differentiation requires that the signal is multiplied by its frequency, so that if the spurious signal is of a higher frequency than the vibration signal then after differentiation the spurious signal may appear to be far more significant. A similar effect occurs with integration when the spurious signal is of a lower frequency than the vibration signal.

## 9.11 Monitoring strategies and standards

Early attempts at establishing acceptance criteria for vibration characteristics of rotating machinery were based upon overall levels of vibration, the transducer output being applied directly to the measuring instrument, so that no account was taken of the presence of different frequencies in the complete signal. The earliest standards were based upon subjective opinions of experienced plant personnel who would indicate whether, in their view, a machine was running 'rough' or 'smooth'. The actual overall vibration level was then measured and correlated with the verdict of the operator. The use of such data, relating only to overall vibration levels, is preferable to having no monitoring criteria at all, but is unreliable in that standards produced in this way become highly specific to one type and design of machine. For example the criteria suggested by Rathbone (1939), based on work with machines mounted on rigid foundations, differ considerably from those suggested by Yates (1949) whose work related mainly to flexibly mounted marine turbines, as shown in Figure 9.19. More recent standards for overall vibration levels are the VDI 2056 criteria (1964), which include different criteria for several groups of machines as shown in Figure 9.20. Despite these limitations, overall (unfiltered) vibration level monitoring is one of the methods most commonly used by plant operators. The method is relatively inexpensive and is usually capable of predicting machine faults when carried out regularly by experienced personnel. The general approach is to collect data on a regular basis and to infer an impending machine defect when vibration levels show significant rising trends. With the help of modern equipment the whole process can be computerized so that measurements are recorded electronically by a small portable instrument, and then played back into a computer for analysis and storage. The major limitation of overall level monitoring is probably that it cannot help in diagnosing a machine fault.

In situations where experience with the type of machine under consideration is not at hand, or where failure of a machine component might be extremely dangerous or costly if not detected at the earliest possible opportunity, regular overall vibration level monitoring does not offer adequate protection against fault development. Furthermore, the acceptance criteria discussed above may be inappropriate; there have been several instances where these criteria would suggest that vibration levels were

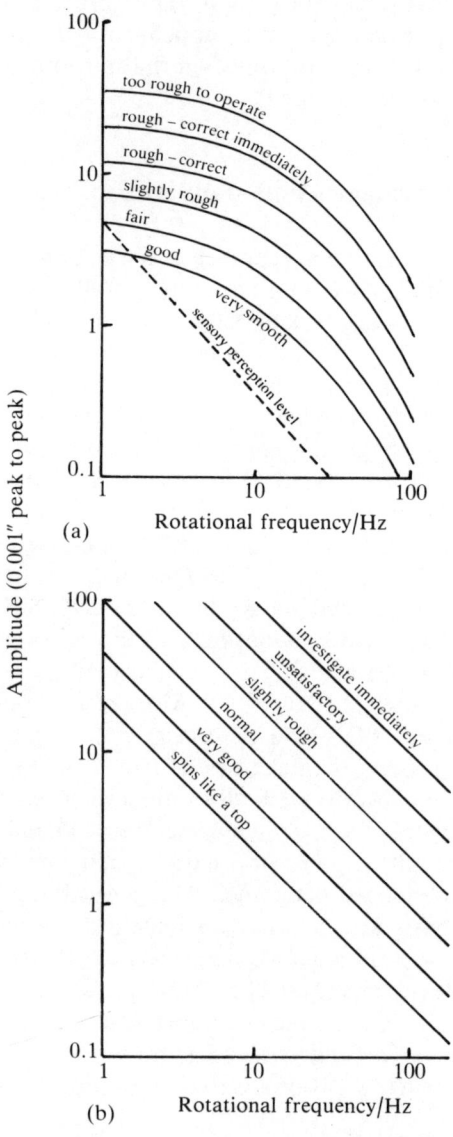

**Figure 9.19** A comparison of the vibration monitoring standards suggested, (a) by Rathbone (1939); (b) by Yates (1949).

## MONITORING

| RMS velocity (mm. s$^{-1}$) | GROUP K — small machines up to 15 kW | GROUP M — medium machines 15–75 kW or up to 300 kW on special foundations | GROUP G — large machines with rigid and heavy foundations whose natural frequencies exceed machine frequency | GROUP T — large machines operating at speeds above foundation's natural frequency |
|---|---|---|---|---|
| 28 | not permissible | not permissible | not permissible | not permissible |
| 16 | | | | |
| 11.2 | | | | just tolerable |
| 7.1 | just tolerable | just tolerable | just tolerable | allowable |
| 4.5 | | | | |
| 2.8 | | | allowable | |
| 1.6 | allowable | allowable | | good |
| 1.12 | | | good | |
| 0.71 | good | good | | |
| 0.45 | | | | |

**Figure 9.20** The VDI 2056 (1964) vibration monitoring criteria.

acceptable when in fact a serious fault condition existed (Downham and Woods 1971). An improvement in the accuracy of monitoring techniques, and in their ability to help in fault diagnosis, is obtained when the signal under consideration is filtered before being measured, and, if considered necessary, when the signal is monitored continually. One way is to use octave band filters and thus obtain overall signal levels within different octave bands; an improvement may be obtained if $\frac{1}{3}$ octave band filters are used, and a further improvement still if an analyser with constant percentage bandwidth filters or narrow-band filters is used. For most precise work,

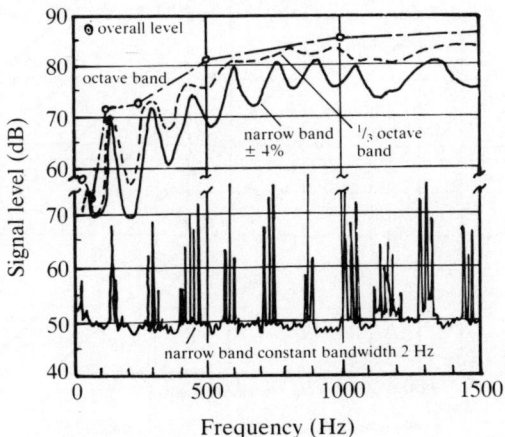

**Figure 9.21** Effect of filter bandwidth on the precision of frequency measurement.

## MEASUREMENTS AND DIAGNOSTICS

modern spectrum analysers are able to provide a frequency resolution to less than 2 Hz. A comparison of the precision provided by each of these filter bandwidths is indicated in Figure 9.21. Clearly, the resolution provided by narrow-band analysers is essential when many close but distinct harmonics and sidebands are sought by the maintenance personnel, whilst a less precise instrument may be adequate in other situations. When signals can be separated into their constituent frequency components a machine vibration severity chart, as suggested by Colby (1978), may be used as a guide in the absence of baseline data and previous experience; the chart shown in Figure 9.22 is intended for machines such as motors, fans,

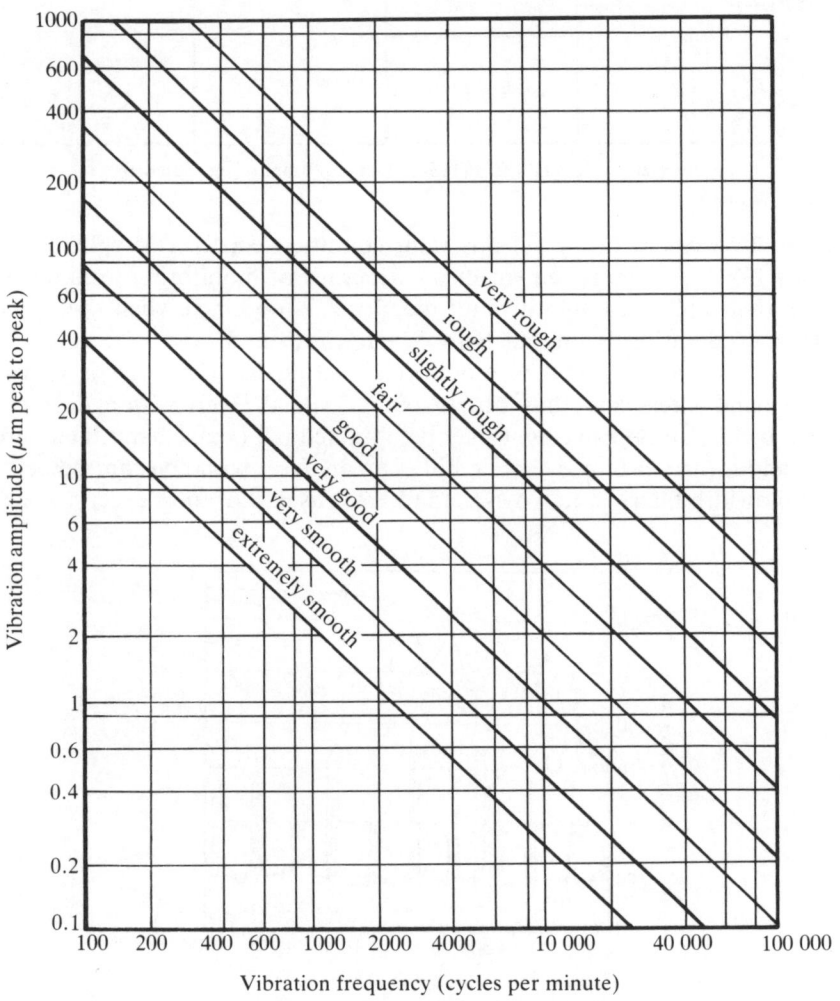

**Figure 9.22** A machine vibration severity chart for assessing vibration levels.

blowers, pumps and other general rotating machinery. In most cases the vibration spectrum is recorded regularly so that changes in magnitude and frequency content are easily recognized. Diagnosis of particular frequencies that are being excited, and hence diagnosis of the fault condition, is sometimes helped if a Campbell diagram indicating excitation frequencies and resonant frequencies of the system is available; an example of such a diagram is shown in Figure 9.23.

A relatively recent development in machine condition monitoring is the use of airborne sound measurements. The analysis equipment used is the same as that described for vibration analysis, except that the electrical signal originates from a microphone instead of a vibration transducer. The method is more difficult in practice because the sound has components associated with all moving parts of the machine, as well as those emanating from other neighbouring machines. Some examples of the method have been reported by Bannister and Donato (1971) and by Borhaug and Mitchell (1973).

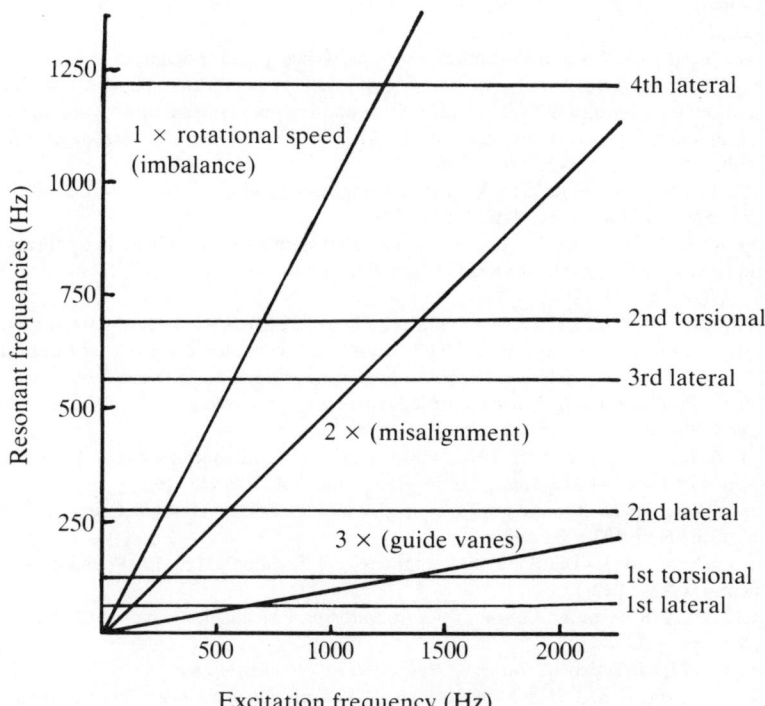

**Figure 9.23** A Campbell diagram for predicting speeds when high vibration levels are likely to occur.

# References

Bannister, R. L. and V. Donato 1971. Signature analysis of turbomachinery. *J. Sound and Vibration*, 5, **9**, 14–21.
Bently, D. E. 1982. Monitoring rolling element bearings, *Orbit* November, 2–15.
Bently, D. E. 1983. Studies reveal physical phenomena of rotor rubs. *Orbit* **4**, **3**, 3–5.
Bently, D. E. 1986. Vibration analysis techniques for detecting and diagnosing shaft cracks. *Orbit* 1, **7**, 18–21.
Bently Nevada 1978. Glitch: definition and methods of correction, including shaft burnishing to remove electrical runout. Applications Note 011, 8/78. Bently Nevada Corp. Minden, Nevada.
Borhaug, J. E. & J. S. Mitchell 1973. Widened frequency range is improving today's machinery vibration analysis. *Power* **117**, 51–53.
Brown, P. J. 1977. Condition monitoring of rolling element bearings. *Noise Control Vibration and Insulation*. February, 40–44.
Cameron, A. 1981. *Basic lubrication theory.* New York: Wiley.
Carmody, T. 1972. The measurement of vibration as a diagnostic tool. *Trans. Inst. Marine Engs.* **84**, 147–159.
Colby, G. A. 1978. Preventive maintenance in rotating machinery using vibration analysis. *Proc. Am. Gas. Assoc. Conf. Montreal.* 8–10 May. Paper 78-T-7. 352–357.
Collacott, R. A. 1979. *Vibration monitoring and diagnosis*. New York: Wiley.
Den Hartog, J. P. 1956. *Mechanical vibrations.* New York: McGraw-Hill.
Dimaragonas, A. D. & S. A. Paipetis 1983. *Analytical methods in rotor dynamics*. London: Applied Science.
Downham, E. 1976. Vibration in rotating machinery: malfunction diagnosis – art & science. *Proc. IMech.E. Conf. Vibrations in Rotating Machinery*, Cambridge. Paper C160 176. 1–6.
Downham, E. & R. Woods 1971. The rationale of monitoring vibration on rotating machinery in continuously operating process plant. *Trans. ASME Conf. Vibrations, Toronto.* September 8–10. Paper No. 71-vibr-96.
Ehrich, F. F. 1969. The dynamic stability of rotor/stator radial rubs in rotating machinery. *Trans. ASME. J. Eng. Ind.*, **91B**, 1025–1028.
Ehrich, F. F. 1972. Identification and avoidance of instabilities and self-excited vibrations in rotating machinery. *Trans. ASME. Design Eng. Conf. and Show.*, Chicago. May 8–11. Paper 72-DE-21.
Eshelman, R. L. 1980. The role of sum and difference frequencies in rotating machinery fault analysis. *Vibrations in rotating machinery*. London: Mechanical Engineering Publications.
Haddad, S. D. & M. J. Corless 1978. Vibration measurements to monitor faults in rotating machines. *Proc. Int. Conf. Noise Control Engineering, Inter-Noise 78.* San Francisco. 8–10 May. 963–966.
Henry, T. A. & B. E. Okah-Avae 1976. Vibrations in cracked shafts. *Proc. I.MechE. Conf. Vibrations in Rotating Machines*, Cambridge. Paper C162/76. 15–19.
Hewlett Packard. 1970. Floating measurements and guarding. Hewlett Packard, Geneva, Applications Note 123, 20 June.
Hewlett Packard 1983. Digital signal analyzer applications. Hewlett Packard, Geneva, Applications Note 243-1.
Hirschfeld, F. 1978. A new technology for monitoring bearing performance. *Design News*, July, 42–48.
Jackson, C. 1975. Practical vibrations. *Hydrocarbon Processing*, **54**.
Probert, S. D., Marsden, J. P. & T. W. Holmes 1971. *Experimentation for engineering students vol. 1: experimental method and measurement.* London: Heinemann.
Rathbone, T. C. 1939. Vibration tolerance. *Power Plant Engineering* **43**, 721ff.

Sadowy, M. 1959. Elementary chatter in machine tools. *Tool Engineer* **43**.

Saugerud, O. T. 1977. Practical condition monitoring of rotating machinery by vibration analysis. *Det Norske Veritas Classification and Registry of Shipping*. No. 101. February.

Shatoff, J. 1976. Using vibration analysis to determine the dynamic health of turbine/generators. *Power* **120**, 5, 23–28.

Spettel, T. 1985. The dynamics of a vertical pump. *Orbit* **6**, 1, 5–8.

Spettel, T. 1986. Shaft position measurements reveal the cause of turbine failure. *Orbit* **7**, 1, 6–9.

Sturrock, D. & B. Hoelscher 1985. Reducing maintenance cost with automated data collection and reduction. *Orbit* **6**, 1, 10–13.

Taylor, J. I. 1979. *Identification of gear defects by vibration analysis*. Vibration Institute, Clarendon Hills, Illinois.

VDI 2056. 1964. Criteria for assessing mechanical vibrations of machines. Verein Deutscher Ingenieure, Düsseldorf, Federal Republic of Germany.

Yates, H. G. 1949. Vibration diagnosis in marine geared turbines. *Trans. North East Coast Inst. Engineering Shipbuilders*. **65**, 225ff.

# Index

*note*: terms in italic lettering indicate figure numbers, terms in bold lettering indicate section numbers

accelerometer 248
acceptance criteria 271
aerodynamic effects 84, **3.8**, 98–100, *3.12*,
    *3.13*, **6.6**, 185–7, 253
  cross-coupling 100, 185–6
  misalignment 99, 186
  seals 99
  stiffness and damping 99
  turbine blades 99
  whirl 100
alignment (of couplings) 255, *9.7*
attitude angle 18, 22, *2.15*, 32, *2.23*

backward precession 258, 259
balancing 3, **7**, 189–214
  balance tolerances **7.6**, 212–13
  Dalby's method 195
  dynamic balance 190, 191
  flexible rotors 189, 204–11
  influence coefficient method 199, **7.5**,
      209–11
  modal balancing **7.4**, 204–9
  mr polygon 191, *7.2*, *7.7*
  mrx polygon 194, *7.6*, *7.7*
  N plane method 205
  N + 2 plane method 206
  overhung shafts 207
  rigid rotors **7.2**, 189–97, **7.3**, 197–203
  static balance 190
  unified method 211
baseline reference measurement 264, 265,
    266, 274
bearing flexibility 2
bearing systems **2**, 5
bearing vibration 2
bistable operation 54, **2.5.3**, *2.38*, 56
Bode diagram 198, 251, 270
boundary conditions 19, 21, 31, 39, 44, 50,
    51, 52, 63, 64, 111

Campbell diagram 275, *9.23*
cascade plot 251, *9.4*
cavitation 25, 30, 51
characteristic equation 170
characteristic matrix 174
coloured noise 228
compressibility number 31, 32, *2.20–4*
cracks **9.7**, 261–2, *9.11*

critical speed 16, 25, 49, 71, 72, 190, 215,
    252
cross-coupling 185–6

diagnosis (of faults) **9**, 248–75
differentiation 270
Dunkerley's formula **4.8**, 138–40
dynamic stiffness 215, 216
dynamic stiffness matrix – *see* 'multi-degree
    of freedom systems'

earth loop 268
eccentricity ratio 16, 22, 27, 28, *2.21*
effective viscosity 19
eigenvalues 174
Euler's method 53, 172

fault diagnosis **9**, 248–75
  cracks **9.7**, 261–2
  gears **9.9**, 266–7
  imbalance **9.3**, 252–3
  loose components **9.6**, 260–1
  misalignment **9.4**, 253–6
  rolling element bearings **9.8**, 262–6
  rubs **9.5**, 256–60
finite difference method 20, 59
flexible foundations **3.6**, 91–5, *3.8*
force transmission 95
flexible rotor 73, (*see also* 'flexible shaft')
flexible shaft
  bearing stiffness and damping coefficients
      (effect of) 84
  deflection at the bearings 86
  deflection at the rotor 87, 89
  flexible bearings **3.5**, 83–91, *3.7*
  force transmission 85, 89
  phase lag angles 89
  rigid bearings **3.2**, 69–74
  stability 169
flow (of lubricant) 25, 27, *2.18*, 37, 40, 45,
    59, 60
force transmission
  rigid shaft 77, 82
free–free conditions 246
frequency domain 251
friction (in fluid film bearings) **2.3.4**, 24–5,
    *2.17*, 32, *2.22*, 58
fundamental mode 204

278

# INDEX

gas bearings 19, **2.4**, 30–48
  attitude angle 32, *2.23*
  boundary conditions 31, 39, 44
  compressibility 30, 35
  compressibility number 31, 32, *2.20–4*, 39
  damping 32, *2.25*, 39, 41, *2.30*, 47, *2.34*
  discharge coefficient 37
  dynamic characteristics 32, *2.25*, 45, *2.33*, *2.34*
  externally pressurized (aerostatic) 30, **2.4.3**, 35–41, *2.26*
  friction 32, *2.22*
  hydrodynamic **2.4.2**, 30–5
  linearization 40
  linearized 'ph' solution 31
  load 32, *2.21*, 36, 45, *2.32*
  orifice types 37, *2.28*
  permeability 43
  porous 30, **2.4.4**, 41–8
  stiffness 32, *2.25*, 36, 39, 41, *2.29*, 47, *2.33*
  squeeze number 39, *2.29*, 2.30, 45, *2.33*, *2.34*
geared systems 253, (*see also* 'torsional vibrations')
  faults **9.9**, 266–7
Gumbel boundary condition 21
gyroscopic effects **3.7**, 95–8, *3.10*, *3.11*, 110, 124

Half-Sommerfeld boundary condition 19, 50, 52
harmonics 258, 260, 264, 266
Hertzian theory 8
hybrid bearings **2.6.4**, 66
hydrodynamic oil film bearings **2.3**, 16–29, 253
  attitude angle 18, 22, *2.15*
  axial groove 17
  boundary conditions 19, 21
  circumferential groove 16, 17, 48
  damping 17, 25, 28, *2.19*
  dynamic characteristics **2.3.6**, 25–9, *2.19*
  finite bearings **2.3.3**, 20–5
  friction **2.3.4**, 24–5, *2.17*
  hole entry 17, 48
  lemon bore 16, 17
  load-carrying capacity 17
  lobed 17
  long bearing approximation 19, 24, 25
  lubricant flow rate 25, 27, *2.18*
  partial arc 16, 17
  plain cylindrical bearing 16, 17
  short bearing approximation 19, 24, 25
  steady-state operation 19, 27
  stiffness 17, 25, 28, *2.19*
  tilting pad 17

hydrostatic oil film bearings **2.6**, 57–66
  boundary conditions 63, 64
  capillary 57, 60
  damping 64, *2.44*
  flow rate (of lubricant) 61, *2.42*
  flow coefficients 59, 60
  friction 58
  hybrid bearings **2.6.4**, 66
  load 61, *2.41*
  orifice 57, 60
  restrictor coefficients 60
  steady-state operation **2.6.2**, 58–61, 62, 63
  stiffness 58, 64, *2.43*
hysteresis damping 179

imbalance 3, 71, 78, **9.3**, 252–3
impedance measurement **8**, 215–47
  impulse testing 238
  multi-frequency testing **8.3.3**, 227–9
  using centrifugal forcing **8.4**, 229–37
  using static loading **8.2**, 217–22
  using transient methods **8.5**
  using vibrators **8.3**, 222–9
impedance matching *4.15*
impedance method – *see* 'multi-degree of freedom systems' and 'torsional vibrations'
impulse 227
impulse testing 238, 267
influence coefficient 70, 219, 220 (*see also* 'multi-degree of freedom systems' and 'balancing')
instability 3, 16, **6**, 167–88, (*see also* 'stability' and 'unstable vibrations')
  aerodynamic effects **6.6**, 185–7
  characteristic equation 170
  characteristic matrix 174
  eigenvalues 174
  flexible shaft 169
  gravity 183
  hysteresis damping 179
  internal friction **6.4**, 178–81
  linearized stiffness and damping **6.2.1**, 168–71
  misalignment 186
  multi-degree of freedom systems **6.2.3**, 173–7
  non-linearity **6.2.2**, 171–3
  oil whirl **6.2**, 167–77
  polar asymmetry **6.5** 181–5
  pulsating torque **6.7**, 187, *6.12*
  resonant whip **6.3**, 177–8
  rigid rotor 169
  Routh–Hurwitz stability criterion 170, 183
  stability map 171, *6.2*
  stability reserve 171

279

# INDEX

time-marching 174
  whirl orbits **6.4**, *6.5*
integration 270
interference fit 179
internal friction **6.4**, 178–81
iterative method 21

jump speed 55, *2.38*

key phasor 248

linearized 'ph' solution 31
Lissajous figure 249
least squares estimator 229, 232, 242
long bearing approximation 19, 24, 25
loose components **9.6**, 260–1
lubricant streamers 25, *2.16*
lumped parameter 58

machine health 3, (*see also* 'faults')
machine vibration severity chart 274, *9.22*
magnetic effects **4.7**, 134–8, *4.16*
  negative stiffness 136
measurements **9**, 248–75
misalignment 186, **9.4**, 253–6, **9.7**
mode shape 157, *5.10*
mode (of vibration) 197, 198, 204, 205
monitoring standards **9.11**, 271–5
multi-degree of freedom systems **4**, 102–44
  boundary conditions 111
  distributed mass 11, 122, 140
  Dunkerley's formula **4.8**, 138–40
  dynamic stiffness matrix method **4.5**, 121–6, **4.6.4**
  flexible bearings and pedestals **4.6**, 127–34
  forced response 113
  free–free impedance 115, 119
  gyroscopic effects 110, 124
  impedance and (receptance) methods **4.4**, 113–21, **5.6**, 130–3, **4.6.3**, 131–3, *4.15*
  influence coefficient method **4.2**, 102–7, *4.1*, 108, **4.6.1**, 127–8, 137
  natural frequencies 106, 107, 113, 132, 138, 140
  Rayleigh's method **4.9**, 140–3
  stability analysis **6.2.3**, 173–7
  transfer matrix method **4.3**, 107–13, **4.6.2**, 128–30
multi-frequency testing **8.3.3**, 227–9

noise **9.10.1**, 267–9
non-destructive test 262
non-linearity 14, 15, 49, **6.2.2**, 171–3
nulling 251, 270
Nyquist diagram 251

octave band filters 273
oil film 16
oil whirl 16, 17
orbit
  squeeze-film bearings 54, **2.37**, 56, 57
  stable and unstable *6.4*, *6.5*
  steady-state 54, 249, **9.2**, 252, *9.5*
  rigid shaft 76, *3.4*, 81, *3.6*
oscilloscope 249

perturbation method 43
phase lag angle 198
pin–pin critical speed 190
polar asymmetry **6.5**, 181–5
pre-load **9.4**, 253–6
proximity transducer 248
pseudo-random binary sequences 228
pulsating torque **6.7**, 187, *6.12*

Rayleigh's method **4.9**, 140–3
receptance **8.3.1**, 223–4
reference measurement 264, 265, 266, 274
reference signal 248, 250, 251
relaxation 21, 60
resonance 73, 177
resonant whip **6.3** 177–8
response
  forced torsional **5.6**, 163–5
  to imbalance 16, 25, 28, 113, 119, 132, 134
Reynolds' boundary condition 19, 21
Reynolds' equation **2.3.2**, 18, 19, 30, 43, 49, 58, 59, 61
  squeeze-film term 62
rigid rotor 73, 75, *3.3*, *3.5*, 169, (*see also* 'rigid shaft')
rigid shaft **3.3**, **3.4**, (*see also* 'rigid rotor')
  bearing stiffness and damping coefficients (effect of) 78, 79, 82
  damping and cross-coupling **3.4**, 77, *3.5*, 79
  flexible anisotropic bearings **3.3**, **3.4**, 75–83, *3.3*, *3.5*
  force transmission 77, 82
  orbits 76, *3.4*, 81, *3.6*
  phase lag 81, 83
rolling element bearings **2.2**, 5–15, table 2.1
  ball bearings 8
  damping 7
  deformation constant 8, 12
  elastic deformation **2.2.2**, 7–12
  faults **9.8**, 262–6, *9.12*, *9.13*, *9.14*
  internal clearance 7
  load zone 11
  roller bearings 8
  stiffness 7, **2.2.3**, 12–15
rotor–bearing interaction **1.2**

280

# INDEX

Routh-Hurwitz stability criterion 170, 183
rubs **9.5**, 256–60
Runge–Kutta method 172
runout 251, 267, **9.10.2**, 269–70

Schroeder phased harmonics 228
screening 268
self-excited vibrations 16, (*see also* 'instability')
shock pulse monitoring 265
short bearing approximation 19, 24, 25, 49, 50
sidebands 264
single mass rotor 3, 69–101
Sommerfeld boundary condition 19
Sommerfeld number 21, 23, 25, 26, *2.19*
sound measurement 275
spectrum analyzer 251, 264, 266, 273
spurious signals **9.10**, 267–71
squeeze-film bearings 19, **2.5**, 48–57, *2.35*, *2.36*
  boundary conditions 50, 51, 52
  cavitation 51
  centralizing springs 49
  damping **2.5.5**, 56–7
  end seals 49
  jump speed 55, *2.38*
  stability 57
  stiffness **2.5.5**, 56–7
  '$2\pi$' film and pressurization effects **2.5.4**, 55–6
squeeze-film term (in Reynolds' equation) 62
stability 2, 25, 28, 57, 226, (*see also* 'instability')
standards – *see* 'monitoring standards'
steady-state
  locus 22, *2.15*, 32, *2.24*
  orbit 54, (*see also* 'whirl orbit')
subharmonics 257, 259
sum and difference frequencies 264

Swift-Stieber boundary condition 21
synchronous vibration 257
system dynamics 5

time domain 249
time-marching 174
torsional vibrations 3, 5, 145–66, 259
  damping 147, **5.5**, 160–3, 165
  equivalent shaft 148, *5.4*
  forced 147, **5.6**, 163–5
  geared systems **5.4**, 154–60, 165
  impedance method **5.6**, 163–5
  multi-degree of freedom systems **5.3**, 150–4, **5.6**, 163–5
  natural frequencies 140, 153, 157, **5.6**, 163–5
  node 147
  simple systems **5.2**, 145–50
  torsional stiffness 145
  transfer matrix method 150, 152, 163
tracking filter 250, 270
transducers 248
transfer matrix method – *see* 'multi-degree of freedom systems' *and* 'torsional vibrations'

unstable vibrations 3, 54, **2.5.3**, *2.38*, 56

velocity transducer 248
vibration standards – *see* 'monitoring standards'

waterfall plot 251, *9.4*
wear 262, 264
whip **6.3**, 177–8
whirl 100
whirl orbit – *see* 'orbit'
whirling speed 71
white noise 228

x on y facility 249